Aurelien Forget
Biofabrication

Also of interest

Materials for Medical Application
Robert B. Heimann (Ed.), 2020
ISBN 978-3-11-061919-5, e-ISBN 978-3-11-061924-9

Single-Use Technology.
A Practical Guide to Design and Implementation
Adriana G. Lopes and Andrew Brown, 2019
ISBN 978-3-11-064055-7, e-ISBN 978-3-11-064058-8

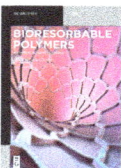

Bioresorbable Polymers.
Biomedical Applications
Declan Devine (Ed.), 2019
ISBN 978-3-11-064056-4, e-ISBN 978-3-11-064057-1

Aurelien Forget

Biofabrication

—

DE GRUYTER

Author
Dr. Aurelien Forget
Institute for Macromolecular Chemistry
Albert-Ludwigs-University Freiburg
Stefan-Meier-Str. 31
79104 Freiburg
Germany
aurelien.forget@makro.uni-freiburg.de

ISBN 978-1-5015-2335-9
e-ISBN (PDF) 978-1-5015-1573-6
e-ISBN (EPUB) 978-1-5015-1582-8

Library of Congress Control Number: 2022936925

Bibliographic information published by the Deutsche Nationalbibliothek
The Deutsche Nationalbibliothek lists this publication in the Deutsche Nationalbibliografie;
detailed bibliographic data are available on the internet at http://dnb.dnb.de.

© 2022 Walter de Gruyter Inc., Boston/Berlin
Cover image: Aurelien Forget
Typesetting: Integra Software Services Pvt. Ltd.
Printing and binding: CPI books GmbH, Leck

www.degruyter.com

Contents

1 Biofabrication history and definition

1.1 Aim of biofabrication

During our life, most individuals will go through injuries or diseases that may lead to the malfunction of vital organs. A dream as old as humanity is to repair injured or damaged tissues. From this dream many myths have emerged over our history, depicting heroes and half-gods that were invincible because of their tissue regeneration capacity. Contrary to the axolotl, lizard, or salamander (figure 1.1) which can regrow limbs or only their tail, the human body has a limited regeneration capacity. Although cells are renewed within organs during our life, for instance, the skin cells are replaced every 39 days, bones are remodeled, and a full turnover of the skeletons takes about 10 years. Conversely, some of our tissues do not regrow past a certain age. As an example, tendons, will not regrow or repair past the age of 10. This means that past this time you will have to keep these tissues, through wear and tears and injury. So, the body regeneration capacity is quite limited, and we are not able to regrow a full organ such as a limb or a finger. Fueled by the salamander's capacity to regrow a fully functional limb, our dream of repairing injury is still alive. But, today, thanks to technology and progress in science that has let us better understand our biology, the dream of invincibility has now shifted from a myth to a technological race to fabricate new organs.

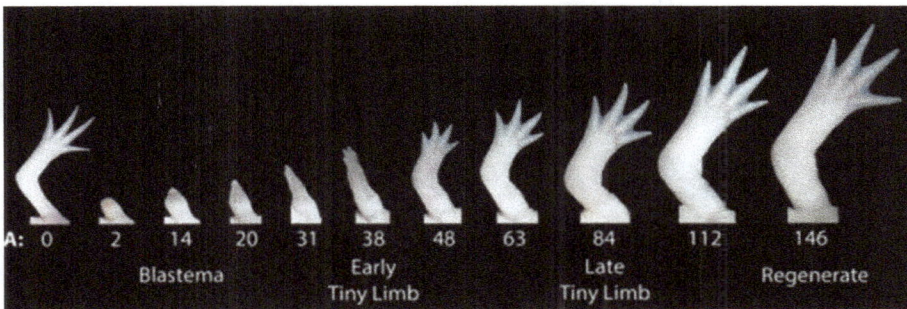

Figure 1.1: The regeneration of the salamander limb after amputation. It takes approximately 4 months for the limb to fully regrow, a dream for humans. Picture from Wells et al. [1].

1.2 Replacement: engineering prosthetics

It is only since the seventeenth century with Robert Hook that we have acquired the notion of cells and only since the 1910s with the work of Burrow that we can reliably culture mammalian cells *in vitro*. So, the first attempt at repairing tissue was strictly an engineered approach, and examples were found in Egypt where artifacts of

https://doi.org/10.1515/9781501515736-001

prosthetics to replace cut toes were found. But the first industrialization of prosthetics was led by Ambroise Paré in the sixteenth century, who is considered as the father of modern prosthetics. Paré lead a hospital in the heart of Paris, where he seeked to heal amputees and soldiers who were coming back from wars. The first idea was to use materials that could be manufactured cost-effectively to be adapted for many patients. This engineering approach used (Figure 1.1a) materials such as wood, leather, and metal hinges to provide articulated prosthetics. Since the beginning, the materials that we use for the replacement of organs play a critically important role and naturally, with the introduction of more effective materials, prosthetics evolved with the emergence of performance materials. The leather-based prosthetics evolved over centuries to incorporate plastic and composite materials that are lightweight and easily customizable. Further development of materials and the introduction of high-performance composites, such as carbon fiber resins, enable the manufacturing of prosthetics that could overcome the natural performance of natural limbs or at least allow prosthetic patients to perform at the highest level. This was particularly illustrated with the 2012 London Olympics performance of the now infamous Oscar Pistorius, a double amputee wearing a prosthetic limb which allowed him to run the 400 m discipline with non-paraplegic athletes [2] (Figure 1.2b). This demonstrates the advances made over the centuries in understanding mechanobiology, the mechanics and movement of the body, and the development of materials that could mimic, and optimize these movements.

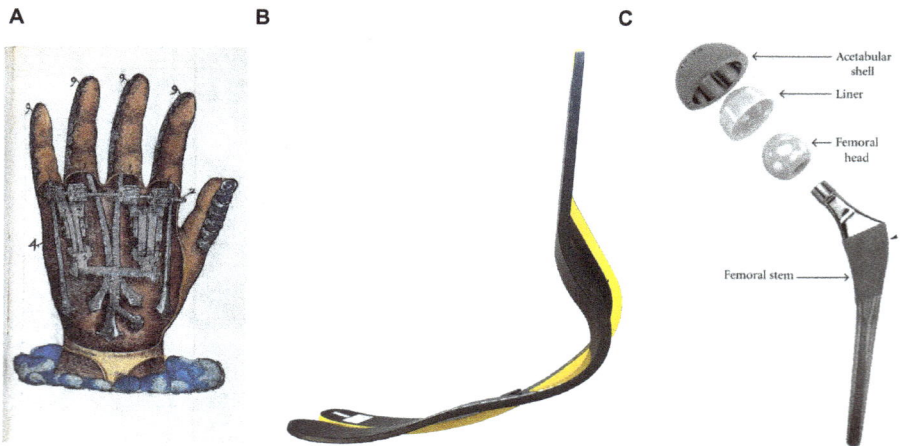

Figure 1.2: Prosthetics through the ages. (A) Illustration of mechanical hand by Ambroise Paré, 1564, (B) Cheetah blade running prosthesis by Össur since 2004 [6]. (C) Total hip replacement components and examples of commercial hip stem implants. Small arrows illustrate porous coating areas using sinter technologies [7].

Along with the development of new materials came the opportunity to develop prosthetic implants, meaning that the prosthesis is not confined outside the body, but incorporated in the body and thus will interact with the immune system and tissues. For that purpose, biocompatible materials are needed. One of the most common implants is articular implants. In the USA alone, about 7 million people are living with prosthetic implants such as the knee, or hip implants [3] (Figure 1.2c). The implants are made in biocompatible metal alloys, such as titanium alloy or zirconia, and are composed of two parts [4]. One part is hammered in the leg and a second part: a ball is hammered in the hip. These two parts coming in contact are made of the same or different materials to create low-wear low-friction prosthetic through metal-on-metal or metal-on-poly(ethylene) interactions. These implants allow full movement of the legs and can last up to 28 years in place [5].

All these prosthetics approaches led to fully mechanical implants to address the loss of mobility, that is, organs such as legs, toes, or hands that are needed for movement of the body. Going further with implants, it becomes clear that for other organs a simple mechanical implant would not be equipped to do the needed work to restore the function of the damaged organ. The heart, which is a muscle that contracts at different frequency during the day depending on the patient's activity, is quite challenging to replace and need a piece of engineering that is more advanced than for a prosthetic limb. The first implant to address heart defect or cardiac insufficiency is known as a peacemaker. It is composed of a sensor that tracks the cardiac rhythm and an electro-mechanics stimulator to correct arrhythmia (Figure 1.3a). Further engineering approaches in repairing organs have led to the development of artificial dental implants, shoulder prostheses, lumbar disc, or cochlear implants for hearing aids. These prosthetics implants are helping patients to live a better and longer life. But these implants were developed within an engineering paradigm that replace an organ with a machine. These organ replacements are made of metallic parts which in some cases are moving but do not have the same performance as their biological equivalents. Metals can corrode in the aqueous environment of the body; plastic sensors can lose efficacy as there are recognized by the immune system as foreign objects. Like any manufactured product, these implants have an expiration date which depends on the complexity of the implant and its function. For instance, the company Carmat developed a mechanical heart that can fully replace a living heart (Figure 1.3b). However, highly advanced mechanical machinery needs constant supervision and revision, which is not possible with implanted equipment. The current operational time of this mechanical heart is 5 years and is therefore used for patients awaiting heart transplant as a provisory measure. Solutions to overcome the limitation of a mechanical or mechatronic engineering approach would lead to a biological equivalent of the defected organ or injured organ.

A   B

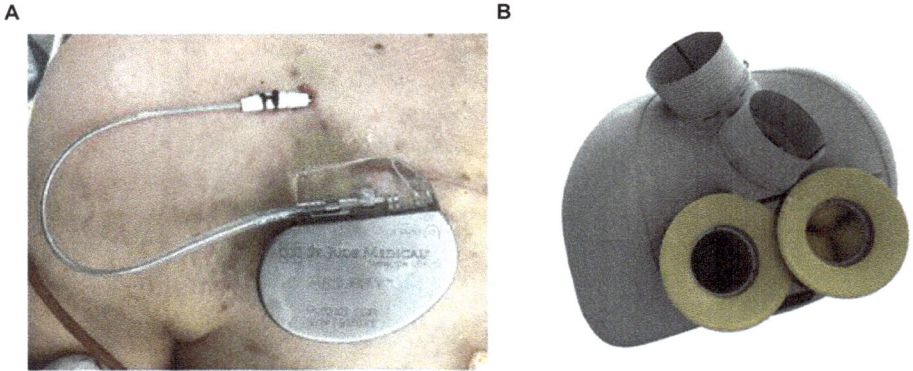

Figure 1.3: Implant for internal organs. (A) Permanent–temporary pacemaker (PTPM): an active-fixation, single-chamber pacemaker lead is fixated to the right ventricle apex or septum and connected to an external pulse generator on the skin [8]. (B) Compliance bag surrounding the prosthesis; atrial suture flanges and ejection conduits [9].

1.3 Replacement: transplanting living tissues

The dream of organ regeneration is as old as our civilization and still alive today. Illustrated by the mythological figures of Prometheus for the Greeks and Tityus for the Romans. Both Prometheus and Tityus's liver was eaten by a crow every day. The liver would regenerate and every night the crow would come again to eat their liver. Althought Romans nor Greek new about the regenerative capabilities of the liver, these myths highlight our ancient desire for regeneration. Combining the advances made by surgery and the development of industrial processes further lead to the establishment of the idea of engineering humans, which is best illustrated by Mary Shelley's Frankenstein considered as the modern myth of Prometheus. Even today, the myth of regeneration is still alive in our culture and often illustrated in movies and comic books where the protagonist is depicted with the capability to regenerate body parts upon injury. But how far are we from realizing this dream?

Early attempts of using biological tissue to regenerate injured tissues have been found in old manuscripts originating from India in the Sushruta Samhita which are dated from 1000 BC [10]. In this document, a surgical procedure to transplant skin flaps from the patient to repair the earlobe is described. Influenced by the Indian literature, Tagliacozzi published the arm-to-nose skin transplantation believed to have been developed by the Branca and Vianeos family (Figure 1.4). This procedure called the Italian method was developed in the fifteenth century [11]. This was commonly used in Italy for the reconstruction of cut noses from fencing accidents. This is considered as the first tissue transplantation of modern medicine bringing the idea of taking tissue from a part of the body that is non-damaged to replace the injured tissue.

Figure 1.4: Illustration from the sixteenth-century technique of rhinoplasty shcwing the transplantation of skin from the arm to the nose. From De Curtorum chirurgia per insitionem (1597) by Gasparo Tagliacozzi.

Further iteration led to taking tissue from cadavers which was first achieved by Girdnerin in the nineteenth century [12]. These were the first transplantations of tissue which eventually led to the first organ transplantation of the thyroid performed by Kocher in 1881 [13]. Decades later, with the understanding of the immune systems and findings of anti-rejection drugs, other organs followed such as kidney, heart, liver, and lungs. Today, organs transplantation is a logistic challenge that requires coordination between multiple medical centres to promptly match donors with a patient. But the establishment of a transplantation network across countries and territories, while increasing the matching success, has not been able to resolve the crgan shortage. Although policies established in Spain for instance, that considers all citizen as donor except if they actively opt-out have allowed decreasing the length of the waiting list, there is still a shortage of organs that will not be resolved soon. Indeed, transplanted organs have a limited lifetime in the patient. Often, organ recipients need to get several transplantations during their life. So, to address the organ shortage, the idea of growing fully functional living transplants has emerged in the late 1990s with the first demonstration of the growing of living cartilage tissue on the back of a mouse was made by

Vacanti (Figure 1.5) The picture made it around the globe, and this started the field of tissue engineering.

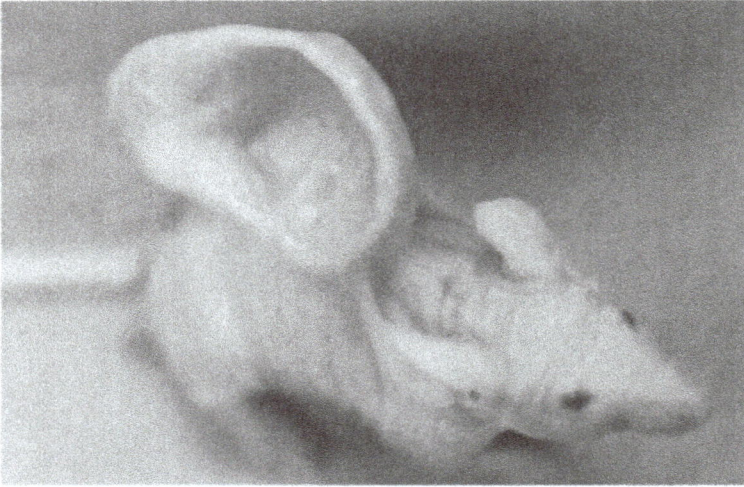

Figure 1.5: The Vacanti mouse is a laboratory mouse that has cartilage tissue grown subcutaneously in what looked like a human ear on its back [14].

1.4 Tissue engineering

Tissue engineering is combining materials science principles with cells biology principles to create living implants. This new approach lets us envision endless possibilities in creating living implants that could replace injured or diseased tissues. Tissue engineering when proposed suggested the possibility to tackle the shortage of organ transplants. If we could make on-demand living implants that are as good as living organs, we would not be limited to cadaver donors to reduce the number of patients on the organ transplantation waiting list. Because tissue engineering proposed to use single cells to grow an organ on a biocompatible material, it brings the possibility to use the patient's cells to create a new organ and create an allograft Table 1.1. Indeed, organ transplantation, because it is using living materials from a foreign system is recognized by the patient immune system as a threat. So, to avoid organ rejection, one needs to match the organ donor with the receiving patient to assure the best compatibility. Additionally, immunosuppressive drugs are needed to trick the patient's immune system into accepting the implant. Using cells that originate from the patient to engineer a replacing tissue opens the possibility to get rid of immunosuppressive drugs after the implantation, thus considerably improving the life of the patient and longevity of the organ. As tissue engineering developed, several successes have demonstrated the relevance of the field: tissue-engineered bladder [15], or nose-to-knee cartilage transplantation [16]. But these successes have been long to develop and remain for the most part an artisanal task that is difficult to bring to scale. If we want to have mass production, industrialization of the tissue engineering processes is needed.

1.A Types of transplantation

It is key to understand the difference between the different types of transplantation. This table has summarized the type of transplantation possible (Table 1.1).

Table 1.1: Types of transplantation.

	Donor			
	Animal	Human	Identical twin	Recipient
Recipient of the transplantation techniques	Xenograft	Allograft	Isograft	Autograft

1.B Organ systems

In this book, we will talk about the different types of biofabrication techniques to create living implants. It is key to understand the type of biofabrication and what can be created with each type of biofabrication technique, for instance, cells, tissue, organ, and organ systems (Figure 1.6).

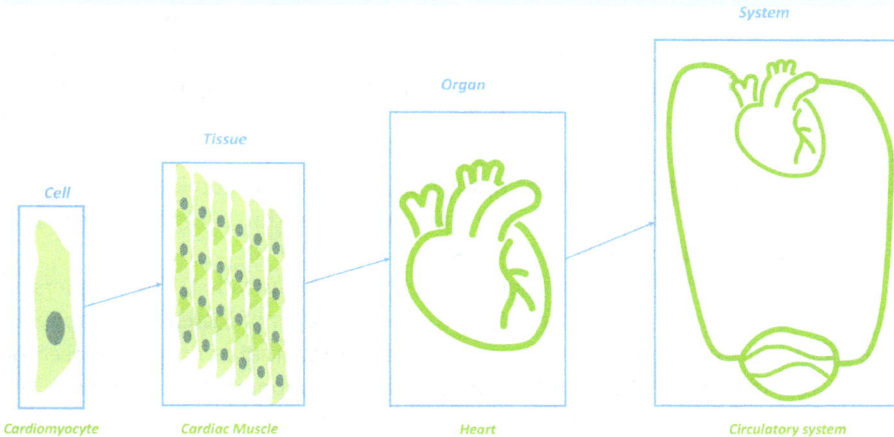

Figure 1.6: The hierarchy of system and cells.

1.5 Engineering in medicine

More and more machines and technical devices have made their way into the operation theaters. Highly engineered and complex tools are now part of surgical procedure: kits containing all the surgical tools to perform a procedure, catheters that are inflatable, analytical instruments such as magnetic resonance imaging or X-ray and bypass machine that can replace kidney, heart, and lung to allow for the surgeon to operate on the organs. The constant development of medical tools has led to more and more advanced instruments. For instance, human-guided robots like the Davinci robots from Intuitive Surgical facilitate minimally invasive surgical approaches, with the potential to perform remote surgery and in the future may be fully automated procedures. Modern medicine heavily relies on the technological advances made in the past decades and is interconnected with engineering.

1.6 Biofabrication

We can see that we have a convergence of engineering, biology, and materials science toward the development of tools and procedures to improve medical care at a mass level. Biofabrication comes at the interface of all these disciplines to propose an industrial approach for the engineering of living implants and biological products. Biofabrication can be defined as the production of complex living and non-living biological products from raw materials such as cells, molecules, extracellular matrices, and biomaterials by using industrial fabrication techniques. That is creating a framework of processes and techniques that will allow for the mass production and standardization of these biological products. The goal of biofabrication is to overcome the limitation of current tissue engineering procedures that have been successful for a single patient but to translate these procedures for the masses and as a therapy that can serve several patients.

In this book, we will present different aspects of biofabrication. The chapters of this book have been built to present methods and concepts of increased complexity in the manufacturing process. Starting with living tissue that is only composed of cells, we will then present methods using scaffolds to support the cell growth. Finally, we will discuss the limitation and challenges associated with the manufacturing of biological products and their translation to the clinical space.

1.7 Quiz

1. When dates the first record of an attempt at medical repair?
2. What is biofabrication?

3. What is tissue engineering?
4. What is the main limitation of organ transplantation?
5. What is the main limitation of prosthetics?
6. In the myth of Prometheus and Tityus, which organ was eaten by crows every night?
7. Which of its organ can the axolotl regenerate?
8. What is an allograft?
9. What is the difference between organ and tissue?
10. What are the requirements of implants?

References

[1]   Wells, K.M.; Kelley, K.; Baumel, M.; Vieira, W.A.; McCusker, C.D. Neural Control of Growth and Size in the Axolotl Limb Regenerate. *Elife* 2021, 10. https://doi.org/10.7554/eLife.68584.

[2]   International Olympic Committee. 4x400m Relay men https://olympics.com/en/olympic-games/london-2012/results/athletics/4x400m-relay-men.

[3]   Maradit Kremers, H.; Larson, D.R.; Crowson, C.S.; Kremers, W.K.; Washington, R.E.; Steiner, C.A.; Jiranek, W.A.; Berry, D.J. Prevalence of Total Hip and Knee Replacement in the United States. *J. Bone Jt. Surgery-American Vol.* 2015, 97 (17), 1386–1397. https://doi.org/10.2106/JBJS.N.01141.

[4]   Merola, M.; Affatato, S. Materials for Hip Prostheses: A Review of Wear and Loading Considerations. *Materials (Basel)* 2019, 12 (3), 495. https://doi.org/10.3390/ma12030495.

[5]   Evans, J.T.; Evans, J.P.; Walker, R.W.; Blom, A.W.; Whitehouse, M.R.; Sayers, A. How Long Does A Hip Replacement Last? A Systematic Review and Meta-Analysis of Case Series and National Registry Reports with More than 15 Years of Follow-Up. *Lancet* 2019, 393 (10172), 647–654. https://doi.org/10.1016/S0140-6736(18)31665-9.

[6]   Taboga, P.; Drees, E.K.; Beck, O.N.; Grabowski, A.M. Prosthetic Model, but Not Stiffness or Height, Affects Maximum Running Velocity in Athletes with Unilateral Transtibial Amputations. *Sci. Rep.* 2020, 10 (1), 1763. https://doi.org/10.1038/s41598-019-56479-8.

[7]   Murr, L.E.; Gaytan, S.M.; Martinez, E.; Medina, F.; Wicker, R.B. Next Generation Orthopaedic Implants by Additive Manufacturing Using Electron Beam Melting. *Int. J. Biomater.* 2012, 2012, 1–14. https://doi.org/10.1155/2012/245727.

[8]   Leong, D.; Sovari, A.A.; Ehdaie, A.; Chakravarty, T.; Liu, Q.; Jilaihawi, H.; Makkar, R.; Wang, X.; Cingolani, E.; Shehata, M. Permanent-Temporary Pacemakers in the Management of Patients with Conduction Abnormalities after Transcatheter Aortic Valve Replacement. *J. Interv. Card. Electrophysiol.* 2018, 52 (1), 111–116. https://doi.org/10.1007/s10840-018-0345-z.

[9]   Mohacsi, P.; Leprince, P. The CARMAT Total Artificial Heart. *Eur. J. Cardio-Thorac. Surg.* 2014, 46 (6), 933–934. https://doi.org/10.1093/ejcts/ezu333.

[10]   Menon, I.A.; Haberman, H.F. Dermatological Writings of Ancient India. *Med. Hist.* 1969, 13 (4), 387–392. https://doi.org/10.1017/S0025727300014824.

[11]   Greco, M.; Ciriaco, A.G.; Vonella, M.; Vitagliano, T. The Primacy of the Vianeo Family in the Invention of Nasal Reconstruction Technique. *Ann. Plast. Surg.* 2010, 64 (6), 702–705. https://doi.org/10.1097/SAP.0b013e3181d9aaad.

[12]   Reis, R.L. *Encyclopedia of Tissue Engineering and Regenerative Medicine*; 2019.

[13] Tan, S.Y.; Shigaki, D. Emil Theodor Kocher (1841–1917): Thyroid Surgeon and Nobel Laureate. *Singapore Med. J.* 2008, 49 (9), 662–663.

[14] Cao, Y.; Vacanti, J.P.; Paige, K.T.; Upton, J.; Vacanti, C.A. Transplantation of Chondrocytes Utilizing a Polymer-Cell Construct to Produce Tissue-Engineered Cartilage in the Shape of a Human Ear. *Plast. Reconstr. Surg.* 1997, 297–302; discussion 303–304. https://doi.org/10.1097/00006534-199708000-00001.

[15] Atala, A.; Bauer, S.B.; Soker, S.; Yoo, J.J.; Retik, A.B. Tissue-Engineered Autologous Bladders for Patients Needing Cystoplasty. *Lancet* 2006, 367 (9518), 1241–1246. https://doi.org/10.1016/S0140-6736(06)68438-9.

[16] Acevedo Rua, L.; Mumme, M.; Manferdini, C.; Darwiche, S.; Khalil, A.; Hilpert, M.; Buchner, D.A.; Lisignoli, G.; Occhetta, P.; von Rechenberg, B., et al. Engineered Nasal Cartilage for the Repair of Osteoarthritic Knee Cartilage Defects. *Sci. Transl. Med.* 2021, 13 (609). https://doi.org/10.1126/scitranslmed.aaz4499.

2 Scaffold-free biofabrication

2.1 The challenges of biomaterials

Biofabrication approaches often use a biomaterial that is processed into a scaffold that supports the growth and organization of cells into functional tissues. However, this approach is not perfect and has some challenges. So, the hypothesis that it is required to have a scaffold for the cells to grow, is not always supported by experiments. Indeed, cells can generate their scaffold by secreting macromolecules which form the extracellular matrix (ECM) in the tissue. If we can induce the formation of natural ECM by the cells *in vitro*, then we do not need biomaterials to support the cells. Biomaterials, while providing support for the cell to grow, generate new challenges. Some biomaterials can induce an immune response and so the scaffold made up of the biomaterials can be seen as a foreign object and can induce an adverse immune response. The second challenge is that the biomaterials' biodegradability needs to be taken into consideration. For some applications, a degradable scaffold might be beneficial, for others not. If a biodegradable scaffold is required, then the degradation rate needs to be optimized for each application. Finally, the selected biomaterials need to be sterilizable by conventional industrial sterilization methods such as autoclave, gamma irradiation, or ethylene oxide. Beyond these challenges, scaffold enables to provide support and a shape to direct the cell growth. This support can be crucial to provide mechanical support for the cells to grow and mature. In the absence of mechanical support, a higher amount of cell is required to create mechanical support based on the secretion of natural ECM macromolecules such as hyaluronic acid, proteoglycan, and collagen.

2.A Extracellular matrix (ECM)

The ECM is a three-dimensional network that is composed of macromolecules, minerals, growth factors, and enzymes (Figure 2.1). The network is created by fibrillar and hydrogel-forming macromolecules such as collagen, other hydrogel-forming macromolecules such as fibronectin, or non-hydrogel forming such as heparin and hyaluronic acid. Also present in the hydrogel is cell-generated enzymes such as matrix metalloproteinase that are metal activated (usually zincs) that can cleave the macromolecules of the ECM. Other cells-generated signaling molecules are present in the ECM such as growth factors that can interact with the macromolecules like heparin. The ECM is created by the cells and it is critical for binding cells together and creating functional and mechanically stable synthetic tissues.

https://doi.org/10.1515/9781501515736-002

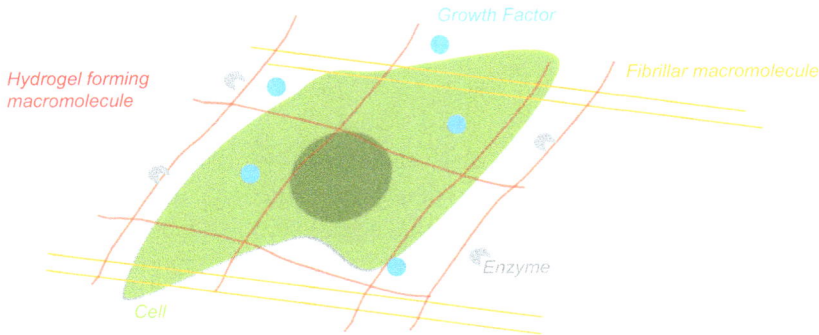

Figure 2.1: The cell and its extracellular matrix.

2.2 Bioreactors

At the laboratory scale, cells are cultivated in sterile containers that provide adhesion for cells. These containers, are coming in different shapes and volumes. For low volumes, cells are generally cultivated in microplates that are constituted of wells capable of holding volumes as low as a couple of microliter up to several milliliters. These plates are made of poly(styrene) that is gamma irradiated to induce cell adhesion and sterilization in a single process. The microplates are designated by the number of wells typically from 96 to 6 wells per plate at the laboratory scales translating in volumes ranging from 100 μL to 10 mL. Then for higher volumes, T-flasks are used. These consist of a trapezoidal shape that can hold volumes from 40 to 400 mL. The cells are cultivated on the bottom of these flasks which can be stacked in the incubator. These containers are mainly used for adherent cells such as fibroblast (skin cells) or endothelial cells (blood vessel cells). For cell conditions where adhesion is not required, one can use non-adherent conical tubes of different volumes. These tubes are made of poly(propylene) and come in two different shapes of tube: Eppendorf's tubes ranging from 0.5 to 2 mL and Falcon's tube from 15 to 50 mL. The conical shape allows for the recovery of cells after centrifugation. These flasks can sustain a centrifugal force of up to 30,000 g and are also used for the recovery of protein and other molecules (Figure 2.3).

In increasing the number of cells, bigger containers are needed, so-called bioreactors. These are also used in a laboratory setup. They are made of poly(propylene) or glass and can range from 125 mL to several liters. Processes optimized with these bioreactors then be scaled up to industrial setup using tanks of several hundred liters. The bioreactor is a cylinders with several opening on the top that allows the changing of the culture media and are equipped with a mechanical stirrer that keeps the cells in suspension. Indeed, without agitation, cells will fall at the bottom of the reactor by gravity. However, some type of cells only grows when they are attached to a support, and it is

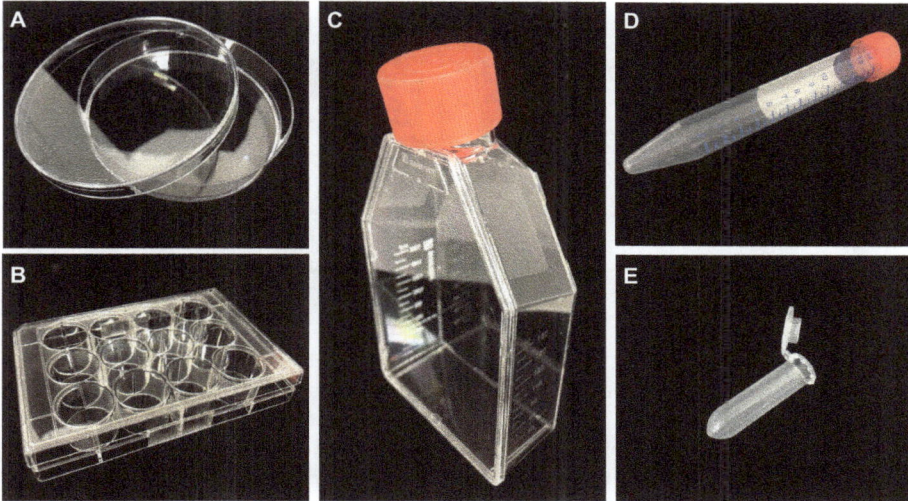

Figure 2.2: Typical plastic ware used in cell biology for the cell culture. (A) Petri dish laboratory and (B) well plate, (C) T flask, (E) Eppendorf's tube and (D) Falcon tubes.

not possible to grow them in suspension. So, to increase the culture surface area from the T-flask planar format but reduce the footprint of the reactors, small particles on which cell scan attached to are added in the bioreactors. The spheres or particles are usually made of polymers of 500 μm diameter to which the cells can attach, thus increasing the surface area. Commercial setup to further increase the available surface area for cell culture consists of fibrous bioreactors. These are presented as cylinders packed with micrometer diameter and parallelly aligned fibers to which the cells can attach (Figure 2.4). Between the fibers, the cell culture media can be circulated, and the product of the cells culture can be extracted, such as antibodies. The fibers can be plain or hollow, thus further increasing the available culture surface. Inherited from biotechnology and vaccine production, these two techniques allow for scaling up the production of cells and obtaining a large number of cells which can be then used for biofabrication.

A   B

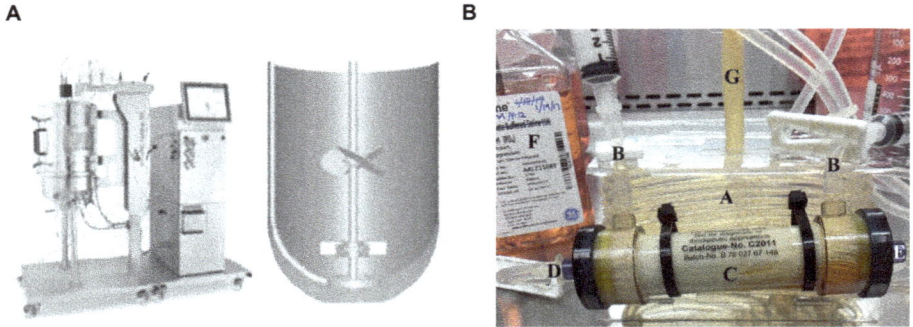

Figure 2.3: Scaling up cell culture. (A) The stirred single-use cell culture bioreactors from BIOSTAT CultiBag STR with a working volume of 50 L [1]. (B) Components of a typical hollow fiber bioreactor setup. The entire setup, including the perfusion pump, is within a temperature-controlled cell culture incubator. (a) gas exchange circulator; (b) Luer-Lock harvest and injection ports (shown with syringes attached); (c) hollow fiber cartridge; (d) circulating media inlet; (e) circulating media outlet; (f) complete media reservoir; (g) tubing interface to perfusion pump (pump not shown) [2].

❗ 2.B Cell surface adhesion mechanism

On tissue culture surfaces such as plastic Petri dishes, the plastic of the surface is treated to be hydrophilic. Molecules present in the cell culture media such as albumin will form a coating on the substrate. Then cells suspended in this media will sediment on top of these coated surfaces and adhere to the adsorbed molecules.

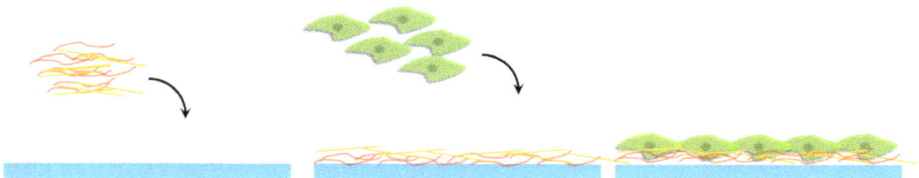

Figure 2.4: Cell-adhesion mechanism. First, protein adsorbs on the cell culture substrate. Subsequently, cells attached to the proteins coating the surface of the cell culture substrate.

2.3 From single cells to tissue

2.3.1 Non-adherent mold

Once a critical number of cells is obtained, we need to organize them into tissue without using a scaffold. Some types of cells can organize by themself through gravitation into a spheroid when they are cultivated on a non-adherent plate. This is the case for chondrocytes which are cartilage cells and many cancer cells. Others need some help and need to be forced into aggregation to form a tissue. To achieve this, one can use molds that are non-adhesive which will force cells to find each other and organize into a predictable shape. To do that the mold will restrict space and guide the gravitation of the cells (Figure 2.5). Typically, these molds can be made of poly(dimethylsiloxane) or hydrogel such as agarose. These molds can be obtained by making a negative shape using typical manufacturing techniques or 3D printed molds. The negative mold is then used to make the mold that will get in contact with the cell. This technique is called replica molding and is widely used across different industries to make a molded part. This approach can be used to create cell aggregates in a shape of a ring by using a doughnut mold. More advanced architecture can be obtained such as honeycomb by using a mold with an adequate shape. However, the cells need to be cultured long enough to have the produced the enough ECM to provide them mechanical strength to sustain their shape once they are removed from the mold. If removed too early, then the cells do not have the time to create the needed ECM macromolecules to maintain the shape and the cell aggregate will collapse.

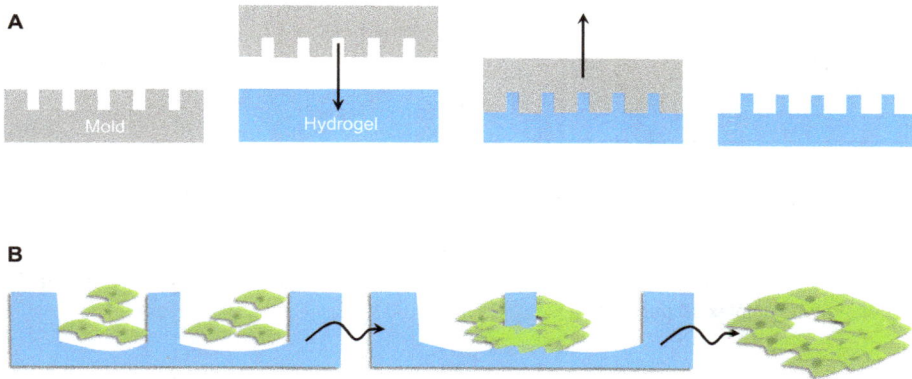

Figure 2.5: Controlling cell organization with a mold. (A) The mold is created using a replica molding technique. A negative mold is manufactured and used to stamp a hydrogel precursor liquid. The hydrogel is set, and the mold is removed leaving its imprint in the hydrogel. (B) Using a circular non-adhesive mold cells (green) can be organized in the shape of doughnuts recreating hollow cellular organization observed in blood vessels for instance.

2.3.2 Acoustic wave

Going a step forward is to get rid of the mold and directly position individual cells precisely without contact. This can be achieved by using acoustic waves to move particles and cells into predictable shapes. This approach is quite interesting, at first, it is highly versatile so the manufacturing can be made instantaneously without the need to prefabricate a mold. Second, it reduces the risk of contamination through contact with the mold or contamination with the contact of the mold materials (Figure 2.6). Emerging from this technology, acoustic wave generators are capable of organizing cells into 3D morphogenic patterns. Such instruments are capable of creating a standing wave that is generated when a wave traveling in one direction is interacting with a wave traveling in the opposite direction. The overlapping wave is called a standing wave which oscillates in time but not in space. By changing the frequency and amplitude of the acoustic waves, the object trapped in the nodes and antinodes can be displaced within the media [3].

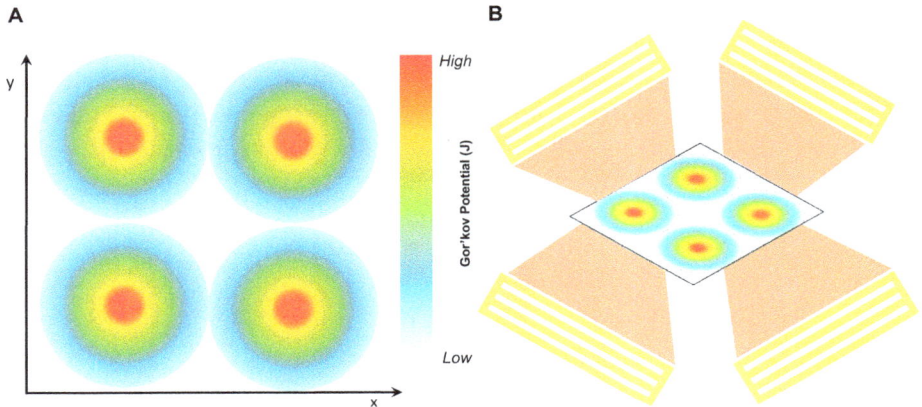

Figure 2.6: (A) The patterns have high and low potentials that permit to precisely position the cells within the nodes of the standing wave. (B) Acoustic wave setup composed of piezo-activated actuator that applies a wave at a set frequency and amplitude. In the container, an acoustic standing wave will be generated by using two opposite piezo generators. The four actuators enable the creation of complex 3D patterns.

2.4 Cell sheet

While 3D architecture is interesting for several types of tissue biofabrication approaches, some therapeutic applications preferably deliver cells as a sheet. As an example, skins replicates can be created and applied as a sheet of cells to cover a wound, but also cardiomyocytes sheets applied as remediation of myocardial infarct. To make cell sheets, one needs to be able to grow cells on a surface, easily achieved using conventional cell culture technique, but to remove the cells as a single mechanically stable living tissue is challenging. One method proposed to do this is by using a temperature-responsive surface coating that lifts the whole-cell sheet together without altering the biology of the cell by using cleaving enzymes such as trypsin.

The material poly(N-isopropyl acrylamide) (poly(NIPAAM)) exhibits a lower critical solution temperature (LCST). This means that above the LCST the polymer is hydrophobic. It will then form a thin coating on a petri dish on to which proteins can attach. Below the LCST, the polymer chains are extended and soluble in water. Which means that the poly(NIPAM) can detach from the tissue culture plate and lift with them the proteins that had attached to it (see knowledge box cell adhesion mechanism). As a result, the cell sheet sheet is lifted with the deposited ECM and can be harvested as a single piece of living tissue (Figure 2.7). Once the cell sheet is obtained, the sheets can be combined to reproduce different cell layers, for instance, the different layers of the skin reproducing the epidermis and the dermis. Using the sheet manufacturing approach applied to cardiomyocyte allows for creating several layers of cardiac cells that have enough mechanical strength to be sutured to the heart.

Going further with the sheet technique of cardiac tissue, one can try to reproduce the striation anatomy of the cardiac muscle. To reproduce the linear pattern of the cardiac tissue, cardiomyocytes can be cultured on a mold that forces the cells to align into natural striation. On poly(dimethylsiloxane), PDMS, molds, the cells cannot adhere and can be then lifted and detached into a mechanically stable sheet of cells that can be sutured into a cardiac patch. In this patch, the cells can be stimulated to beat like a cardiac muscle.

Another application for cell sheets is skin sheets. The technique consists in creating layers of fibroblast, keratinocytes, and melanocytes by alternative immersion of the cell sheets. The dermal layer is made by seeding fibroblast on a flat surface that is immersed in cell culture media. Once matured, the sheet is raised and the second layer constituted of keratinocyte is added on top of the fibroblast layer. The multicell culture sheets are then immersed in a culture media for maturation of the cells by exposing the layers to the air–liquid interface. Developed in the 1990s, these skins cells are used for cosmetic testing, Indeed, since 2009 it is forbidden to use animals for cosmetic testing, and the artificial skins are now the industry standard for testing the safety of molecules used in cosmetic formulations. Current research aims at developing more advanced models including hair and nerve into the model.

A At 37°C

B At 10°C

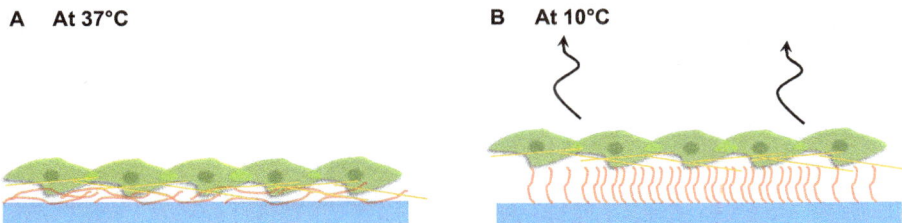

Figure 2.7: Temperature-actuated coating for cell sheet manufacturing. The petri dish is coated with poly(NIPAAM) polymer. At the cell incubation temperature (37 °C), the poly(NIPAAM) polymer forms a thin sheet coating. As the temperature is lowered, below the LCST, the polymer becomes hydrophilic and detached from the surface allowing to lift the cell layer with its ECM. The cell sheet can be detached and further used.

2.C Trypsinization

Once the cells have attached to the bottom of the tissue culture substrate, they will secrete macromolecules of the ECM and form strong bonds with the surface. To remove them, one needs to cut these bonds. The enzyme trypsin digests proteins by cutting the bonds between the amino acids. The coated proteins being cut becomes inefficient and the cells will detach and float in the media suspension. This process is called trypsinization.

Figure 2.8: Action of the enzyme trypsin on the adsorbed molecules.

2.5 Biofabrication of spheroids

Spheroids are cells organized into spherical shapes. They are particularly interesting for the simulation of cancer tissues and cartilages [4]. To manufacture these spheroids, you need to force the cells to come together into a sphere. Naturally, cells will fall by gravity when suspended in a cell culture media, but not every cell type can form spheroids.

2.5.1 Hanging drop

Spontaneous aggregation into spheroids is a well-characterized and established methodology due to its ability to generate reproducible spheroids and their similarity to near-native tissue [5]. This methodology of spheroid aggregation has been exploited by the technique of hanging drop. In this method, a small droplet of cell suspension (single cell type or multiple cell type systems can be used) is pipetted onto the lid of a multiwell plate that due to the surface tension between the lid material and the water holds the droplet in place following its inversion (Figure 2.9). Density and type of cell suspension can be varied based on desired spheroid size or type. Following pipetting and as a function of time, the cells within the droplet accumulate at its tip, self-aggregate at the liquid-air interface, and eventually proliferate. During this technique, moisture/humidity within the plate is maintained to prevent drying of the droplet containing the cell suspension. This technique has the advantage of being simple, inexpensive, and reproducible in terms of generating a single spheroid per droplet and results in tightly packed spheroids of cells.

2.5.2 Liquid overlay technique

Figure 2.9: The hanging drop method is where individual cells are suspended in a droplet of media that is positioned on an inverted petri dish lid. The gravity forces the cells to come together into a spheroid.

As an alternative to the hanging drop methods, one can use a non-adhesive or attachment limiting substrate like agarose or non-adhesive petri dish to force cell aggregation. The cells added on these surfaces are forced to generate cell–cell attachment because they cannot generate cell–substrate attachment. This is a two-step technique: first, the cells are grown on non-adhesive tissue culture plates, wherein the cells migrate toward each other resulting in cell–cell aggregation and spheroid formation. In the second stage, the cellular aggregates grow. Several non-adhesive materials such as poly(2-hydroxyethyl methacrylate), polyethylene glycol, and agar can be used. While this is a simple technique, it suffers from several disadvantages including heterogeneity in size and cell number.

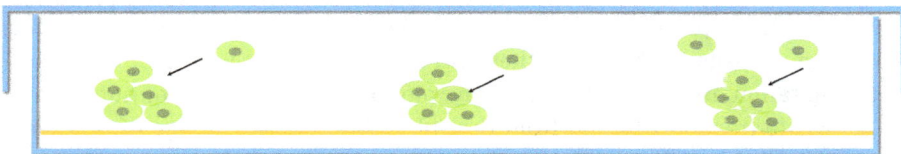

Figure 2.10: Using a non-adhesive coating (yellow), the cells cannot attach to the bottom of the cell and are forced to aggregate together.

In some cases, the coating can be a temperature-responsive material such as poly(NI-PAAM). As previously described, the cells can adhere to it and form a uniform sheet. When the temperature is lowered, the polymer solubilizes and the cells detach. Either the cells will stay together and form a sheet of cells or they will role and aggregated into spheroids. This depends on the type of cells and incubation conditions.

Figure 2.11: Temperature-sensitive poly(NIPAAM) forms an adhesive layer at 37 °C. At low temperature, the polymer solubilizes in water and the cells detached and aggregate into spheroids.

To obtain spheroids of homogeneous size, one can combine the coating technique with microwells that will limit the spheroid growth and size. For some cell types that naturally aggregate, microwells are sufficient to direct the formation of spheroids. But for some other cell types, it is challenging to induce the formation of spheroids. For instance, when the cell–cell attachment is not favored over the cell–substrate adhesion. In this case, non-adherent coating of the microwell is needed. This has the advantage to overcome the limitation of using only the non-adherent coating and enabling to control the spheroid size [6].

Figure 2.12: (left) Microwells that restrain the spheroid size and (right) same microwells that are coated with a non-adherent substrate that forces the cell aggregation.

2.5.3 Active rotational method

The methods presented above are classified as passive methods. But active methods are often needed to obtain spheroids reliably at a large scale. To this effect, spinner flask and rotary cell culture system (RCCS) are two of the agitation techniques used for growing cells in suspension culture (Figure 2.13). In the spinner flask, the cells are placed in a bioreactor that contains an impeller that maintains the cells in suspension. The cells then will come in contact with each other during the agitation and form spheroids. The motion provides a flux of nutrients as well as waste transport to and from the spheroids. The advantage of this technique is that it is easily scalable to the industrial scale of spheroid production, but one must take into consideration the shear forces generated by this technique which may alter the cell physiology. However, this method can result in the inconsistent size of spheroids [7].

The RCCS was introduced by NASA in 1992 to mimic microgravity [8]. This consists of a vessel that rotates on its horizontal axis and like in the spinner flask the constant rotation prevents the cells from adhering to the chamber walls. The culture vessel is filled with culture medium, inducing low fluid turbulence as well as low shear forces. Low fluid turbulence is provided in RCCS in comparison to the spinner flask as it reduce the shear forces induced to the cells.

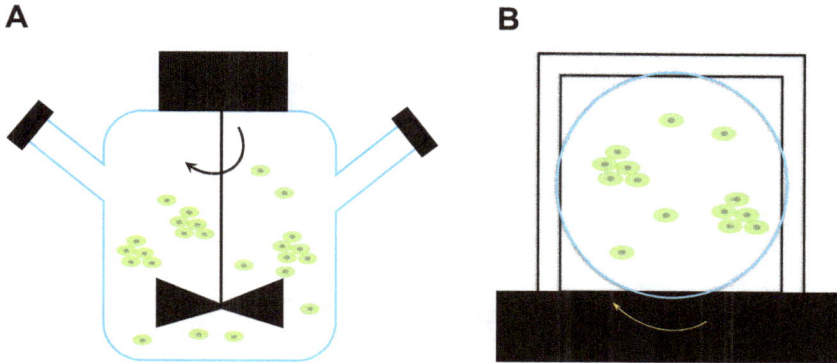

Figure 2.13: The two rotational techniques (**left**) are the spinning flask and (**right**) the rotary cell culture system (RCCS).

2.5.4 Microcarrier beads

Formation of cell aggregate can be obtained by centrifugation of a cell suspension in a Falcon tube or Eppendorf's tube. But this leads to a non-control spheroids size (Figure 2.14). Alternatively, microcarriers are spherical beads of diameter around 500 µm that provide a defined surface area for the culture of a high density of cells in smaller volumes. This technique is particularly adapted for cells that do not aggregate spontaneously [9]. Microcarrier beads can be either solid or porous. Once the cells adhere, it is followed by cell growth. Some microbeads can degrade leading to the formation of a cellular aggregate. Non-degradable beads can also be used, but then the beads remain within the spheroid core. Microcarrier beads can also be coated with cell adhesion-promoting materials like gelatin, collagen, or laminin.

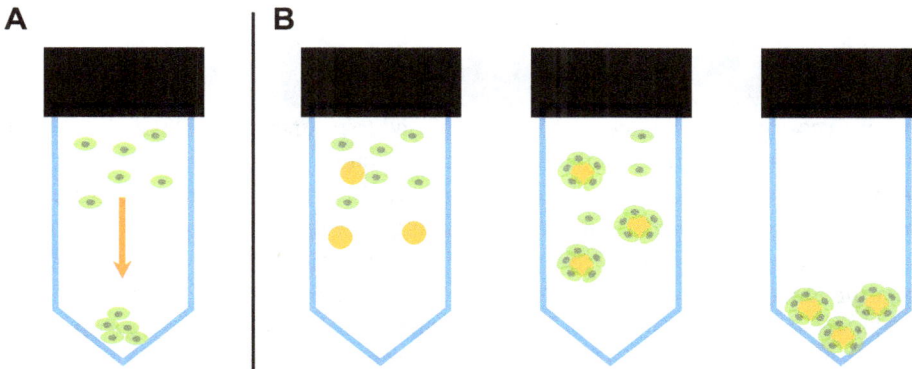

Figure 2.14: (**left**) Cells are forced to aggregate by centrifugal force. (**right**) Use of cell-adhesive microcarrier that aggregates the cells into spheres of equal size.

2.D Cell lineage

The cell lineage is the developmental history of a cell that can be traced back to the cell from which it comes. From the fertilized embryo, four main lineages are created which give the different cells observed in the adult organism. The current focus of research aims at isolating cells from abundant tissue that can be easily isolated such as fat and skin and manipulating the cells to create a different type of cells. However, biological limitations restrain the jump between different primary germ layers. To date, it is not possible to create a skin cell from a muscle cell (Figure 2.15).

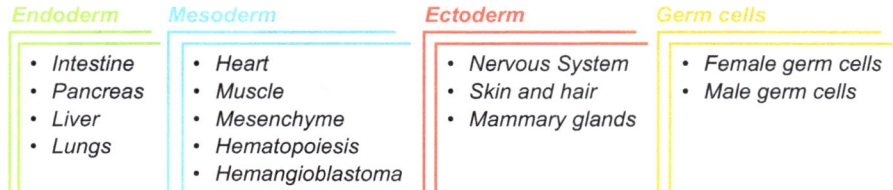

Endoderm	Mesoderm	Ectoderm	Germ cells
• Intestine	• Heart	• Nervous System	• Female germ cells
• Pancreas	• Muscle	• Skin and hair	• Male germ cells
• Liver	• Mesenchyme	• Mammary glands	
• Lungs	• Hematopoiesis		
	• Hemangioblastoma		

Figure 2.15: The four different cell lineage.

Mesenchymal stem cells are cells that can be harvested from adult individuals and differentiated into several types of cells of the mesoderm. They are isolated from different tissues including bone marrow, umbilical cord, and adipose tissues. Because mesenchymal stem cells can multiply without differentiation, these cells open the possibility to generate a high number of cells that can then be differentiated into a particular tissue (Figure 2.16).

Mesenchymal Stem Cells (MSCs)

Cell	Osteoblast	Chondrocyte	Muscle cell	Epithelial cell	Adipocyte
tissue	bone	cartilage	muscle	skin	fat

Figure 2.16: The lineage of mesenchymal stem cells.

2.5.5 Magnetic levitation

In some cases, magnetic nanoparticles can be used to magnetize a cell population and then use magnetic force to manipulate the cells, Figure 2.17 [10]. This technique has been used for the generation of 3D tumor spheroids that are reminiscent of tumors *in vivo*. Usually, nanoparticles made of iron oxide (Fe_3O4)-encapsulated in poly(lactic-co-glycolide) polymer are internalized by the cells, which become magnetic. Apart from the monoculture of cells using the magnetic levitation method, it has also been used for co-culturing to force cells of different origins to come together. While this technique utilizes magnetic nanoparticles that could potentially affect cell viability, the use of magnetic nanoparticles has been approved for imaging agents. Nevertheless, the use of iron oxide as 3D spheroid models, in drug testing as well as tissue engineering, needs to be on a case-by-case toxicologically assessed.

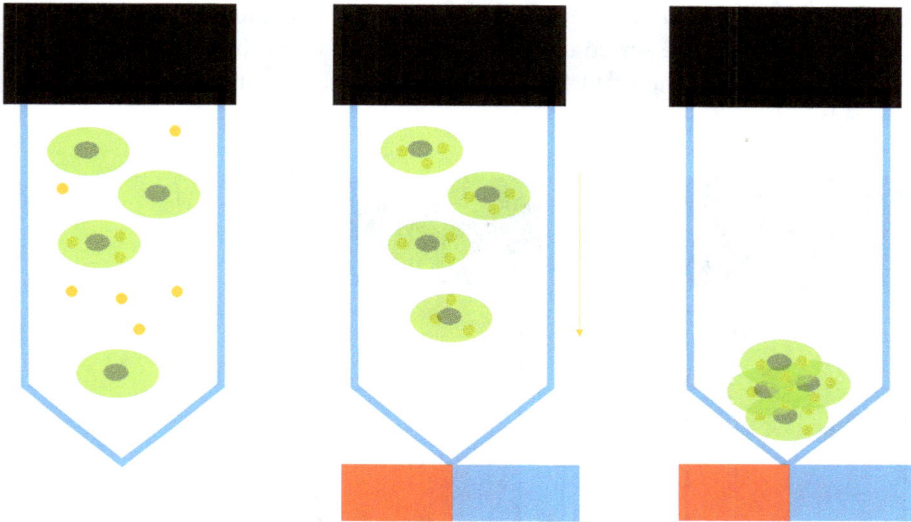

Figure 2.17: Magnetic nanoparticles (yellow) are incubated and uptaken by the cells. Upon exposure to a magnetic field, magnetized cells are attracted together to form a spheroid.

2.5.6 Acoustic aggregation

We have reviewed in the previous section that acoustic waves can be used to organize single cells into morphogenic structures. Similarly, spheroids can be created by forcing the cells to aggregate into gravity holes formed by an acoustic wave. Starting from a single cells suspension, an acoustic wave is generated that consists of peaks and depression in the aqueous culture media. Cells will then concentrate in the valleys and have the chance to aggregate into spheroids. Once these spheroids are created they can be further manipulated to form more complex structures using a different acoustic wave that will move the freshly created spheroids into a new pattern that will force the aggregation into tissue-like structures [11].

2.5.7 Emulsion

Emulsions are used in many industrial processes from polymer synthesis to food manufacturing. This technique uses the dispersion of water in oil and permits the creation of round structures that are isolated from each other. Applied to cell culture, cells can be dispersed into a water phase that is processed into an emulsion in oil. Of course, in this application, the oils need to be non-cytotoxic and easily removable after the process to recover the cells. Cells are then encapsulated into the water phase and are forced to meet each other as they cannot escape into the oil phase. This emulsion creates several droplets like the hanging drop technique but in a process that can be scaled up in bioreactors. The water droplet size can be adjusted by optimizing the emulsion thus permitting the creation of homogeneous spheroids [12].

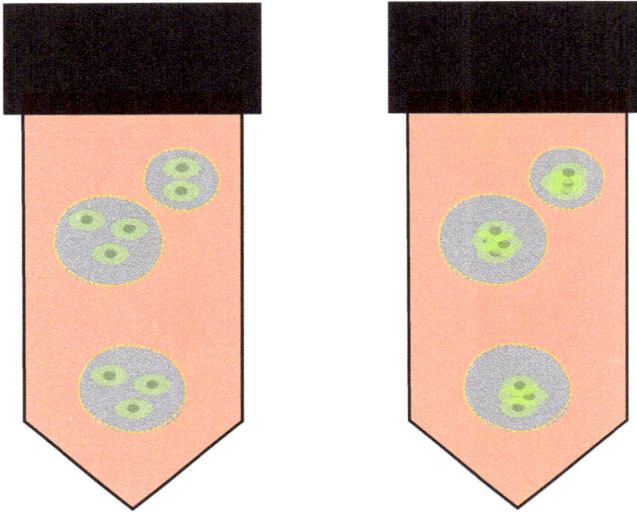

Figure 2.18: Emulsions of water (**blue**) in oil (**red**) are created which confines the cell in a set diameter.

2.5.8 Applications of spheroids

From the generation of organoids for drug testing to the formation of biological transplants, spheroids have a broad range of applications. One of the major challenges during drug development is to assess the liver toxicity of a drug. Therefore, being able to test the drug on a cellular organism that could accurately predict the human response *in vitro* would be of great benefice. Spheroid constituted of Kupffer cells, stellate cells, and biliary cells organized into spheroids have been shown to stimulate the effect of drugs on the liver [13]. In cancer studies, spheroids have played a critical role as they enable the reproduction of the necrotic center observed in tumors. The lack of oxygen in cancer tumors leads to the death of cells or a change in physiology. This can be reproduced *in vitro* using cancer tumor spheroids that are grown to a critical size [14]. Cancer cell spheroid when they reach a critical diameter, the core of the spheroid lacks of oxygen and becomes necrotic.

Apart from drug testing, spheroids can further be used as a building block to create a more complex or bigger tissue. For instance, spheroids are used as a building block for 3D printing using the needle array, called the Kenzan needles to form tubular structures. Spheroids are prepared in microwells and then strung on needles that are arranged in the shape of the attended tissue. For instance, the needles are organized into a cylinder and on each needle, spheroids are placed layer after layer creating first a circle and then a cylinder [15] (see Chapter on Bioprinting). The technique permits to rapidly create tissue with a high number of cells. The spheroids can be further manipulated using a 3D bioprinter that will position the spheroids into a particular shape. Using different spheroids complex tissues can be formed by the association of the organoids together. This technique was developed by Organovo and was attended to make liver replicates and other types of tissue to be used for drug testing and development.

2.6 Quiz

1. What is the major drawback of biomaterials for cell culture?
2. What is the ECM?
3. Why are spheroids important for cell structure?
4. Which additive manufacturing technique uses spheroids?
5. What are the four cell lineages?
6. What is trypsin and its role in cell culture?
7. What is the behavior of a polymer that has a LCST?
8. What is an example of polymer with LCST and one of its conditions applications?
9. Which spheroid manufacturing technique can simulate space conditions?
10. Which polymer can be used for a non-adherent coating?

References

[1] Löffelholz, C.; Kaiser, S.C.; Kraume, M.; Eibl, R.; Eibl, D. Dynamic Single-Use Bioreactors Used in Modern Liter- and M3- Scale Biotechnological Processes: Engineering Characteristics and Scaling Up. 2013, 1–44. https://doi.org/10.1007/10_2013_187.

[2] Yan, I.K.; Shukla, N.; Borrelli, D.A.; Patel, T. *Use of a Hollow Fiber Bioreactor to Collect Extracellular Vesicles from Cells in Culture*; 2018; pp 35–41. https://doi.org/10.1007/978-1-4939-7652-2_4.

[3] Guex, A.G.; Di Marzio, N.; Eglin, D.; Alini, M.; Serra, T. The Waves that Make the Pattern: A Review on Acoustic Manipulation in Biomedical Research. *Mater. Today Bio.*, **2021**, 10, 100110. https://doi.org/10.1016/j.mtbio.2021.100110.

[4] Arya, N.; Forget, A.; Sarem, M.; Shastri, V.P. RGDSP Functionalized Carboxylated Agarose as Extrudable Carriers for Chondrocyte Delivery. *Mater. Sci. Eng. C* **2019**, 99, 103–111. https://doi.org/10.1016/j.msec.2019.01.080.

[5] Tung, Y.C.; Hsiao, A.Y.; Allen, S.G.; Torisawa, Y.S.; Ho, M.; Takayama, S. High-Throughput 3D Spheroid Culture and Drug Testing Using a 384 Hanging Drop Array. *Analyst* **2011**, 136 (3), 473–478. https://doi.org/10.1039/c0an00609b.

[6] Valdoz, J.C.; Jacobs, D.J.; Cribbs, C.G.; Johnson, B.C.; Hemeyer, B.M.; Dodson, E.L.; Saunooke, J.A.; Franks, N.A.; Poulson, P.D.; Garfield, S.R.; et al. An Improved Scalable Hydrogel Dish for Spheroid Culture. *Life* **2021**, 11 (6), 517. https://doi.org/10.3390/life11060517.

[7] Hirschhaeuser, F.; Menne, H.; Dittfeld, C.; West, J.; Mueller-Klieser, W.; Kunz-Schughart, L.A. Multicellular Tumor Spheroids: An Underestimated Tool Is Catching up Again. *J. Biotechnol.* 2010, 148 (1), 3–15. https://doi.org/10.1016/j.jbiotec.2010.01.012.

[8] Li, S.; Ma, Z.; Niu, Z.; Qian, H.; Xuan, D.; Hou, R.; Ni, L. NASA-Approved Rotary Bioreactor Enhances Proliferation and Osteogenesis of Human Periodontal Ligament Stem Cells. *Stem Cells Dev.* **2009**, 18 (9), 1273–1282. https://doi.org/10.1089/scd.2008.0371.

[9] Jong, B.K.;; Three-Dimensional Tissue Culture Models in Cancer Biology. *Semin. Cancer Biol.* 2005, 15 (5 SPEC. ISS.), 365–377. https://doi.org/10.1016/j.semcancer.2005.05.002.

[10] Lewis, N.S.; Lewis, E.E.L.; Mullin, M.; Wheadon, H.; Dalby, M.J.; Berry, C.C. Magnetically Levitated Mesenchymal Stem Cell Spheroids Cultured with a Collagen Gel Maintain Phenotype and Quiescence. *J. Tissue Eng.*, **2017**, 8, 204173141770442. https://doi.org/10.1177/2041731417704428.

[11] Chen, K.; Wu, M.; Guo, F.; Li, P.; Chan, C.Y.; Mao, Z.; Li, S.; Ren, L.; Zhang, R.; Huang, T.J. Rapid Formation of Size-Controllable Multicellular Spheroids via 3D Acoustic Tweezers. *Lab Chip* 2016, 16 (14), 2636–2643. https://doi.org/10.1039/C6LC00444J.

[12] Chan, H.F.; Zhang, Y.; Ho, Y.-P.; Chiu, Y.-L.; Jung, Y.; Leong, K.W. Rapid Formation of Multicellular Spheroids in Double-Emulsion Droplets with Controllable Microenvironment. *Sci. Rep.* 2013, 3 (1), 3462. https://doi.org/10.1038/srep03462.

[13] Bell, C.C.; Hendriks, D.F.G.; Moro, S.M.L.; Ellis, E.; Walsh, J.; Renblom, A.; Fredriksson Puigvert, L.; Dankers, A.C.A.; Jacobs, F.; Snoeys, J.; et al. Characterization of Primary Human Hepatocyte Spheroids as a Model System for Drug-Induced Liver Injury, Liver Function and Disease. *Sci. Rep.* **2016**, 6 (1), 25187. https://doi.org/10.1038/srep25187.

[14] Arya, N.; Forget, A. *Biomaterials Based Strategies for Engineering Tumor Microenvironment*; 2017; Vol. 66. https://doi.org/10.1007/978-981-10-3328-5_8.

[15] Arai, K.; Murata, D.; Verissimo, A.R.; Mukae, Y.; Itoh, M.; Nakamura, A.; Morita, S.; Nakayama, K. Fabrication of Scaffold-Free Tubular Cardiac Constructs Using a Bio-3D Printer. *PLoS One* **2018**, 13 (12), 1–17. https://doi.org/10.1371/journal.pone.0209162.

3 Organic and inorganic materials for biofabrication

3.1 What is a biomaterial?

A biomaterial is usually defined as a natural or synthetic material that is suitable for introduction into living tissue especially as part of a medical device. With this definition, we can conclude that biomaterials are non-cytotoxic and can be of different origins. But one must be careful on how to qualify a biomaterial. Indeed, a material can be biocompatible for a specific application but toxic for another application. The qualification is dependent on the usage and site of insertion. Following this principle, regulatory agencies do not qualify biomaterials for implantation, but medical devices made of biomaterials for a specific application. Hence comes the question of what sort of biomaterials can be used for biofabrication. One of the main uses of biomaterials in biofabrication is to make scaffolds that can be used to direct the organization of cells into tissues. So, biomaterials can be processed into scaffold, hydrogel, or surfaces that can be used as a cell substrate for cell culture. Biomaterials are classified into three major families: polymer, ceramic, and metal (figure 3.1). Each of these materials is used for implant or medical devices either as a degradable or non-degradable implant.

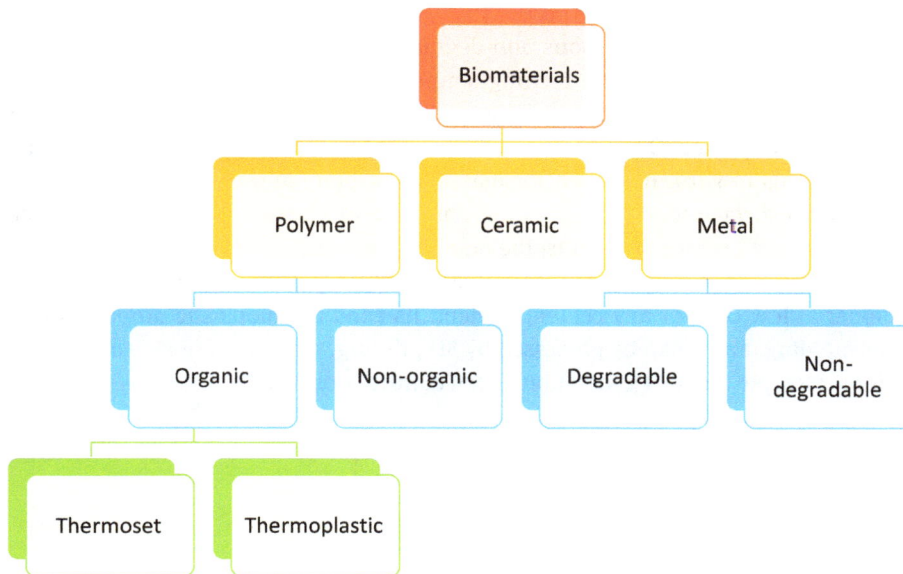

Figure 3.1: The subfamilies of biomaterials.

https://doi.org/10.1515/9781501515736-003

3.2 Organic polymers

3.A Polymer lexicon

Glass transition (T_g): The glass transition is the reversible transition in an amorphous or semi-crystalline polymer from a hard and brittle state into a viscous or rubbery state by increasing the temperature. The reverse transition, from a viscous liquid into the glass state, is called vitrification.

Melting temperature (T_m): The melting point of a polymer is the temperature at which it changes from solid state to liquid and all the chains are in a flow motion.

Thermoplastic: is a plastic polymer material that becomes pliable or moldable at a certain elevated temperature and solidifies upon cooling. As such, the polymer can be molded by pressure and heat. These materials are particularly suitable for mass production as they can precisely and rapidly process molded parts. Once molded the polymer part can be reproduced and sometimes recycles into other parts.

Thermoset: is a polymer that is obtained by the irreversibly hardening or curing of a soft polymer or viscous liquid prepolymer often called a resin. The resin or flowing polymer is introduced in a mold. Then, the curing, hardening, or setting of the polymer resin can be induced by heat or radiation or promoted by high pressure or mixing with a catalyst. Once the materials are cured, the thermoset polymers cannot be remolded.

3.2.1 Poly(styrene)

Poly(styrene) (PS) is an amorphous non-degradable thermoplastic with a glass transition (T_g) around 100 °C and a melting temperature of about 240 °C (figure 3.2). It is obtained by an anionic radical polymerization that allows getting high molecular weight polymer chains. Over the years, PS has become the material of choice for the culture of cells *in vitro*. It is used for making well plate and other plastic ware. It can undergo surface treatment in either chemical or plasma to modify its surface chemistry which in turn modulates the adhesion of cells. It can be manufactured at low cost, sterilized in bulk by gamma radiation, which permits to guarantee a sterile container at low cost for any cell experiment. Its good thermoplastic properties permit its molding, and it can be processed by 3D printing, thus opening new opportunities for prototypic the next generation of cell culture substrates.

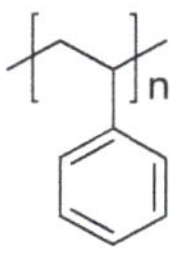

Figure 3.2: Chemical structure of poly(styrene).

3.B Anionic radical polymerization of styrene

The polymerization of styrene occurs through an anionic radical polymerizaticn where two active sites are built on a chain and the polymer chain growth by both extremities (figure 3.3). (1.) The polymerization is activated with naphthalene, (2) the anionic radical is then transferred to the styrene monomer, and (3) the chain is grown, and the polymerization is propagated.

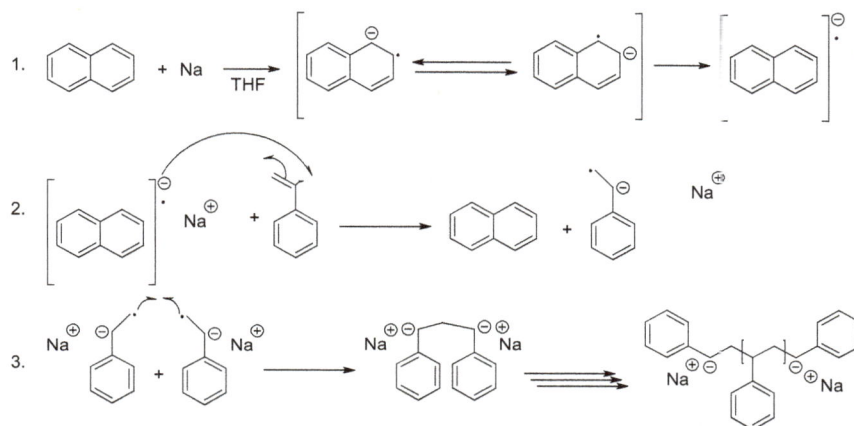

Figure 3.3: Radical anionic polymerization mechanism of poly(styrene).

3.2.2 Poly(ethylene)

Poly(ethylene) (PE) is the most abundant plastic material on earth. It is used for diverse applications from plastic bags to a container for biomedical applications including syringes and other medical devices. PE was originally synthesized by a German chemist, Hans von Pechmann, in 1898. But it was only after the Second World War that production ramped up with the discovery of new synthesis routes. The typical synthesis process for the polymerization of PE begins using ethylene, a gaseous hydrocarbon (C_2H_4) produced from petrochemical sources with a catalyst such as titanium (III) chloride, developed by Ziegler and Natta. PE obtained is semi-crystalline with a T_g around -125 °C for the low-density PE. It is a thermoplastic with a melting temperature ranging from 110 °C to 140 °C depending on the chain structure. Indeed, depending on the synthesis route one can obtain a partially branched polymer or linear polymer. The more the PE is linear, the higher will be the density. In addition to the branching, the length of the polymer chain can also vary depending on the catalyst used. There are therefore different types of PE polymer with different properties and domains of application. In medical devices such as implants, the ultra-high molecular weight high-density poly(ethylene) (UHMWPE) is mostly used (Table 3.1). It gives tough materials that are self-lubricant and is used for the contact part for hip implants between the limb and the hip. Thus, UHMWPE acts as a load barring replacement for articular cartilage.

Table 3.1: Reaction scheme of the free radical polymerization of poly(ethylene) and a table summarizing the different types of poly(ethylene) and their polymer chain structures. I being a radical initiator.

Type of PE	Name	Density	Chain
Low density	LDPE	0.88–0.91	
Linear low density	LLDPE	0.91–0.925	
Medium density	MDPE	0.926–0.94	
High density	HDPE	0.941–0.965	
Ultra-high molecular weight	UHMWPE	>0.966	

3.2.3 Poly(propylene), (PP)

PP is a semi-crystalline thermoplastic with a glass transition around 0 °C and a melting transition ranging from 130 °C to 171 °C depending on the crystallinity degree. Because of the methyl side chain, PP can only be synthesized with metal catalysts such as Ziegler–Natta. The catalyst permits the control of the molecular structure of the obtained materials and the orientation of the methyl group on the chiral carbon. Indeed the methyl group of the PP can be positioned in a different position depending on the monomer incorporation in the polymer chain. For methyl that is all on the same side, it is an isotactic (all chiral carbons are R or S) PP. If there is alternating R and S chiral carbon, the PP is syndiotactic, and if there is no organization, the PP is atactic Table 3.2. PP is used in many biomedical applications for making non-degradable implants such as hernia mesh, but it is often associated with a foreign-body response. Due to PP good mechanical proper-ties, it is used for sutures and has a mesh to maintain the organ in place after trauma or surgery. It is also used to make many devices such as syringes and other medical containers used for holding solutions for injection.

Table 3.2: Reaction scheme of the free radical polymerization of poly(propylene). I being a radical initiator.

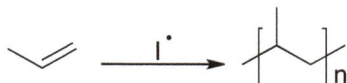

Molecular structure	Type	Crystallinity	T_g	T_m
	Atactic	Amorphous	−20 °C	0 °C
	Syndiotactic	Semi-crystalline	−8 °C	160 °C
	Isotactic	Semi-crystalline	0 °C	184 °C

3.C Ziegler–Natta catalyst

Ziegler–Natta mechanism for the polymerization of poly(propylene) (PP) and high molecular weight PE (figure 3.4). Where L is an unspecified ligand, M is a metal such as Al, Li, Mg, or Zn and R is an alkane chain. First, the catalyst is activated and coordinated with an alkene mono-mer. A rearrangement occurs and the chain starts to grow. The coordination rearrangement is repeated as many times as the monomer is integrated into the polymer.

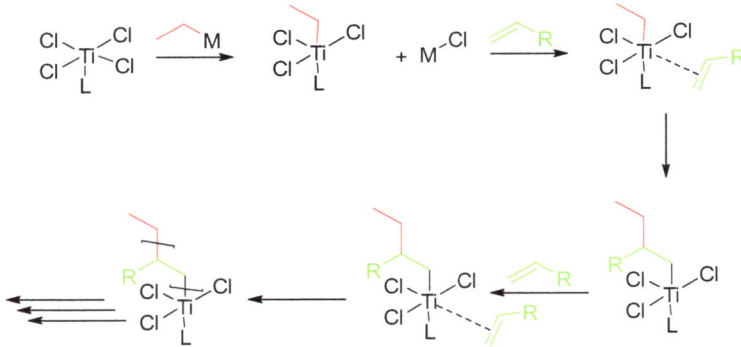

Figure 3.4: Ziegler–Natta mechanism.

3.2.4 Poly(tetrafluoroethylene)

Poly(tetrafluoroethylene) PTFE is a semi-crystalline polymer with a T_g of 115 °C and a melting temperature T_m of 327 °C (figure 3.5). However, the melting temperature is close to its degradation temperature, and it is therefore challenging to process it in the melted state. The polymer is obtained by radical chain polymerization from the gaseous monomer tetrafluoroethylene. Due to its fluorine atoms, PTFE has low friction, is anti-fouling, and is highly hydrophobic which makes it a material of choice for certain biomedical applications. For example, membrane filtration, as it permits to avoid adhesion of the molecules being filtrated. But the exceptional thermo-mechanical character of PTFE (thermal isolation and low flow) makes it hard to process like the other thermoplastics and so PTFE is usually machined to create parts. However, because it releases small debris, it cannot be used for low friction applications like for articular joints. For these applications, UHMWPE is preferred. The expanded form of PTFE (Gore-Tex™) is used for vascular prostheses, sutures, and some cardiac patches. The expansion process makes the materials porous, waterproof but allow for the exchange of gas.

Figure 3.5: Chemical structure of poly(tetrafluoroethylene).

3.2.5 Poly(amide), Nylon

Nylon is usually obtained by polycondensation of sebacoyl chloride with diamine monomers [1]. It is a semi-crystalline material with a T_g of about 130 °C and a melting temperature from 178 °C to 265 °C depending on the length of the alkane chain (figure 3.6). Because amine groups can form interchain h-bounds with the oxygen atoms, polyamide (PA) forms naturally long strands of materials that are easily processed as a fiber. Therefore, one of the main applications in the biomedical field for nylon was found for sutures. Another application for PA was for the manufacturing of balloon catheters as it provides a material with a high strength that allows the balloon to be bloated with high pressure.

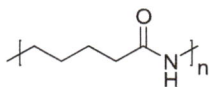

Figure 3.6: Chemical structure of poly(amide).

3.2.6 Poly(carbonate)

Poly(carbonate) (PC) is an amorphous polymer with a T_g of 147 °C and with a melting temperature between 230 °C and 260 °C (figure 3.7). PC is obtained by polycondensation of bisphenol A (BPA) and phosgene. However, due to safety concerns, intensive research is focusing on finding solutions to remove the use of BPA. That is BPA is reported to have hormone-mimicking properties and is therefore said to be a chemical to which humans should not be exposed. Because of its amorphous properties, PC gives a transparent material. This property is quite important for many applications where the samples need to be monitored in a container. For instance, when handling blood samples, monitoring flow, color, and coagulation.

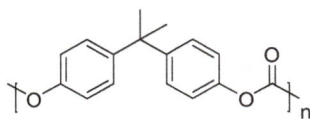

Figure 3.7: Chemical structure of poly(carbonate).

3.2.7 Poly(vinyl chloride)

Poly(vinyl chloride) (PVC) is the third most-produced polymer with a T_g of 82 °C and melting temperature of 260 °C; PVC is usually a rigid material at room temperature (figure 3.8). However, by the addition of plasticizers such as phthalate, citrate adipates, or azelates the plasticity of PVC can be tuned [2]. Its low cost and ease of manufacturing makes it a candidate of choice for the cost-effective manufacturing of high-volume devices and their packaging. Additionally, PVC can be sterilized with many methods from autoclave, ethylene oxide, and chemical sterilization processes. But the main concern of PVC is the addition of plasticizers that could be leaking from the bulk materials and contaminate the surrounding environment. In cause, the phthalate, which after many studies over the past decades, there is to date a lack of data showing harmful effects of phthalate that would prevent the use of PVC for medical devices. About one-third of disposable medical devices are manufactured from flexible PVC. It is mostly used in medical devices that interact, both directly and indirectly, with human body fluids or tissues (e.g., blood or intravenous storage bags and/or tubing). It is widely employed in the storage of blood derivatives.

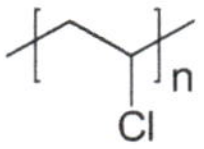

Figure 3.8: Chemical structure of poly(vinyl chloride)

3.2.8 Poly(ethyl vinyl acetate)

Poly(ethyl vinyl acetate) (PEVA or EVA) is a copolymer that is obtained by the polymerization of vinyl acetate with ethylene in any proportion (figure 3.9). It is a semi-crystalline thermoplastic. The ratio of each monomer will affect the properties and thus resulting in diverse applications for the copolymer. An increased proportion of vinyl acetate in the copolymer induces an elastomer type of behavior which is favored for providing resistance to shock, transparency, and gas permeability. These enhanced properties are achieved to the detriment of the resistance to traction, heat, and melting point. Typical PEVA will have a T_g of about $-25\,°C$, and a melting temperature between $45\,°C$ and $105\,°C$ depending on the content of vinyl acetate [3]. EVA at industrial scale is usually polymerized in bulk, through an exothermic process above $100\,°C$ under pressure. But medical-grade polymerization occurs in solution at low pressure and usually requires the use of a buffer to maintain the pH and reduce the hydrolysis of acetate. The main use of PEVA is for plastic packaging for food and agricultural application as it provides an adhesive film due to the capabilities of the acetate moieties to adhere to polar substrates such as paper. PEVA is used in several controlled drug delivery devices for small and large molecular weight drugs. These devices are fabricated by using casting and freeze-drying methods and are mainly used for subcutaneous implantation. Varying the content of vinyl acetate copolymer in the bulk affects the permeability of the device and thus the delivery profile of the hormone. One of the major successes of PEVA implants is the estrogen delivery rod that permits a sustained delivery and contraceptive effect for up to 3 years. The low melting temperature and the possibility to mix it with active pharmaceutical ingredient makes

PEVA a great candidate for biofabrication applications as it can be processed by 3D printers [4, 5].

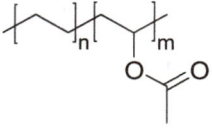

Figure 3.9: Chemical structure of the copolymer poly(ethylene vinyl acetate). Adjusting the ratio of vinyl acetate can be used to precisely tune the properties of the resulting polymer.

Drug release mechanism

There are three main mechanisms of drug release for a drug carrier (figure 3.10). The carrier loaded with its payload will release its cargo by diffusion of the molecule through its polymer network from the inside to the outside. This is the main mechanism in the hydrogel type of carrier. In a bioresorbable polymer system, the payload will be released as the carrier degrades, and here there are two main degradation mechanisms. These polymers are degrading through aqueous hydrolysis. So, the type of degradation erosion of the polymer will be dictated by the ratio between water diffusion and degradation rate of the polymer. If the water diffuses to the polymer bulk faster than its degradation, then the carrier will have a bulk degradation or erosion leading to a porous material. If the polymers degrade faster than the water can diffuse in the polymer then the carrier will have surface erosion, here only the surface of the materials is degraded. In turn, we will have two different release

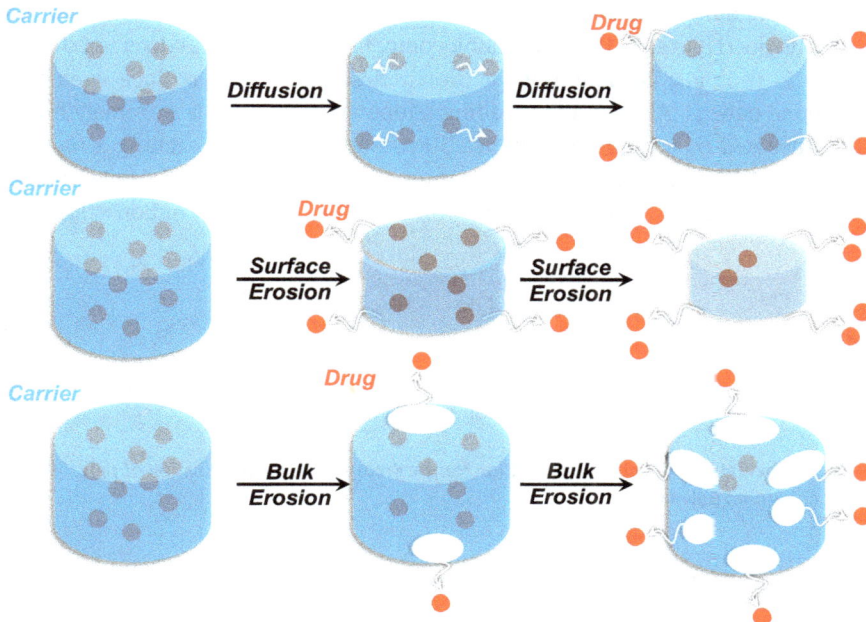

Figure 3.10: The three different drug release mechanisms of drug-loaded polymer are: diffusion of the active pharmaceutical ingredient out of the polymer matrix, erosion of the polymer matrix from the surface, and erosion of the bulk polymer matrix.

kinetic profiles. A surface erosion rate will give a near-zero-order kinetic release and the release will be directly proportionate to the surface area of the materials. Conversely, bulk erosion is more difficult to control, and a more complex kinetic order is achieved. Here the degradation rate is proportional to the volume of the implant and so as the implant erodes, its volume declines and its release rate as well. Therefore, the materials for the drug delivery system need to be meticulously chosen for each drug delivery application. Typically, polyester polymers such as polycaprolactone (PCL), poly(lactic acid) (PLA), and polyglycolic acid (PGA) undergo bulk erosion. Polyanhydride and poly(orthoester) are conversely, considered as surface eroding [31].

3.2.9 Poly(ether ether ketone)

Poly(ether ether ketone) (PEEK) is a polymer parts of the larger family of poly(aryl ether ketones) that comprise several variations of the ether ether ketones and ether ketone ketone polymeric architecture (figure 3.11). PEEK is a semi-crystalline amorphous polymer with a glass transition that occurs around 143 °C and a melting temperature around 345 °C. Its high-performance mechanical properties exhibit an elastic modulus up to 4 GPa and can be reinforced by different phases such as glass fibers, hydroxyapatite, and metallic oxides to match bone elastic modulus (18 GPa). The material is obtained by a step-growth polymerization process through the reaction of 4, 4 difluorobenzophenone with hydroquinone sodium salt at temperatures above 300 °C in aprotic solvents (Figure. 3.11). PEEK is a biocompatible material that does not degrade *in vivo*, it is thus a great material for prosthetics implants to replace even load-bearing bone defects. Although the melting temperature of PEEK is extremely high, it can be processed by a high-performance 3D printer capable of reaching a temperature above 390 °C, above the flow temperature of the polymer. This permits the creation of custom-made polymeric implants for reconstructive surgery of bones [6, 7].

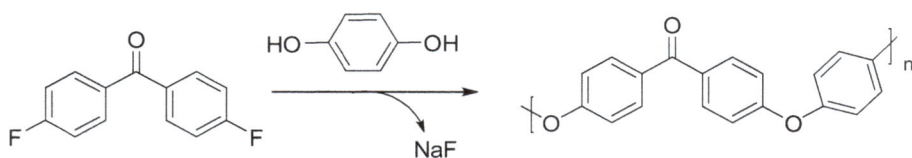

Figure 3.11: Schematic of the chemical reaction for the step-growth polymerization of poly (ether ether ketone) polymers.

3.2.10 Poly(glycolic acid)

Poly(glycolic acid), (PGA) is a thermoplastic semi-crystalline polymer that has a glass temperature of 40 °C and a melting temperature of 230 °C. This material is used to make biodegradable surgical sutures (figure 3.12). PGA undergoes biodegradation through hydrolysis of the ester bond between the repeating units of the polymer chain. This degradation is catalyzed by an alkaline and acidic environment. This material can also be processed as a mesh for wound dressing. Because it is thermoplastic, it is suitable for its processing through melt extrusion 3D printing. PGA can be synthesized through different polymerization routes, but the most efficient one is through ring-opening polymerization (ROP) of glycolide catalyzed by tin octoate, a catalyst approved in polymer used in food and medical application (Figure. 3.12). The polymerization of lactone by $Sn(Oct)_2$ is undergone through a coordination–insertion mechanism. The polymer obtained through ROP is of high molecular weight (Mn > 100 k g.mol^{-1}) and high crystallinity of about 50%. PGA is considered a hydrophilic polymer. Due to its good wettability, its degradation rate is relatively fast. For instance, in-vivo PGA can lose about 50% of its weight in a couple of weeks. The addition of a co-monomer through polycondensation allows for the precise tuning of the physical properties of the polymers such as degradation rate, thermal and mechanical properties. Because of its thermoplastic properties, PGA can be processed by fuse deposition modeling 3D printing to make biodegradable scaffolds [8, 9].

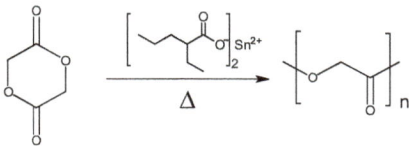

Figure 3.12: Ring-opening polymerization of glycolide to obtain PGA using tin(II) catalyst.

3.2.11 Poly(lactic acid)

Poly(lactic acid), (PLA) is a thermoplastic semi-crystalline polymer that has a glass transition at 60 °C and a melting temperature of 150 °C (figure 3.13). However, it can be processed into a transparent amorphous film by casting the PLA in a cold mold without a nucleation agent (Figure 3.13). PLA is a fully biobased polymer that is obtained by ROP of lactide extracted from corn starch. The starch is depolymerized through acid or enzymatic treatment to obtain glucose. The glucose undergoes then glycolysis through an enzyme treatment to give lactide. The process is achieved by fermentation of the starch using microorganisms such as the *Lactobacillus* bacteria. The fermentation process dates back to the nineteenth century, but with the recent advances in biotechnology, the process has been improved to obtain biobased-monomers with cost competing with oil-based monomers. Once the lactide is obtained, it is polymerized through a ROP using thin octanoate following the same mechanism as for PGA. In contrast to PGA, PLA has asymmetrical methyl groups which convey its hydrophilic properties and longer degradation rate. However, both lactide and glycolide monomers can be polymerized following the same process, they can then be polymerized together to form a copolymer poly(lactic-co-glycolic acid) (PLGA).

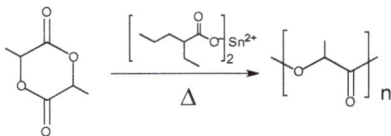

Figure 3.13: Synthesis of poly(lactic acid) from lactide monomer using heat with the tin(II) octanoate catalyst.

3.2.12 Poly(lactic-co-glycolic acid)

Poly(lactic-co-glycolic acid), (PLGA) is also a thermoplastic copolymer with a melting temperature of about 250 °C and glass temperature above physiological temperature, usually about 40 °C (figure 3.14). Controlling the ratio of PGA and PLA in the PLGA copolymer allows for the precise optimisation of the degradation rate of the polymer and its melting and hydrophobic properties. Because PLA contains an asymmetric methyl group, PLA is more hydrophobic than PGA. Therefore, a higher content of PLA in the copolymer will increase its hydrophobicity. The increased hydrophobicity will in turn decrease the interaction of the polymer with the aqueous milieu and reduce its degradation rate. For PLGA, the degradation rate can be controlled by two mechanisms: the molecular weight of each block of the polymer and the ratio of PGA to PLA (Figure 3.14). The possibility to precisely control the degradation rates of PLGA permits designing drug release systems in which the cargo is released at a constant rate. The drug is dispersed into the polymer matrix, the polymers degrade thus releasing the drug. The PLGA copolymer can be prepared as a block copolymer or as a random co-polymer. The block copolymer allows forming micelles or liposomes nanoparticles whereas the random copolymer is more suited for the manufacturing of eluting implants and microparticles. The PLGA is used for drug delivery in the oral and parenteral formulation. Alternative synthesis routes of PLGA are being investigated to remove the need for a tin catalyst, despite the tin octoate being approved in polymer used in food and certain biomedical applications.

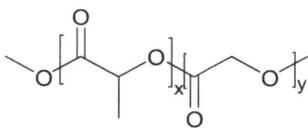

Figure 3.14: The chemical structure of poly(lactic-co-glycolic acid) shows the block copolymer structures that allow to precisely tune the hydrophobicity and degradation rate of the polymer by changing the ratio of lactic acid to glycolic acid.

3.2.13 Polycaprolactone

Polycaprolactone, (PCL) is a thermoplastic polymer with a glass transition of – 60 °C and depending on its degree of crystallinity a melting temperature ranging from 59 °C to 64 °C (figure 3.15). Due to its degradation by hydrolysis and ease of manufacturing, PCL has been for a long time a polymer of choice for engineering medical devices.

Due to its low melting point, it is particularly suitable for advanced manufacturing techniques such as 3D printing and melt electrowriting, further described later in this book. High molecular weight PCL is usually achieved by ROP of ε-caprolactone. Several catalysts can be used for the polymerization of PCL including metal-based, organic, inorganic, and enzymatic systems. While metal-based catalysts are predominantly used, enzymatic systems are being intensively studied. The catalyst is mainly chosen for an application and the desired reaction conditions. For biomedical applications, the tin octanoate catalyst is mainly used due to its history in food and biomedical space. PCL degrades slower than PLA, PGA, and copolymer thereof. up to several years for PCL to totally degrades *in vivo*. The addition of copolymer can modify the degradation profile of the resulting materials and thus allow for the design of drug delivery devices with degradation profiles adapted to a specific application. In the human body, degradation is a two-step process. During the first year, ester groups are hydrolytically cleaved. Next, intracellular degradation occurs in the second phase beginning when the polymer is more highly crystalline and of low molecular weight, typically below 3,000 g. $\cdot$mol^{-1}. At this stage, different cells can degrade in about two weeks the fragments such as macrophage and giant cells. But the degradation is ultimately dependent on the molecular weight and crystallinity of the polymer or composite.

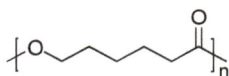

Figure 3.15: Polycaprolactone is polymerized through ring-opening polymerization of ε- caprolactone. Several polymerization strategies are available including enzymatic, anionic, and cationic.

3.D Coordination–insertion mechanism for ring opening polymerization (ROP)

Many polymers discussed in this chapter are synthesized by ROP. The most common mechanism for the ROP is the catalyzed coordination–insertion mechanism with tin octanoate (figure 3.16). Here, the metal catalyst coordinates with the lactone to open the ring and insert one of its alkoxy groups. Here, the oxygens are colored to highlight the mechanism.

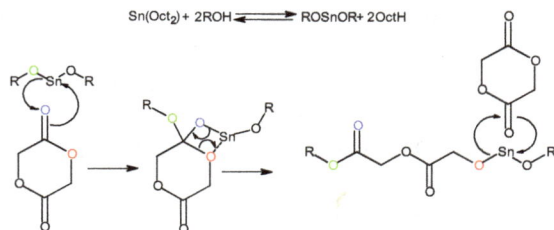

Figure 3.16: Ring-opening polymerization mechanism.

3.2.14 Poly(vinyl alcohol)

Poly(vinyl alcohol) (PVA) is a semi-crystalline water-soluble polymer with a glass transition of 80 °C and a melting temperature of 200 °C (Figure 3.17). Because the monomer, vinyl alcohol, is thermodynamically unstable due to its tautomerization to acetaldehyde, PVA cannot be directly synthesized by radical polymerization. PVA is obtained by hydrolysis of poly(vinyl acetate), through a base-catalyzed transesterification with ethanol. Because of its alcohol functional groups, PVA can be processed as a physical hydrogel through the formation of complexes with divalent cations such as calcium or cupper, and as chemical hydrogels using crosslinkers such as boric acid, glutaraldehyde (see Chapter 4 for hydrogels). PVA hydrogel can be obtained by the crystallization of the polymer in a DMSO/water solution which upon cooling below room temperature will induce the crystallization of the polymer and its crosslinking. Exchanging the DMSO solvent with water gives a transmitting hydrogel of high tensile strength. One of the applications of PVA chemically crosslinked hydrogel is used to make degradable contact lenses. PVA is completely degradable by microorganisms and the mechanism involved enzymatic dehydrogenases and secondary alcohol oxidases. But the main application of PVA is to form fibers for the textile industry and as an additive, in many formulations of composites and cement as they can serve as a hydrating phase.

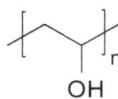

Figure 3.17: Chemical structure of the biodegradable thermoplastic PVA. Obtained through the saponification of polyacetate.

3.2.15 Polyanhydride

Polyanhydrides were first investigated by Hill and Carothers at the beginning of the twentieth century, but their low aqueous stability is a great limitation for their industrial application (Figure 3.18). With the development of tissue engineering and drug delivery system, their degradation behavior became of interest for sutures and the engineering of diverse medical devices. Polyanhydride is synthesized by dehydrating carboxylic acid difunctional monomers to form corresponding polyanhydride linkages. The usual industrial route involves the use of sebacic chloride to reach a typical molecular weight of about 15–20 kDa. Like any anhydride functional group, the polymer chain is highly sensitive to moisture and quickly degrades in water through a hydrolysis mechanism to give the original carboxylic acid monomers. Because sebacic acid naturally occurs in human physiology, it is a good monomer to use in a biodegradable polymer. Like the other polymer systems presented earlier, their degradability rate can be tuned by introducing hydrophobic groups such as aromatics which will reduce the water access to the polymer chain and thus the degradation rate. Additionally, their molecular weight and degree of crystallinity will affect their degradation rate, the higher the crystallinity and the higher the molecular weight, the slower the polymer will degrade. Typical degradation occurs in a couple of weeks. Depending on the monomers polyanhydride will have different glass transition temperatures (T_g). For instance, an aromatic repeat unit will give a polymer with a T_g above physiological temperature while having a solubility in organic solvent, and a melting temperature typically between 50 °C and 90 °C. Thus, these thermochemical properties make it a polymer easy to process.

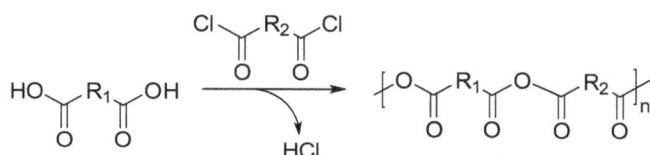

Figure 3.18: Polymerization of a polyanhydride by the reaction of a dicarboxylic acid with diacyl chloride monomers. The polymer quickly degrades in the presence of water.

3.2.16 Poly(propylene fumarate)

Poly(propylene fumarate) (PPF) is an amorphous thermoplastic polymer with T_g dependent on the molecular weight and is typically below physiological temperature (10–30 °C) (Figure 3.19). It is usually synthesized in an organic solvent by step-growth polymerization using the various metal catalyst available including chromium (III) complex, or manganese dietanoate from maleic anhydride and propylene oxide. Other synthetic routes involve a multistep method using diethyl fumarate with propylene oxide and a catalyst such as zinc chloride. The interesting properties of PPF are that it can be crosslinked by radical polymerization using a UV photoinitiator. This has particular interest for the manufacturing of bone or dental implants as the polymer can be molded into the defect space and then harden using UV crosslinking. Additional crosslinking molecules can be used such as poly(ethylene glycol) terminated with methyl acrylate, for which the molecular weight can be varied to the obtained resin of diverse mechanical properties up to the range of MPa elastic modulus reproducing the toughness of bones. The resin is then degraded through hydrolysis, and it is highly impacted by the molecular weight of the polymer, its crosslinking density, and crosslinker type. As we have reviewed for the previous polymers, higher molecular weight and denser resin will decrease the degradation rate [10, 11].

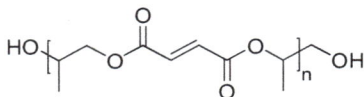

Figure 3.19: The chemical structure of poly(propylene fumarate) PPF can be further crosslinked by radical polymerization into a resin due to its unsaturated carbon bond.

3.2.17 Polyurethane

Polyurethane (PU) is synthesized by step-growth polymerization to obtain a semi-crystalline polymer. PU is an interesting polymer that can be obtained as foam, thermoset materials but also thermoplastic materials. These differences can be obtained by making either homopolymer (thermoset) or block copolymer made of segments that contain long linear blocks made of hard and soft segments. The soft segments are built with polyols that provide flexibility to the polymer chains and the hard segments are built with diisocyanate monomers that are not flexible (Figure 3.20). As a thermoset, PU can be processed as foam either with close cells or open cells geometry. In an open-cell system, the pores created by the foaming reaction are interconnected, this is usually generated by the addition of extra gas. Whereas in a close cell system, the pore generated by the foaming process is not interconnected. For biomedical applications and implants, it is generally preferred to have open cells where cells can penetrate and migrate within the implant. Once the implant is colonized by the cells, they can generate their matrix and create artificial tissue. PU can undergo degradation *in vivo* through the hydrolysis of the urethane bonds. In thermoplastic PU, the degradation usually occurs at the intersection of hard and soft segments [12].

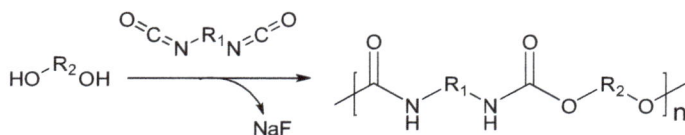

Figure 3.20: Scheme of the chemical reaction for the step-growth polymerization of the polyurethane.

can also alter their folding on the surface. In turn, cells from the immune systems will experience different types of molecules and molecule folding at the polymer surface. As a result, different types of cells will then adhere to the surface. Alternatively, the same progenitor cells could adhere to the surface but depending on the information obtained on the polymer surface will differentiate into two distinct cell lineages and secrete different chemokine which will, in turn, convey a signal of "isolation" or "integration" of the implanted materials. We can see by this mechanism that selecting the right polymer for a specific application is paramount for the good integration or isolation of the implanted material. Some applications require the implant to stick with the tissue to have integration such as implants used for the replacement of load-bearing bones. But other applications such as drug delivery rode that need to be removed once the cargo is delivered after several years, the polymer cannot be integrated into the tissue as it would limit the removal of the implant.

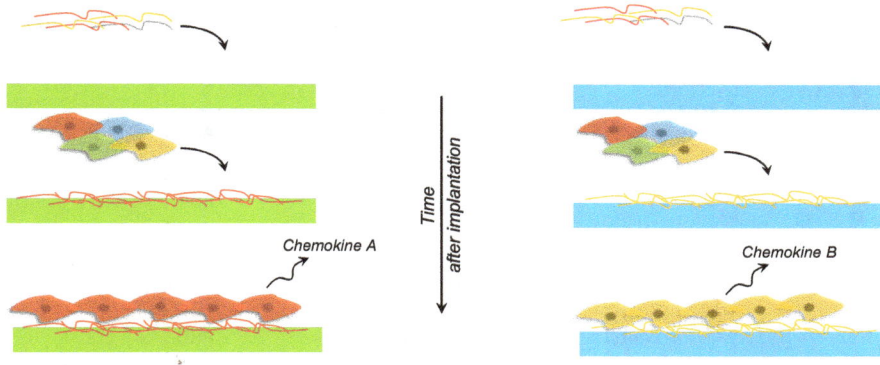

Figure 3.21: Comparison of implant behavior after implantation. The green material adsorbs the red protein, which in turn recruits the red cell that will set a biological signal using a chemokine. Conversely, the blue material adsorbs the yellow protein which recruits the yellow cell that will send a different signal. Thus, green and blue materials will not induce the same biological response after implantation.

3.2.18 Poly(hydroxy alkanoates)

Poly(hydroxy alkanoates) (PHAs) are semi-crystalline thermoplastic polymers with a T_g varying between $-50\ °C$ and $5\ °C$ and a melting temperature of comprised between $60\ °C$ and $180\ °C$ depending on their molecular structure (Figure 3.22). These polymers are obtained by bacteria biosynthesis and are used by the bacteria as energy storage materials. The longer the side chain R is the higher is the T_g and T_m. As a polyester, PHA can be degraded by hydrolysis but also by bacterial depolymerase. The PHA usually degrade in 4 years *in vitro* under physiological conditions. Due to their tunable properties, these materials have been used for the biofabrication of a broad range of soft and hard tissue by changing the R side chain chemistry. Other thermoplastic PHAs can be formulated for drug delivery applications and be processed as fibers and microparticles. Due to their slow degradation rates, the drug delivery applications are limited to long-term ones. But they still suffer from several drawbacks mainly the production cost and the melting, the melting temperature is often close to the degradation temperature. The current use of PHA is directed toward blends with other natural organic polymers. Finally, the modification of the polymer required genetic engineering of the bacteria, and this is an expensive and long process. Additionally, the needed to obtain pure materials are tedious and costly as for biomedical application one needs to remove pyrogenic materials issue from the bacteria wall because lipopolysaccharides can cause inflammation once implanted [13–15].

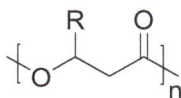

Figure 3.22: Scheme of the chemical structure of poly(hydroxy alkanoate) polymers and their nomenclature.

	Alkyl (R)	Name Poly(3hydroxy-)	Abb.
	H	Propionate	PHP
	CH_3	Butyrate	P3HB
	CH_2CH_3	Valerate	PHV
	Propyl	Hexanoate	PHHx
	Butyl	Heptanoate	PHH
	Pentyl	Octanoate	PHO
	Hexyl	Nonanoate	PHN

3.2.19 Poly(ether sulfone)

Poly(ether sulfone) (PES) is an amorphous non-degradable thermoplastic polymer that has a glass transition around 188 °C and can be sterilized by many processes, which makes it a good candidate for biomedical applications. These polymers can be made either by polycondensation or ROP (Figure 3.23). The resulting polymer is transparent and can be used for medical devices that can withstand many steam sterilization cycles without deformation. Thus, it can be used to make surgical tools as an alternative to stainless steel. PES starts to flow at a temperature above 340 °C but still can be processed by a high-performance 3D printer. It can be made as a composite with an inorganic calcium phosphate phase for dental implants.

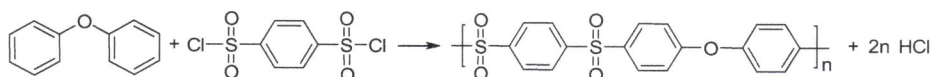

Figure 3.23: Scheme of the polymerization reaction of poly(ether sulfone).

3.3 Synthetic inorganic biomaterials

According to IUPAC, an inorganic compound is defined by a compound that is not organic. An organic molecule is a molecule that contains carbon bound to hydrogen. As such carbon dioxide is not only an organic molecule. Counter-intuitively to the popular belief that living objects are composed of organic molecules, mammalians are predominantly composed of inorganic materials. Indeed, the mineral phase in mammalian tissue accounts for about 70% of the whole body. While the organic phase is composed of the majority of cells and extracellular matrix (collagen being the most abundant polymer in the tissues), the bones, teeth, and salts found in the aqueous phase (Mg, Na, K, Ca, F, and Cl) account for most of the total body mass.

3.3.1 Calcium phosphate

The inorganic phase of the bone is composed of a calcium phosphate crystal that has a specific Ca/P ratio of 1.67 called hydroxyapatite (HAp). But reproducing this exact ratio *in vitro* and at an industrial scale is quite challenging. Indeed, several calcium phosphate crystals can be generated. Some of them reproduce naturally occurring crystals. Each different ratio of calcium to phosphate leads to a different crystal structure that has then different biological activity (Table 3.3). These materials are used to support the regeneration of bones by providing a natural environment for osteointegration. The HAp can also be used for reinforcing polymer melts. Successfully, this has been used to make a composite with PEEK and provide a better osteointegration of the implant. HAp is usually synthesized following three different routes: dry, wet, and high temperature. Dry methods use calcium and phosphate precursors that are either heated up or ground and compressed under heat to obtain a crystal. The wet methods have precursor solutions containing calcium and phosphate ions that are then precipitated through precise pH adjustment followed by aging, filtration, and drying. Often the obtained crystal undergoes a heat treatment or sintering around 1,000 °C. Other wet methods included hydrothermal (heated solution), hydrolysis. High-temperature approaches used high-temperature reaction at 500 °C and to induce the formation of HAp crystals then sintering above 1,000 °C of solution or solid salts [16–18].

Table 3.3: The different phases of calcium phosphate and their occurrence in mammalian tissues.

$$Ca^{2+} \quad O^- \overset{\overset{\displaystyle O}{\|}}{\underset{\underset{\displaystyle O}{|}}{P}} - OH$$

Ratio	Name	Formula	Occurrence
0.5	MCPM Monobasic calcium phosphate monohydrate	$Ca(H_2PO_4)_2 \bullet H_2O$	
1.0	DCPA/monetite Dicalcium phosphate anhydrous	$CaHPO_4$	
1.0	DCPD/brushite Dibasic calcium phosphate dihydrate	$CaHPO_4 \bullet 2H_2O$	
1.33	OCP Octacalcium phosphate	$Ca_8(HPO_4)_2(PO_4)_4 \bullet 5H_2O$	Dental calculi and urinary stones
1.5	α-TCP α-Tricalcium phosphate	$\alpha\text{-}Ca_3(PO_4)_2$	
1.5	β-TCP β-Tricalcium phosphate	$\beta\text{-}Ca_3(PO_4)_2$	Dental calculi, arthritic cartilage, soft tissue deposits
1.2	ACP Amorphous calcium phosphate	$Ca_xH_y(PO_4)_znH_2O$ $n = 3\text{--}4.5$	Heart calcification, kidney stones
1.5	CPPD Calcium prophosphate	$Ca_2P_2O_7 \bullet 2H_2O$	Pseudo-gout deposits in synovium fluids
1.5–1.67	CDHA Calcium deficient hydroxyapatite	$Ca_{10-x}(HPO_4)_x(PO_4)_{6-x}$ $(OH)_{2-x}$ $0 < x < 2$	
1.67	HAp Hydroxyapatite	$Ca_{10}(PO_4)_6(OH)_2$	Enamel, dentin, bone, dental calculi
2.0	Hilgenstockite Tetracalcium phosphate	$Ca_4(PO_4)_2O$	

3.3.2 Bioactive glass

In the search for an alternative to early metal implants that had a poor osteointegration, meaning a poor binding to the bone, Hench investigated different formulations of glass and their behavior in animal bone defects. The first bioactive glass melt was obtained with a composition of 46.1 mol%, SiO_2, 24.4 mol%, Na_2O, 26.9 mol% CaO, and 2.6 mol% P_2O_5. Based on this pioneering work, further studies have aimed at enlarging the library of bioglass but only two compositions have been cleared for human use and shown better performance compared to usual metal implants. *In vitro* studies have been able to underlie the mechanism behind the integration of the bioglass in bone and other tissue. Once implanted, the glass will release some of its phosphate and calcium components which will stimulate the formation of hydroxyapatite on its surface (Figure 3.24). This inorganic biolayer formed at the surface of the implant will then allow for cells to attach, proliferate, and create a natural extracellular matrix which will create a strong physical linkage between the implanted bioglass and the surrounding tissue. This was particularly flagrant in the original work of Hench when the surgeon couldn't move the implanted bioglass, whereas the metal implant was loose and could be removed [19].

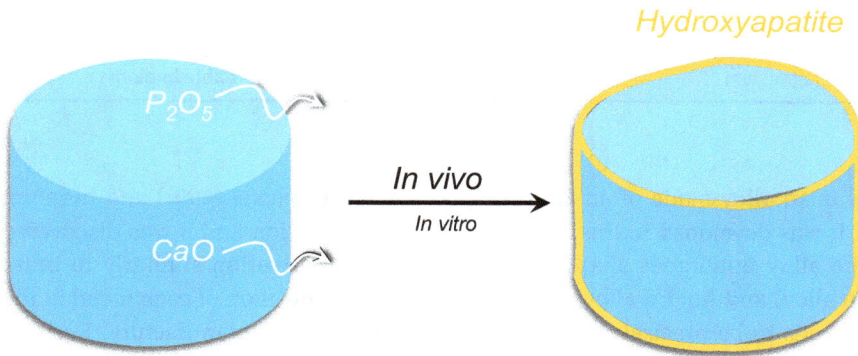

Figure 3.24: Composed of a precise ratio of silica, phosphate, calcium, and sodium, upon implantation the bioglass will release some phosphate and calcium in the surroundings which will favor the attachment of protein and the growth of the hydroxyapatite layer around it, thus favoring the osteointegration.

3.3.3 Metal

Metal has been used in biomedical applications ever since Ambroise Paré's first prosthetics in the sixteenth century. Naturally, once the surgical techniques have allowed placing implants in tissues, metal implants have emerged. For a metal implant tso show a favorable response in a given biological environment in a particular function, it depends on the corrosion resistance and cytotoxicity of corrosion products. This means that the loss of metallic ions from the metal surface to the surrounding environment needs to be well understood as this can lead to degradation of the implant leading to its failure but also an allergic reaction in patients. Over the years, several alloys resisting the harsh and corrosive aqueous and salty environment of biological tissues have been developed (Table 3.4). The recent development of metal 3D printing has opened research and development efforts in manufacturing patient-specific implants made of these biomedical metal alloys [20].

Table 3.4: Composition of the most used metal alloys for biomedical applications.

Name	Type	Composition (%)	Application
NiTiNOL	Nickel–titanium	Ni(49) Ti(51)	Shape memory stents
Ti6AL4V	Titanium	Ti(89) Al(6) V(4)	Bone implant
316L	Stainless steel	Fe(60–67) Cr(17–20) Ni(12–14) Mo (2–4) Mn(2) C(0.03)	Surgical tools, temporary implants and stents
CoCrMo	Cobalt–chromium	Co(58.9–69.5) Cr(27–30) Co(58–59) Cr(26–30) Co(45–56) Cr (19–21)	Dental, hip implant, and stents
AE21	Magnesium	Mg(97) Al(2) Rare Earth(1)	Resorbable implants

Nickel–titanium alloy (NiTiNOL)

Nitinol is an alloy made of an almost equal amount of nickel (49%) and titanium (51%). It was developed in the 1960s for military application and it was discovered that the alloy undergoes shape memory upon heat activation similarly to brass (Cu–Zn alloy) and Au–Cd alloys. But because the concentration of each metal in the alloys needs to be precisely controlled, it is challenging to manufacture. Further, due to its thermo-mechanical properties, it is also difficult to process. But for medical applications that are high added value applications, nitinol has found its main application as self-deployable stents. The stent can be folded into a small shape and introduced in the artery by a catheter and upon heat opens and pushed the artery to restore vascular flow. This technique avoids the use of inflating balloons for the deployment of the stent. The second concern for the current use of nitinol in a biomedical application is the potential allergic reaction to nickel and its carcinogenic property. To remedy to these drawbacks, nitinol stent can be oxidized to have a titanium oxide surface or can be coated with a polymer that can be further formulated with a drug to slowly elute the drug as the polymer degrades [21–24].

Cobalt–chromium alloy (CoCrMO)

Cobalt–chromium alloys possess superior mechanical properties with high corrosion resistance. They are usually supplemented with molybdenum and tungsten. These alloys have shown good biocompatibility and blood compatibility. They are used in various orthopedic, dental, and cardiovascular implants and devices and are classified into four types: ASTM F75 alloy, ASTM F799 alloy, ASTM F90 alloy, and ASTM F562 alloy with a different ratio of cobalt and chromium. The cobalt–chromium alloys are the most common alloys for dental application and hip joint replacement.

316L stainless steel

Stainless steel has been commonly used for surgical medical tools as it can be easily sterilized in an autoclave and the tools can be reused. For this application, stainless steel needs to be resistant to corrosion. The addition of 2% of molybdenum to make the 316 stainless steel improves the resistance to acids and localized corrosion caused by chloride ions. The 316L stands for low-carbon stainless steel to avoid corrosion problems caused by welding. In addition, this metal is inexpensive to manufacture and can be easily shaped by conventional forming techniques. Stainless steel implants are usually used for temporary bone fixation that is removed once the bone has healed.

Ti-based alloy (Ti–6Al–4V)

Titanium has been used for many applications due to its lightweight, resistance to corrosion, and biocompatibility. Commercially pure (CP) titaniums are classified in different grades depending on the quantity and type of impurity. The most used titanium for biomedical implants is the CP titanium grade 5 Ti–6Al–4V. However, the presence of aluminum and vanadium particles has raised some concerns as they might contribute to medical issues such as Alzheimer's and neuropathy [25].

Bioresorbable metals

Current trends in surgery aim to reduce the number of procedures and the invasiveness of these procedures. For a procedure that requires a metal implant because of the strength and load-bearing properties that are needed the development of a metal temporary implant that could be bioresorbable would avoid a second intervention to remove the supporting implants a bioresorbable. As such, magnesium, a metal naturally present in tissues has been investigated as metallic implants. But pure magnesium degrades too quickly and can release hydroxide chloride gas through reaction with water and sodium chloride present in the biological environment. To improve its corrosion resistance, alloy of magnesium-containing aluminum and rare earth have been developed. The second bioresorbable metallic materials proposed are iron implants. But the magnetic properties of irons limit its application as it is not compatible with magnetic resonance imaging techniques. Currently, these bioresorbable materials are used to make stents and screws for bone repair [26].

3.3.4 Inorganic polymers

Poly(dimethylsiloxane)

Beyond metallic and crystalline phases, polymers can also be made as inorganic materials. The most use of these inorganic polymers is Poly(dimethylsiloxane) (PDMS) which is also called silicon. It is a silicone-based polymer that is biocompatible and non-degradable. It has been used for decades for the manufacturing of microfluidic chips and is now used in biofabrication for the fabrication of organ-on-a-chip modules. The silicone polymer is obtained through ROP catalyzed in organic solvents such as acetone by based such as potassium hydroxide and in dichloromethane with acidic sulfate and hydroxide chloride (Figure 3.25).

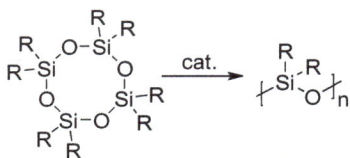

Figure 3.25: Ring-opening polymerization of Poly(dimethylsiloxane) using an alkaline or acidic catalyst in organic solvents.

The obtained polymer has a large molecular weight with high rotation potential through the silicone centers. It results in a flowing polymer that can penetrate thin topography, which makes it the perfect candidate for replica molding applications used to make microfluidic chips. Once the polymer is constrained within its mold, it can be crosslinked to form a strong resin. With the massive development of PDMS and variety of usage, there are many strategies for the crosslinking of PDMS. Some are water catalyzed and can be crosslinked by simple exposure to ambient air, other requires the use of metal catalyst and heat (Figure 3.26). The most frequently used crosslinkers are titanium and tin catalyst that form crosslinked resin upon heat curing and these approaches are used to make microfluidic chips, tubing, and other manufactured goods (Figure 3.27). Variation in the heat treatment time and temperature permits obtaining transparent resin of different mechanical properties and flexibility [27].

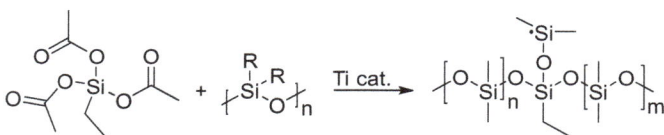

Figure 3.26: Titanium-catalyzed crosslinking of polydimethylsiloxane.

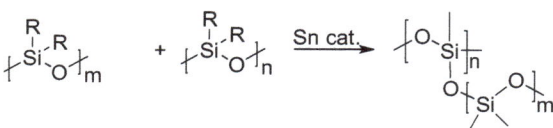

Figure 3.27: Tin-catalyzed heat curing of polydimethylsiloxane resin.

The obtained resin is an inert material that can be implanted in tissue with a limited inflammatory response. The PDMS can be processed into a different shape with micro- and nanotopography that can be further used to modulate the immune response of the implanted material. Once the resin is obtained, it can be further functionalized using silane chemistry. Silane functionalization is obtained by oxidizing the PDMS surface with sodium hydroxide or through an oxygen plasma treatment. Then, silane can react on the freshly oxidized PDMS surface. However, as we have discussed earlier, the PDMS polymer chain is long and highly flexible. This means that the chains that are oxidized on the surface will then migrate within the bulk and the activation will not be stable. There is now a broad variety of silane that can be used for the functionalization of PDMS. The most used one is 3-aminopropyltriethoxysilane (Figure 3.28). Silanes with click chemistry moieties, epoxy, acrylate, or thiol functional groups are now available and this list is growing every year. It is thus now possible to create microfluidic chips with highly precise chemical definition and functionalization. The main application of PDMS in biofabrication is for the manufacturing of organs on-chip. These are PDMS microfluidic chips that are populated with cells or organoids that can be designed to reproduce some physiological environment for drug testing applications.

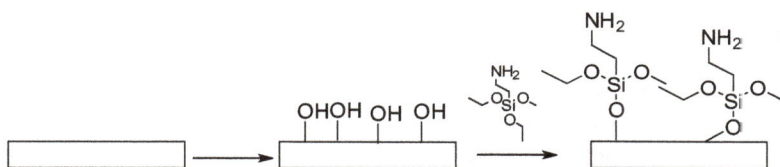

Figure 3.28: Functionalization of a PDMS resin by first oxidizing the surface with sodium hydroxide or plasma activation followed by the reaction of the hydroxyl groups with silane (here 3-aminopropyltriethoxysilane).

Polyphosphazenes

Polyphosphazenes (PPP) is an inorganic polymer that was developed in the 1960s. This is a fully inorganic polymer that degrades into amine and phosphate which are two build blocks of biochemistry. The glass transition of these polymers can be tuned by selecting different side-chain substituents and can vary between $-105\ °C$ and $30\ °C$. The PPP can be synthesized by ROP which is induced either thermally at $250\ °C$ or lower temperature (about $200\ °C$) by using a catalytic amount of Lewis acid such as aluminum chloride. The polymerization can also be conducted at room temperature using trichlorophosphoranimine and phosphate pentachloride through a mechanism of living cationic polymerization (Figure 3.29). The living polymerization approach enables, like with organic polymers, to achieve advanced polymer architectures such as block copolymer and graft polymers with comb or star architecture. The polymer is of particular interest for biomedical applications because it degrades into a building block of biochemistry amine and phosphate. The degradation rate of the polymer can be tuned by selecting different side groups. Likewise, the mechanical properties of the polymer will be determined by its organic sidechain. For instance, a thermoplastic material can be made with trifluoromethoxy side chains, a hydrogel-forming polymer can be obtained with an ethyleneoxy side chain that can be crosslinked with gamma radiation. Their degradation into non-toxic molecules made them particularly interesting materials for drug delivery application and bioresorbable implants with a degradation rate that can occur between weeks and months depending on the side chain [28–30].

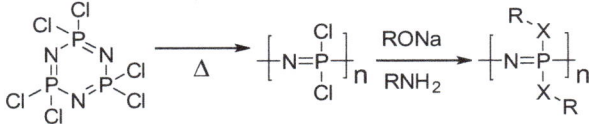

Figure 3.29: With X = O or N, the reaction scheme of the basic thermal activated polymerization reaction of polyphosphazene.

3.4 Quiz

1. What is the main mechanism for ROP?
2. Which polymer can be obtained by ROP?
3. What are the two main biocompatible inorganic polymers?
4. Why bioglass can have a better tissue integration compared to metal implants?
5. To have a bulk erosion, should the water diffusion be greater than the degradation rate of the polymer or vice versa?
6. What is the Ca/P of hydroxyapatite?
7. Which of these polymers is more hydrophobic poly(lactide acid) or poly(glycolide acid)?
8. Which one of the following polymers are made by bacteria: poly(lactide acid), PDMS, or poly(3-hydroxybutyrate)?
9. Which one of these alloys can be used for making thermally actuated metallic implants: cobalt–chromium, nickel–titanium, or titanium–aluminum–zirconium?
10. What is the definition of biomaterials?

References

[1]   Shakiba, M.; Rezvani Ghomi, E.; Khosravi, F.; Jouybar, S.; Bigham, A.; Zare, M.; Abdouss, M.; Moaref, R.; Ramakrishna, S. Nylon – A Material Introduction and Overview for Biomedical Applications. *Polym. Adv. Technol.* **2021**, 32 (9), 3368–3383. https://doi.org/10.1002/pat.5372.

[2]   Chiellini, F.; Ferri, M.; Morelli, A.; Dipaola, L.; Latini, G. Perspectives on Alternatives to Phthalate Plasticized Poly(Vinyl Chloride) in Medical Devices Applications. *Prog. Polym. Sci.* **2013**, 38 (7), 1067–1088. https://doi.org/10.1016/j.progpolymsci.2013.03.001.

[3]   Wang, K.; Deng, Q. The Thermal and Mechanical Properties of Poly(Ethylene-Co-Vinyl Acetate) Random Copolymers (PEVA) and Its Covalently Crosslinked Analogues (CPEVA). *Polymers (Basel)* **2019**, 11 (6), 1055. https://doi.org/10.3390/polym11061055.

[4]   Fu, Y.; Kao, W.J. Drug Release Kinetics and Transport Mechanisms of Non-Degradable and Degradable Polymeric Delivery Systems. *Expert Opin. Drug Deliv.* **2010**, 7 (4), 429–444. https://doi.org/10.1517/17425241003602259.

[5] Schneider, C.; Langer, R.; Loveday, D.; Hair, D. Applications of Ethylene V nyl Acetate Copolymers (EVA) in Drug Delivery Systems. *J. Control. Release* **2017**, 262, 284–295. https://doi.org/10.1016/j.jconrel.2017.08.004.

[6] Panayotov, I.V.; Orti, V.; Cuisinier, F.; Yachouh, J. Polyetheretherketone (PEEK) for Medical Applications. *J. Mater. Sci. Mater. Med.* **2016**, 27 (7), 118. https://doi.org/10.1007/s10856-016-5731-4.

[7] Verma, S.; Sharma, N.; Kango, S.; Sharma, S. Developments of PEEK (Polyetheretherketone) as A Biomedical Material: A Focused Review. *Eur. Polym. J.* **2021**, 147, 110295. https://doi.org/10.1016/j.eurpolymj.2021.110295.

[8] Sanko, V.; Sahin, I.; Aydemir Sezer, U.; Sezer, S. A Versatile Method for the Synthesis of Poly (Glycolic Acid): High Solubility and Tunable Molecular Weights. *Polym. J.* **2019**, 51 (7), 637–647. https://doi.org/10.1038/s41428-019-0182-7.

[9] Samantaray, P.K.; Little, A.; Haddleton, D.M.; McNally, T.; Tan, B.; Sun, Z. Huang, W.; Ji, Y.; Wan, C. Poly(Glycolic Acid) (PGA): A Versatile Building Block Expanding High Performance and Sustainable Bioplastic Applications. *Green Chem.* **2020**, 22 (13), 4055–4081. https://doi.org/10.1039/D0GC01394C.

[10] Wang, S.; Lu, L.; Yaszemski, M.J. Bone-Tissue-Engineering Material Poly(Propylene Fumarate): Correlation between Molecular Weight, Chain Dimensions, and Physical Properties. *Biomacromolecules* **2006**, 7 (6), 1976–1982. https://doi.org/10.1021/bm060096a.

[11] Kasper, F.K.; Tanahashi, K.; Fisher, J.P.; Mikos, A.G. Synthesis of Poly(Propylene Fumarate). *Nat. Protoc.* **2009**, 4 (4), 518–525. https://doi.org/10.1038/nprot.2009.24.

[12] McCarthy, S.J.; Meijs, G.F.; Mitchell, N.; Gunatillake, P.A.; Heath, G.; Brandwood, A.; Schindhelm, K. In-Vivo Degradation of Polyurethanes: Transmission-FTIR Microscopic Characterization of Polyurethanes Sectioned by Cryomicrotomy. *Biomaterials* **1997**, 18 (21), 1387–1409. https://doi.org/10.1016/S0142-9612(97)00083-5.

[13] Rodriguez-Contreras, A.; Recent Advances in the Use of Polyhydroyalkanoates in Biomedicine. *Bioengineering* **2019**, 6 (3), 82. https://doi.org/10.3390/bioengineering6030082.

[14] Li, Z.; Yang, J.; Loh, X.J. Polyhydroxyalkanoates: Opening Doors for a Sustainable Future. *NPG Asia Mater.* **2016**, 8 (4), e265–e265. https://doi.org/10.1038/am.2016.48.

[15] Raza, Z.A.; Abid, S.; Banat, I.M. Polyhydroxyalkanoates: Characteristics, Production, Recent Developments and Applications. *Int. Biodeterior. Biodegradation* **2018**, 126, 45–56. https://doi.org/10.1016/j.ibiod.2017.10.001.

[16] Dorozhkin, S.; Calcium Orthophosphates in Nature, Biology and Medicine. *Materials (Basel)* **2009**, 2 (2), 399–498. https://doi.org/10.3390/ma2020399.

[17] LeGeros, R.Z.; Formation and Transformation of Calcium Phosphates: Relevance to Vascular Calcification. *Zeitschrift für Kardiol.* **2001**, 90 (15), III116–III124. https://doi.org/10.1007/s003920170032.

[18] Mohd Pu'ad, N.A.S.; Abdul Haq, R.H.; Mohd Noh, H.; Abdullah, H.Z.; Idris, M.I.; Lee, T.C. Synthesis Method of Hydroxyapatite: A Review. *Mater. Today Proc.* **2020**, 29, 233–239. https://doi.org/10.1016/j.matpr.2020.05.536.

[19] Hench, L.L.; Splinter, R.J.; Allen, W.C.; Greenlee, T.K. Bonding Mechanisms at the Interface of Ceramic Prosthetic Materials. *J. Biomed. Mater. Res.* **1971**, 5 (6), 117–141. https://doi.org/10.1002/jbm.820050611.

[20] Prasad, K.; Bazaka, O.; Chua, M.; Rochford, M.; Fedrick, L.; Spoor, J.; Symes, R.; Tieppo, M.; Collins, C.; Cao, A.; et al. Metallic Biomaterials: Current Challenges and Opportunities. *Materials (Basel)* **2017**, 10 (8), 884. https://doi.org/10.3390/ma10080884.

[21] Ölander, A.; An Electrochemical Investigation of Solid Cadmium-Gold Alloys. *J. Am. Chem. Soc.* **1932**, 54 (10), 3819–3833. https://doi.org/10.1021/ja01349a004.

[22] Stoeckel, D.; Pelton, A.; Duerig, T. Self-Expanding Nitinol Stents: Material and Design Considerations. *Eur. Radiol.* **2004**, 14 (2), 292–301. https://doi.org/10.1007/s00330-003-2022-5.

[23] Ahlström, M.G.; Thyssen, J.P.; Wennervaldt, M.; Menné, T.; Johansen, J.D. Nickel Allergy and Allergic Contact Dermatitis: A Clinical Review of Immunology, Epidemiology, Exposure, and Treatment. *Contact Dermatitis* **2019**, 81 (4), 227–241. https://doi.org/10.1111/cod.13327.

[24] Kasprzak, K.; Nickel Carcinogenesis. *Mutat. Res. Mol. Mech. Mutagen.* **2003**, 533 (1–2), 67–97. https://doi.org/10.1016/j.mrfmmm.2003.08.021.

[25] Tharani Kumar, S.; Prasanna Devi, S.; Krithika, C.; Raghavan, R. Review of Metallic Biomaterials in Dental Applications. *J. Pharm. Bioallied Sci.* **2020**, 12 (5), 14. https://doi.org/10.4103/jpbs.JPBS_88_20.

[26] Chakraborty Banerjee, P.; Al-Saadi, S.; Choudhary, L.; Harandi, S.E.; Singh, R. Magnesium Implants: Prospects and Challenges. *Materials (Basel)* **2019**, 12 (1), 136. https://doi.org/10.3390/ma12010136.

[27] Cypryk, M.; Polymerization of Cyclic Siloxanes, Silanes, and Related Monomers. In *Polymer Science: A Comprehensive Reference*; Elsevier, **2012**; 451–476. https://doi.org/10.1016/B978-0-444-53349-4.00112-6.

[28] Kim, J.I.; Chun, C.; Kim, B.; Hong, J.M.; Cho, J.-K.; Lee, S.H.; Song, S.-C. Thermosensitive/Magnetic Poly(Organophosphazene) Hydrogel as a Long-Term Magnetic Resonance Contrast Platform. *Biomaterials* **2012**, 33 (1), 218–224. https://doi.org/10.1016/j.biomaterials.2011.09.033.

[29] Gleria, M.; De Jaeger, R. Polyphosphazenes: A Review. **2005**; pp 165–251. https://doi.org/10.1007/b100985.

[30] Rothemund, S.; Teasdale, I. Preparation of Polyphosphazenes: A Tutorial Review. *Chem. Soc. Rev.* **2016**, 45 (19), 5200–5215. https://doi.org/10.1039/C6CS00340K.

[31] Burkersroda, F.; Schedl, L.; Göpferich, A. Why Degradable Polymers Undergo Surface Erosion or Bulk Erosion. *Biomaterials* **2002**, 23 (21), 4221–4231. https://doi.org/10.1016/S0142-9612(02)00170-9.

4 Hydrogels for biofabrication

4.1 What are hydrogels?

There are different definitions for hydrogels either based on their chemical properties or their physical properties. One definition that is widely accepted defines hydrogels as water-soluble polymers that form a tridimensional network of chains connected at crosslinking points. Therefore, if we can control the crosslinking points, we can control the hydrogels. In the following sections, we will discuss how we can create such a tridimensional network. Hydrogels have several fields of applications including food for the making of slurry or as a thickening agent, for agricultural application for the delivery of water and nutrient to the plants by reducing the evaporation rate, in biomedical applications for DNA or protein separation by electrophoresis and for wound dressing or ointment to provide hydration. In the case of biofabrication, hydrogels are highly sought after as they reproduce the natural mechanical environment of natural tissues. Overall, the worldwide market for polymer gel represents several billions USD yearly and covers many fields of application. In cosmetics, hydrogels are used to make hydrating cream. In crude oil extraction, hydrogels due to their swelling properties are used to remove water and bind hydrophilic particles. Since crosslinking points are key to controlling the hydrogel properties, we classify them by their type of crosslinking: chemical or physical (Figure 4.1).

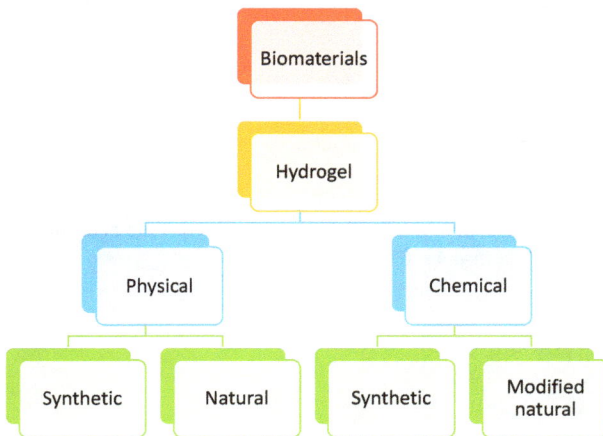

Figure 4.1: Classification of hydrogels by their mode of gelation and origin of the polymers.

https://doi.org/10.1515/9781501515736-004

4.2 Chemical hydrogels

Chemical hydrogels are formed by chemical crosslinking, meaning the crosslinking points are made through covalent bonds. These hydrogels are considered irreversible, meaning the crosslinking points cannot be removed and reformed. To form a hydrogel, we need to form a tridimensional network between the polymer chains. To do so, the polymerization needs to incorporate monomers with a functionality higher than two. Using a monomer with a functionality greater than two will allow two functional groups to be engaged in the chain growth and the remaining third functional group will be engaged in bindings between the polymer chains (Figure 4.2).

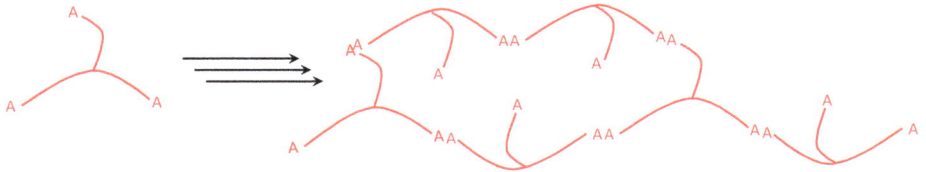

Figure 4.2: Radical polymerization approach where a monomer or monomer pool having total functionality higher than 2 can react to form a three-dimensional network.

The crosslinking of the polymer chains can be formed through two different strategies. One is to grow the polymer chain and then have a second step to create chemical bonds between the polymer chain. A second strategy consists in crosslinking the polymer chains during the chain growth.

4.A Swelling

Hydrogels composed majorly of water can swell as they can uptake water after their formation. By doing so, they will increase volume and mass using the equation below. The difference of mass can be measured, and from this measurement the swelling ratio of the hydrogel can be calculated. This is of particular importance superabsorbents such as a poly(acrylic acid) polymer that can uptake many times their weight in water (Figure 4.3).

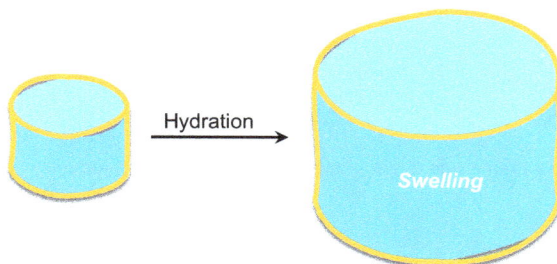

Figure 4.3: Swelling of hydrogel.

$$sw\% = \frac{(m_{swollen} - m_{not\ swollen})}{m_{not\ swollen}}$$

4.2.1 Poly(acryl amide)

Poly(acryl amide) (PAA) are hydrogels that are prepared by radical polymerization in water, following the first strategy presented above. Here, a total functionality of 3 is brought by using a combination of monomers: one monofunctional acrylamide and a bifunctional bisacrylamide (Figure 4.4). The polymerization is initiated by the addition of ammonium persulfate and the base tetramethylene diamine (TEMED). TEMED catalyzes the decomposition of the persulfate to give a free radical that will then transfer to the acrylamide and bis-acrylamide monomers. Adjusting the ratio of acrylamide and bisacrylamide allows the creation of hydrogels of different pore sizes and mechanical properties. This polymer is used for making hydrogel for electrophoresis but also to

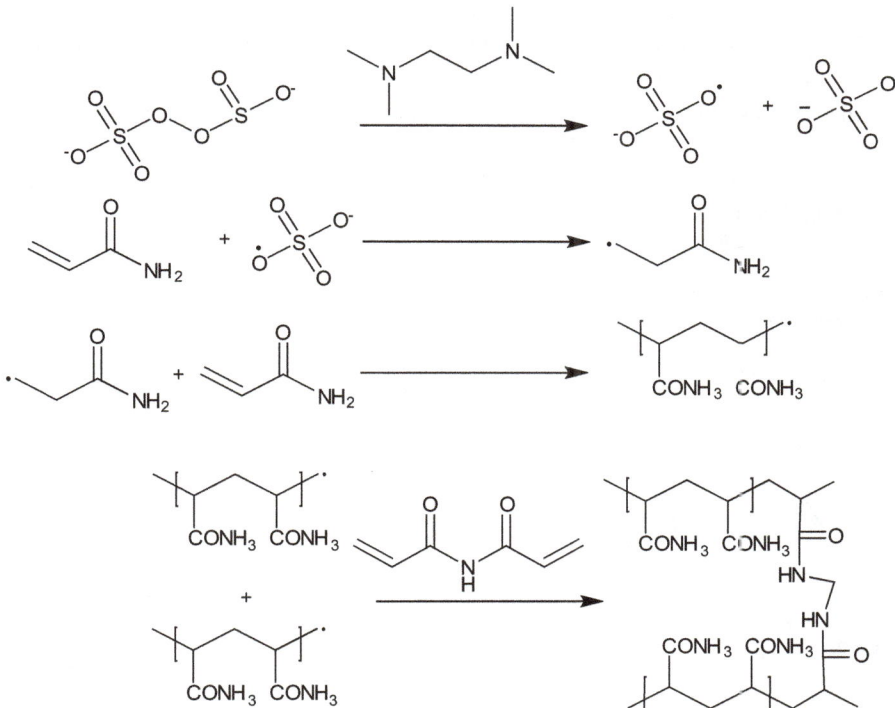

Figure 4.4: Polymerization of poly(acryl amide) three-dimensional hydrogel through radical polymerization.

make surfaces of various elastic moduli to grow cells on it. However, as for many polymers, acrylamide monomers are toxic and thus are not compatible with an in situ reaction. Meaning, all the monomers need to have reacted and need to be removed before having the hydrogel in contact with any living tissue or organism.

4.2.2 Poly(ethylene glycol)

Poly(ethylene glycol) (PEG) is a water-soluble polymer that can form a thick viscous solution at high polymer concentration. Crosslinked PEG hydrogels can be formed by reacting monomers with functionality greater than two. Using monomers that have three, four, or five branches or arms each of the arm terminated with a reactive end groups such as thiol or vinyl sulfone. For instance, mixing monomers with 4-arms that are terminated with vinyl sulfone with monomers with 4-arms that are terminated with thiol functional groups leads to a chemically crosslinked hydrogel through the reaction of thiols with vinyl sulfone (Figure 4.5). This approach can be applied with different functional groups that have different kinetic reaction rates, such as maleimide with thiol or carboxylic acid with an alcohol. Using coupling chemistry that is not toxic and that can occur in aqueous media under physiological conditions allows for the polymerization of these hydrogels in situ or with cells upon addition of the second pre-polymer (far instance (PEG–VS), the two precursors react with each others forming an hydrogel that encapsulate the cells. Meaning, cells can be suspended into a hydrogel precursor solution that contains one of the PEG reactive pre–polymer into the three-dimensional hydrogel network.

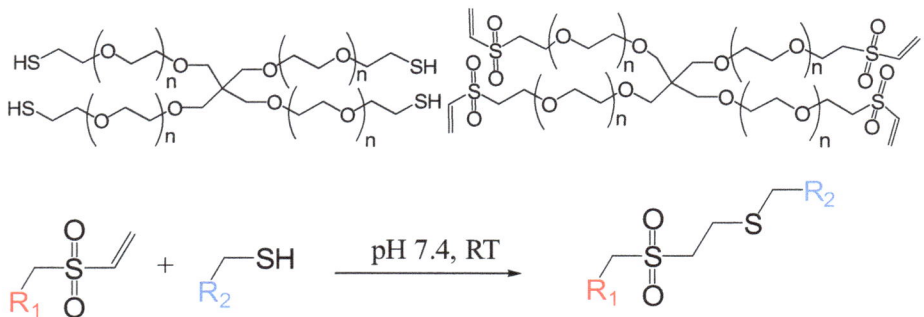

Figure 4.5: Polymerization of PEG hydrogel through a step-growth polymerization as presented in Figure 4.2.

4.2.3 Gamma irradiation

For a medical device or an implant to be used for biomedical applications, be it *in vivo* or *in vitro*, one must sterilize the objects or instrument before its use. Several methods are available including chemical (ethylene oxide) and physical (heat, ultraviolet light). The method of sterilization must be carefully chosen to make sure that the device is not impacted by the sterilization technique. For instance, using steam sterilization (121 °C for several minutes) on a thermoplastic material such as polycaprolactone, which has a T_m of 60 °C, will destruct the device. One of the techniques of choice for sterilization of industrial equipment after their manufacturing and packaging is γ radiation. Gamma rays are generated by a radioactive source and unlike α and β radiation, γ radiation can only be stoped by thick dense materials such as lead or concrete. The γ radiation forms by the radioactive decay of atomic nuclei and it is the shortest electromagnetic wave of about 8 MeV. But for this type of sterilization to be successfull, the materials be sterilized must be able to sustain irradiation. During the gamma irradiation process, radicals are formed that then can induce the polymerization or crosslink of polymer chains (Figure 4.6). Conversely, this can be used as an advantage where a hydrogel must be crosslinked for instance for the crosslinking of PVP hydrogels. In the PVP case, the final step of crosslinking and sterilization occurs in a single step after manufacturing and packaging. Sterilization facility using γ radiation can accommodate whole palettes per process. This is particularly efficient for the poly(styrene) plasticware commonly used in biology laboratories where a whole shipment can be sterilized in their packaging by gamma radiation.

Figure 4.6: Scheme of the chemical reaction occurring between two polymer chains forming radical by γ radiation.

Poly(vinyl pyrrolidone)

Poly(vinyl pyrrolidone) (PVP) is usually crosslinked by gamma radiation. It can then be processed as a firm hydrogel at the same time as undergoing sterilization. Its main application is for binders in pharmaceutical applications. The most commercially successful use of PVP is for its formulation with iodine to form the disinfectant betadine. To form hydrogel PVP has been often used as a blend with poly(vinyl alcohol) (PVA) or polysaccharides such as chitosan and cellulose. It is now used in many wound dressing formulations for its ease of production due to the final sterilization that is combined with the croslinking process.

Poly(acrylic acid)

Poly(acrylic acid) (PAA) is a polyelectrolyte type of polymer which makes it particularly suitable for holding water. It is the main polymer used for superabsorbent applications in diapers and hygienic products. It is not biodegradable and is polymerized by free radical polymerization with initiators such as azobisisobutyronitrile or potassium persulfate. The resulting polymer is biocompatible, water soluble, and has a glass transition around 100 °C. PAA can forms hydrogel by absorbing water, but it often requires to be crosslinked to reinforce the polymer network. This can be achieved by gamma irradiation of the polymer in solution as presented above [8].

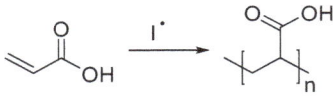

Figure 4.7: Scheme of the free radical polymerization chemical reaction to make poly(acrylic acid).

Further formulations of the acrylic polymers can be made with similar monomers including methacrylic acid and 2-hydroxyethyl methacrylate (HEMA). During the polymerization, the chains can undergo crosslinking instead of doing cyclization and thus form a three-dimensional network. Furthermore, HEMA can undergo photocrosslinking when used in conjecture with a photoinitiator and thus has found application in 3D printing.

4.2.4 Protein crosslinking

Not only synthetic polymers can be chemically and irreversibly crosslinked. Natural polymers can also be irreversibly crosslinked. Proteins are a perfect example. Made of amino acids with side chains made of reactive functional groups, these can be reacted with molecules with a functionality of two to crosslink the polymer chains. The main reactive functional groups in proteins are amine, which reacts rapidly with aldehyde and acetal.

Glutaraldehyde

Compose of an alkane chain terminated by two aldehyde functional groups, each molecule of glutaraldehyde can react with two polymer chains by a reaction with an amide (Figure 4.8). However, glutaraldehyde is cytotoxic and thus cannot be used for the crosslinking of hydrogel-containing cells, nor can it be used without proper removal of the residual unreacted glutaraldehyde. It is usually used to crosslink gelatine or collagen-based hydrogel or sponges. For the crosslinking of gelatine sponges, the gelatine hydrogel is dehydrated by freeze-drying and then crosslinked with glutaraldehyde. But the main usages of glutaraldehyde remain for the fixation of biological samples for staining in histology.

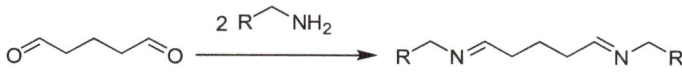

Figure 4.8: Scheme of the chemical reaction of glutaraldehyde with two molecules containing primary amine.

Genipin

Found in the fruit extract of *Genipa americana*, also called *Jagua azul* in Mexico, the genipin molecule can react with two primary amines. As such, it can be used to crosslink proteins by reacting with the amine side chains. It is much less toxic than glutaraldehyde and thus can be used for biomedical applications with less stringent precautions (Figure 4.9).

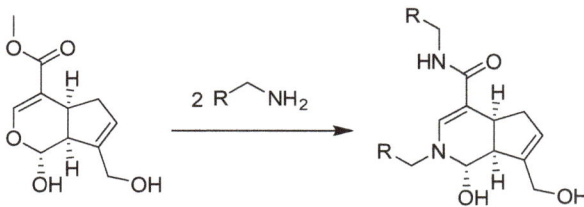

Figure 4.9: A. Scheme of the chemical reaction of the crosslinking of two protein molecules using genipin, a naturally occurring compound.

4.2.5 Photocrosslinking

Another method to induce the formation of radicals is the use of a photoinitiator derived from free radical polymerization. To date, there are several photoinitiators available that upon illumination with the correct wavelength decompose into two radical molecules. The radical once created can then propagate to the polymer or reactive species to form radical that can polymerize and crosslink into a hydrogel. One of the most well-known photopolymers used in biofabrication is gelatine methacrylate (Figure 4.10) (GelMA). It is a gelatine hydrogel that has been modified with methacrylate functional groups. This can be achieved by reacting the gelatine polymer that has an amine side group from the amino acid with anhydrous methacrylate. Beyond gelatine, other natural polymers have been methacrylated using the same process with anhydrous methacrylate: hyaluronic acid, alginate, chitosan, and collagen.

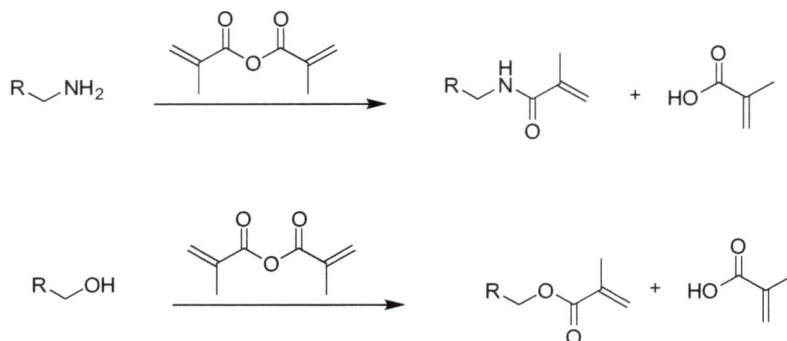

Figure 4.10: Scheme of the chemical reaction of methacrylate of an amine and alcohol functionalized polymer.

A photoinitiator is needed to create a radical. The most used photoinitiators are Irgacure 2959 (LAP) which is activated by light at 365 nm, and lithium phenyl-2,4,6-trimethylbenzoylphosphinate (LAP) at 405 nm (Figure 4.11).

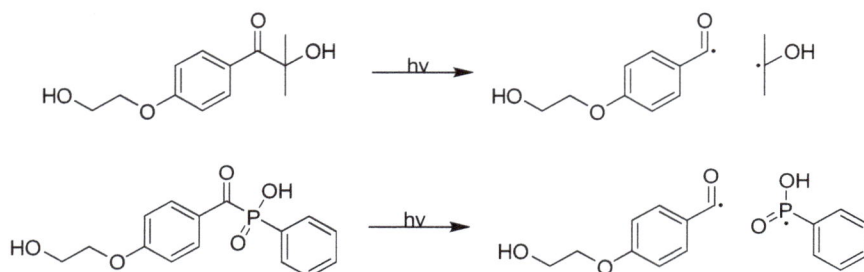

Figure 4.11: Mechanism of a radical formation of a photoinitiator Irgacure (up) and LAP (down).

Once activated, the photoinitiator propagates the radical to methacrylate groups which will, in turn, react with each other thus crosslinking the polymer chains (Figure 4.12). These sorts of systems are used for bioprinting. GelMA can be extruded in a bioprinter and then crosslinked post-extrusion to fix the printed hydrogel. In stereolithography bioprinting, a similar system can be used. However, here, one

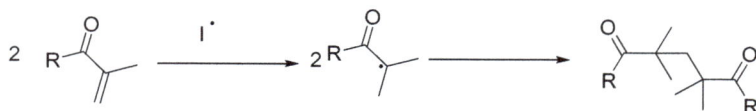

Figure 4.12: Crosslinking of a methacrylate polymer with a photoinitiator.

must be aware of the propagation of the radical beyond the illumination volume, called "shadow cure," and this can lead to a loss of resolution. These bioprinting processes will be discussed in more detail in the following chapters.

Further development of the photocrosslinking of hydrogel leads to the implementation of more complex systems. Great efforts have been focused on creating photoinitiator systems excited by visible light. Among them, Eosin-Y, riboflavin, and camphorquinone are good candidates to substitutes the conventional UV-activated systems. Indeed, UV irradiation is possibly harmful to the encapsulated cells. However, in comparison to UV light photoinitiators, visible light photoinitiators are rapidly cured under normal lighting conditions, as the energy level of the visible light is lower than that of UV light [1]. Another system, the two-photon polymerization (2PP) has been promising for the photocrosslinking of hydrogels to increase the printing resolution. 2PP utilizes high power near-infrared or infrared femtosecond lasers with initiators that generate radicals after absorbing two photons, unlike UV-vis-based photopolymerization where initiators produce radicals after one-photon absorption. If sufficient light power is provided at the focal point, a free radical can be initiated. This means that in 2PP, excitation only occurs where the laser beam is sharp which allows the precise spatial crosslinking. 2PP uses a laser between 700 and 1,050 nm in combination with initiators that are highly conjugated such as rose Bengal, benzophenone, or flavin adenine [2].

4.2.6 Click chemistry

Alternative crosslinking methods that rely on click chemistry reactive functional groups have emerged in the past decades. Click chemistry was defined by Sharpless in 2001 as reactions that are high yielding, stereospecific, simple to perform in a benign solvent, and create only by-products that can be removed without chromatography. Further development of these chemistries has led to the creation of a wide library that is constantly growing as summarized in Table 4.1. Click chemistry reactions are attractive for the crosslinking of hydrogels because the reactions can take place in water and do not have by-products, offering the possibility to create a system where two pre-polymers are mixed to form a hydrogel. These moieties have been extensively used for the functionalization of branched PEG. Each of these systems has a different reaction kinetic. Also, because these moieties are extremely reactive, some of these functional groups can have limited stability under physiological conditions. For instance, the system maleimide–thiol (Michael addition) reacts in seconds, but the maleimide group is stable only a few minutes at pH 7.4. Other systems like the inverse electronic demand Diels–Alder are reversible and have been used to make self-healing hydrogels. The hydrogel cut in two can be reformed into one piece of the hydrogel by rearrangement of the polymer chains and reaction of the diene with the dienophile [23]. Some of the cycloaddition moieties such as the strand

Table 4.1: Click chemistries systems for the crosslinking of polymers to form hydrogels.

	Name	Reaction mechanism
Cycloaddition	SPAAC	
	CuAAC	
	IEDDA	
	Diels–Alder	
Thiol reactive	Michael addition	Thiol–maleimide
		Thiol–acrylate
		Thiol–vinyl sulfone
		Thiol–yne

Table 4.1 (continued)

	Name	Reaction mechanism
Aldehyde	Oxyme	
	Hydrazone	

promoted are highly reactive but require multistep reaction to obtain the final product and, thus are now expensive and limited for highly valuable applications that needs only a limited amount of reagent such as for microscopy labeling experiments. Further development of the click chemistries systems and new synthesis routes for these now expensive moities will lead to reducing cost and the opportunity to use these for hydrogel formation. Another use of these chemistries is for the functionalization of hydrogel with biological molecules. For instance, azide functionalized proteins can be reacted with alkyne functionalized hydrogel with a copper catalyst to form biologically defined hydrogels.

4.3 Physical hydrogels

Synthetic polymers can form reversible hydrogels through non-covalent crosslinking or non-reversible hydrogels upon chemical crosslinking of the polymer network. Other polymers such as PVA form physical hydrogels through hydrophobic interactions, and has been used for the generation of 3D cartilage microenvironments [5]. A distinct advantage of synthetic polymers over naturally sourced biopolymers is the relative ease with which the polymers can be engineered for a dedicated application by modifying their chemical structure, and in turn, physical properties. Likewise, natural polymers can be crosslinked chemically as described earlier in this. But most of the natural polymer forming hydrogel occur through physical crosslinking.

4.3.1 Synthetic physical hydrogels

Poloxamer
These are a family of block copolymers formed from poly(ethylene oxide) and poly(propylene oxide) (PPO) marketed under the name of Pluronic. To date, there

exists a great variety of these polymers as many combinations of and PPO can be made. Indeed, the molecular weight of each block can be varied to modulate the gelling properties of the polymer. Their main use is in food and texture modifiers for the cosmetic industry but they also seen great interest for biomedical applications.

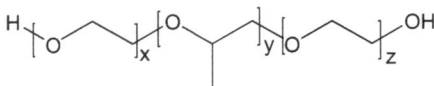

Figure 4.13: Pluronic block copolymer chemical structure made of poly(ethylene oxide) and poly(propylene oxide).

So, the PPO-PEO combination called F127 gives a block copolymer with a lower critical solution temperature (LCST) meaning that at low temperature the polymer is in solution and at high temperature, a gel is formed. Depending on the concentration of the polymer in the solution other organizations of the polymer chains can be obtained including micelles. This organization is due to the aggregation of the PPO chains that are more hydrophobic than the PEO chains (Figure 4.14). An interchain hydrophobic organization that is driven by the temperature and the concentration of the polymer in solution leads to the micellar solution or gel organization of the polymer. Typical hydrogel formation of Pluronic F127 is obtained for solutions concentration above 30% w/w at a temperature above 20 °C, transition from gel to solution occurs at 4 °C or by boiling off the gel.

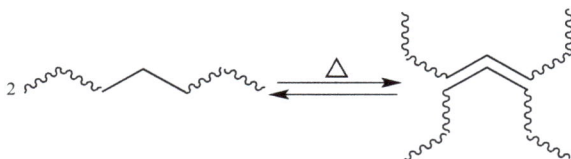

Figure 4.14: Pluronic undergo a thermally-controlled gelation by the aggregation of the hydrophobic block into a crosslinking point when the polymer is solubilized in water.

4.3.2 Natural physical hydrogels

Natural polymers that have been used as synthetic ECM scaffolds are hydrogel-forming biomaterials originating from animals, plants, or microorganisms. These polymers are either polysaccharides or proteins and some of them can be efficiently sourced and obtained in high purity. However, because of their origin, one must be aware of the potential batch-to-batch reproducibility which can be dependent on the source of raw materials or external factors. For instance, algae-extracted polysaccharides can have seasonal variability and their properties can depend on their harvesting area.

4.3.3 Polysaccharide-based (vegetal)

Cellulose

Cellulose is a naturally occurring polysaccharide composed of a repeat unit of β-D-anhydroglucose and can be obtained from plants, algae, and bacteria (Figure 4.15). Hydrogels are made from nanocellulose colloids, which are solid particles dispersed in water. Nanocellulose is obtained through the chemical depolymerization of the raw polysaccharide. The obtained colloids behave like a fluid in the diluted state and like a gel at higher concentrations. The obtention of a hydrogel from the colloidal dispersion of nanocellulose is mainly dependent upon the aspect ratio and volume fraction of the colloids. The colloidal dispersion confers these hydrogel shear-thinning properties. Such properties enable the hydrogel to be injected through small needles and this is a highly attractive property as a rheology modifier in 3D printing inks. However, in contrast to other polysaccharide gels, such as agarose, cellulose hydrogels are not highly stable and can be redispersed in water by shearing. Therefore, the formulation of pure nanocellulose hydrogel is limited and it is often combined with another polymer-forming hydrogel.

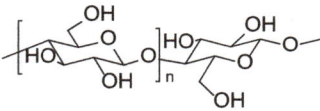

Figure 4.15: Chemical structure of cellulose.

Alginate

One of the most used polysaccharides in biofabrication is alginate, which gels upon complexation with divalent cations. The polysaccharide structure varies greatly depending on the seaweed growth environment, algae species, and extracted algae tissue. It is composed of two saccharides, the mannuronate (M) and guluronate (G) arranged in sequences of M- and G-block regions and randomly inserted M and G units (MG-blocks) (Figure 4.16). The M/G ratio is defined by the ratio of mannuronate to guluronate.

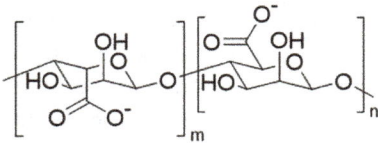

Figure 4.16: Chemical structure of alginate.

Alginate forms a gel in presence of divalent cations such as Ca^{2+}, following the egg-box model. The divalent cations complexes with the carboxylate groups of the G-blocks, while the M-blocks have a lower affinity to complexation (Figure 4.17). Because cations have a higher affinity of divalent cation toward the G-blocks, the gel properties will greatly depend on the M/G ratio and the G-blocks length. Alginate with higher G content and a low M/G ratio will therefore produce a stiffer gel with higher gel strength than alginate with a high M/G ratio. Therefore, it is critical to know the M/G ratio to have reproducible rheological properties of the formed hydrogel. Finally, the nature of the cation will also affect the hydrogel strength. For instance, Ca^{2+} exhibits stronger interactions with the alginate than Mg^{2+}.

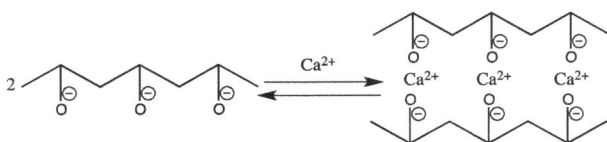

Figure 4.17: Alginate crosslinking mechanism through the complexation of divalent cations, here Ca^{2+}.

Agarose

Extracted from red seaweed, agarose is isolated from agar, a mixture of polysaccharides including agaropectin, a non-gelling polymer. Agarose's backbone is composed of β-D-galactose and 3,6-anhydro-α-L-galactose (3,6 AG) (Figure 4.18). Changes in the composition and structure of agarose polysaccharides such as the presence of α-L-galactose and other minor substituents (sulfate, methyl ether, pyruvic acid) [9] are known to occur depending on the species [9, 10] and seasons [11, 12].

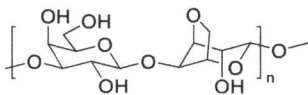

Figure 4.18: Chemical structure of agarose.

The composition of agarose controls the formation of secondary structures of the polysaccharide which then impacts its gelation mechanism that occurs through a phase separation mechanism, involving the formation of double helixes in the polymer backbone and aggregations of these helices into crosslinking points creating a 3D hydrogel network (Figure 4.19). Agarose is a thermoreversible hydrogel that has been widely used for electrophoresis. It is an interesting hydrogel for biomedical applications because it exhibits hysteresis. The transition from solution to gel occurs at a lower temperature (below 37 °C) than the transition from gel to a solution (above 85 °C). It is therefore stable at physiological temperature and forms a translucent hydrogel. There are to date several extraction methods for agarose and each type of extraction can lead to a polysaccharide with different gelling temperatures. Agarose can be obtained as a low melting and high melting temperature hydrogel. Due to its availability and specific mechanical properties, agarose has been a material of choice for the generation of 3D cell microenvironments for the culture and differentiation of chondrocytes [13].

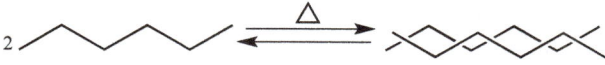

Figure 4.19: Agarose crosslinking mechanism through the aggregation of secondary structure that can be disrupted by heat.

κ-carrageenan
Carrageenans are sulfated polysaccharides extracted from red seaweed. Their backbone is composed of β-D-galactose and 3,6-AG with sulfate groups on the galactose units [14] Figure 4.20. They form hydrogel through an alpha-helical aggregation, like agarose. The resulting hydrogel is strong and brittle. Like other algae-extracted polysaccharides, factors such as species seasons growth conditions, and extraction conditions are influencing the 3,6-AG and sulfate content which in turn alters the helix formation leading to different gel properties [21].

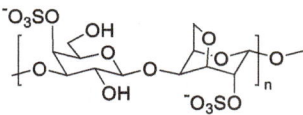

Figure 4.20: Chemical structure of the repeat unit of κ-carrageenan.

Unlike agarose, κ-carrageenan goes through a coil-to-helix transition upon addition of cations, leading to the formation of double helices. In κ-carrageenan, the helix formation is followed by further helix aggregation [22]. The type of cations used to induce the gel formation will impact the mechanical properties of the hydrogel. For instance, κ-carrageenan forms a stronger gel with K^+ than with Na^+ [25]. Not only cations but also some anions such as I^- and SCN^- have been reported to bind to the helix influencing the gelation mechanism by impeding helix aggregations and gelation.

4.3.4 Polysaccharide from microorganism

Dextran
Beyond plants and algae, microorganisms also produce polysaccharides that can form hydrogels, and these can be used for biomedical applications. Dextran is obtained from lactic acid bacteria such as *Leuconostoc mesenteroides* or mutants *Streptococcus sp.*, and is widely used in medicine since the early 1960s as a blood volume expander. Low molecular weight Dextran is also used to increase blood flow and reduce blood viscosity. Dextran neutral polysaccharide chain is composed of α-glucan with glucopyranoside side chains (Figure 4.21). It is freely soluble in water but can be formulated as

a hydrogel at a higher concentration. With the use of dextran for medical applications came the establishment of a supply chain of medical-grade materials. Naturally, dextran has fuelled some hydrogel-based innovation for drug delivery systems or cell encapsulation. But being synthesized by bacteria with many side chains, in comparison to linear polymers, it is often difficult to have a precise characterization of the polysaccharide and all its side chains. This made it rather challenging to chemically modified for further applications.

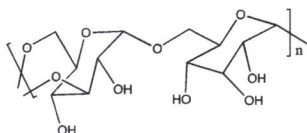

Figure 4.21: Chemical structure of the main linear chain of Dextran.

Gellan gum

The negatively charged polysaccharide gellan gum (GG) is produced by *Sphingomonas elodea* found in the lily plant. Initially used as a replacement for agar for microbiological culture, it is mainly used as a food thickener. The backbone repeat unit is composed of glucose, glucuronic acid, and rhamnose which allow for the easy functionalization of the backbone through reaction with the carboxylic acid of the glucuronic acid (Figure 4.22). The hydrogel formation is thermoreversible and the gel is believed to aggregate through the formation of a double helix and subsequent coil aggregation. The addition of cation to the solution reinforces the hydrogel which is necessary to reach a high gel strength. It has seen research application in the field of tissue engineering for cartilage repair and bone by incorporating inorganic phases such as hydroxyapatite [28]. Due to its use as a food additive, research has been conducted to use it in the formulation of drug delivery systems. Because it can be produced in high quantity in a reproducible manner, GG could be a good candidate for biofabrication application.

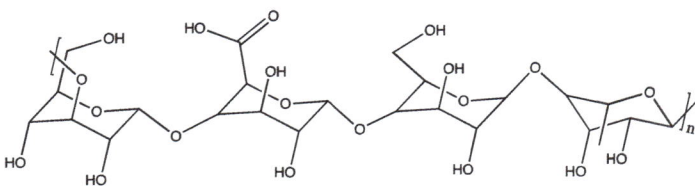

Figure 4.22: Chemical structure of Gellan Gum.

Xanthan gum

A widely used food stabilizer and formulation thickening extracted from the bacteria *Xanthomonas campestris* is xanthan gum (XG). The repeat unit is quite complex and is composed of a glucose backbone, mannose, and glucuronic acid side chain attached to every other glucose [29]. The organization of the backbone chain impacts the gel-to-sol transition temperature. It is a versatile polysaccharide for biomedical applications [30]. The hydrogel formation can be triggered by divalent cation through the complexation with the glucuronic acid sugar. Alternatively, crosslinking is possible through inorganic sodium triphosphate or after chemical modification of the backbone by an organic crosslinker such as methacrylation. The XG formulated with hydroxyapatite phase was proposed for a bone implant and blends of XG with other polysaccharides were investigated for tissue engineering applications *in vivo* [31]. Due to its low cytotoxicity and low immune response, XG is a material to be taken into consideration for biofabrication.

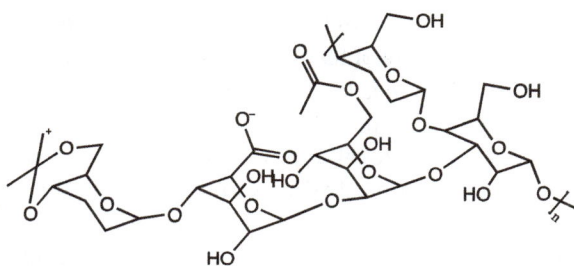

Figure 4.23: Chemical structure of Xanthan polysaccharide.

4.3.5 Polysaccharide from animal

Chitosan
An example of a polysaccharide sourced from animals is chitin, which is extracted in its deacetylated from chitosan, material forming crustacean shells, a by-product of the food industry [32] Figure 4.24. Chitosan-based hydrogels are stable under physiological conditions and have been shown to possess antimicrobial properties, as well as blood clotting properties [33]. It is now possible to obtain chitosan as a medical-grade material produced according to good manufacturing procedures. Thus, it is to expect that new products incorporating chitosan in their formulation will come to the market. It is the only polysaccharide in this list being positively charged and having an amine on its repeat unit. This is of particular importance as it allows for the easy functionalization of the polysaccharide through alternative synthetic routes as with the other polysaccharide listed above. Chitosan hydrogel can be formed by various routes either physical through the blending with glycerophosphate and chemical crosslinking by glutaraldehyde reaction with the amine groups for instance.

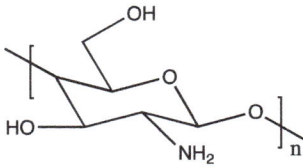

Figure 4.24: Chemical structure of chitosan repeat unit.

4.3.6 Protein-based hydrogels

Fibroin, Silk

The most common producer of fibroin is the *Bombyx mori* worm which produces the silk used in textile (Figure 4.25). It produced the protein fibroin that can be processed into silk fibers. The organization of the protein into anti-parallels β-sheets makes it a resistant fiber that was used for the manufacturing of textile. Once extracted, the protein can be processed into a hydrogel [35]. Fibroin is known to be safe, non-cytotoxic and many companies are now scaling up the production of medical-grade materials. In its regenerated solution, fibroin will form hydrogel at physiological temperature. However, the physical hydrogel is regarded as a weak hydrogel for many applications in tissue engineering. Indeed, the resistant fibers need to be dissociated to be solubilized and this can alter its structure. It is therefore often further crosslinked using chemical methods. Alternatively, the hydrogel can be processed as a carrier for drug delivery application, as a thickening agent, or be processed as a porous sponge.

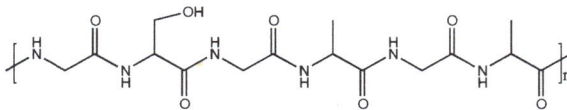

Figure 4.25: Chemical structure of the repeat unit of fibroin protein.

Gelatin

Gelatin is a transparent colorless protein derived from collagen extracted from bone and cartilage of animal tissues. It is obtained by chemical or physical treatment (heat) of bones, which partially hydrolyze collagen present in the tissue. The protein obtained is not able to reform the full triple-helix distinctive of collagen proteins. Gelatine forms a gel at room temperature and is used in many cosmetic formulations, food additives, and more recently as a material of choice for biofabrication. However, the hydrogel formed by gelatine is quite weak. Therefore, gelatine is mostly used in its chemically modified methacrylate form (see paragraph above) which permits chemically crosslinking of the hydrogel by photopolymerization of the methacrylate functional groups. Although its application for medical application raises some concern in terms of the safety profile of the animal extracted materials, it has found a niche application in the bioprinting of tissue models for *in vitro* application. These model tissues are not intended for implantation and thus GelMA is adequate.

4.B Viscosity

It is the measure of the resistance of fluid or a solution of polymer to deformation. It is the physical expression of the thickness of a solution. For instance, mixing water and plain flour would give a thick solution as the ratio of flour increases compared to water, the solution becomes thicker and thicker until it flows extremely slowly that it is not visible to the naked eye. Similarly, one can say that ketchup is thicker than water. We would say that the viscosity of water is lower than this of ketchup. The viscosity will correlate with the capacity of a solution or blend of the solution to flow. If we apply a force to a viscose material, this material will move in the direction of the force (Figure 4.26).

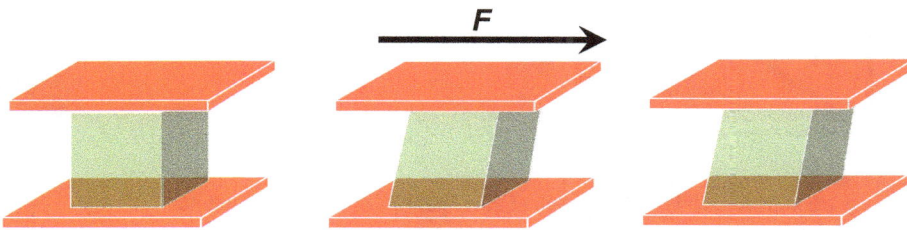

Figure 4.26: Hydrogel viscosity. The force applied to the material irreversibly deforms the hydrogel.

4.C Elasticity

The elasticity is the ability of a deformable body to resist the distorting effect of an applied force. An elastic material that has an elastic behavior will regain its original force upon the application of a force. If we take a hydrogel and apply a shearing force to start its flowing, if the hydrogel has an elastic behavior it will not move, but deform and regain its original shape (Figure 4.27).

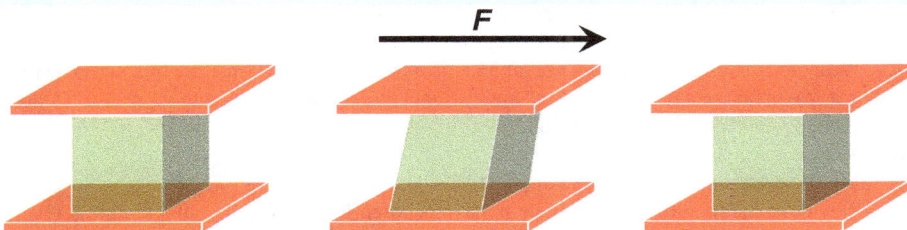

Figure 4.27: Hydrogel elasticity. The force applied to the material deforms the hydrogel that comes back to its initial shape.

4.D Viscoelastic behavior

Hydrogel, being a solid that has high content of water, has some of the properties of water and some of a solid. Meaning that they have a viscous behavior (from the water) and an elastic behavior (from a solid). As such, we say that they have a viscoelastic behavior or that they are viscoelastic solids. But to better describe these materials, one wants to know if they are more viscous than elastic. These means are they flowing or are they staying as a solid and do not flow. To define this, the complex shear modulus is used (Figure 4.28). The complex shear modulus describes the elastic (G') and viscous (G'') components of the material into a vector. The phase shift reflects the ratio of a viscous component with an elastic component.

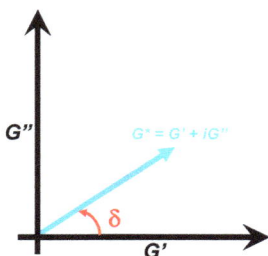

Figure 4.28: Mathematical representation of the shear (G') and loss (G'') modulus.

4.4 Quiz

1. What is the difference between physical and chemical hydrogels?
2. What is the minimum functionality required to polymerize a crosslinked polymer network?
3. What are the two components used to describe the mechanical behavior of hydrogels?
4. What are the three possible sources of hydrogels?
5. Which sugar unit is responsible for the chelation with cation in alginate?
6. Which functional group is used to create photocrosslinkable hydrogel from gelatine?
7. We measure the mass of a hydrogel at 34 g. In a container, we submerged this hydrogel for 2 h. We measure again the mass of the hydrogel and it has increased to 45 g. What is the swelling percentage of the hydrogel?
8. What is the LCST?
9. Give an example of a polymer having an LCST
10. Why does agarose has a hysteresis behaviors when it comes to its sol–gel and gel–sol temperature?

References

[1] Choi, J.R.; Yong, K.W.; Choi, J.Y.; Cowie, A.C. Recent Advances in Photo-Crosslinkable Hydrogels for Biomedical Applications. *Biotechniques* **2019**, 66 (1), 40–53. https://doi.org/10.2144/btn-2018-0083.

[2] Ciuciu, A.I.; Cywiński, P.J. Two-Photon Polymerization of Hydrogels – Versatile Solutions to Fabricate Well-Defined 3D Structures. *RSC Adv.* **2014**, 4 (85), 45504–45516. https://doi.org/10.1039/C4RA06892K.

[3] Weber, L.M.; He, J.; Bradley, B.; Haskins, K.; Anseth, K.S. PEG-Based Hydrogels as an in Vitro Encapsulation Platform for Testing Controlled Beta-Cell Microenvironments. *Acta Biomater.* **2006**, 2 (1), 1–8. https://doi.org/10.1016/j.actbio.2005.10.005.

[4] Peyton, S.R.; Raub, C.B.; Keschrumrus, V.P.; Putnam, A.J. The Use of Poly(Ethylene Glycol) Hydrogels to Investigate the Impact of ECM Chemistry and Mechanics on Smooth Muscle Cells. *Biomaterials* **2006**, 27 (28), 4881–4893. https://doi.org/10.1016/j.biomaterials.2006.05.012.

[5] Taguchi, T.; Kishida, A.; Akashi, M. Hydroxyapatite Formation On/in Poly(Vinyl Alcohol) Hydrogel Matrices Using a Novel Alternate Soaking Process. *Chem. Lett.* **1998**, 711–712. https://doi.org/10.1246/cl.1998.711.

[6] Sarem, M.; Moztarzadeh, F.; Mozafari, M. How Can Genipin Assist Gelatin/Carbohydrate Chitosan Scaffolds to Act as Replacements of Load-Bearing Soft Tissues?. *Carbohydr. Polym.* **2013**, 93 (2), 635–643. https://doi.org/10.1016/j.carbpol.2012.11.099.

[7] Forget, A.; Christensen, J.; Ludeke, S.; Kohler, E.; Tobias, S.; Matloubi, M.; Thomann, R.; Shastri, V.P. Polysaccharide Hydrogels with Tunable Stiffness and Provasculogenic Properties via -helix to -sheet Switch in Secondary Structure. *Proc. Natl. Acad. Sci.* **2013**, 110 (32), 12887–12892. https://doi.org/10.1073/pnas.1222880110.

[8] Nho, Y.-C.; Park, J.-S.; Lim, Y.-M. Preparation of Poly(Acrylic Acid) Hydrogel by Radiation Crosslinking and Its Application for Mucoadhesives. *Polymers (Basel)* **2014**, 6 (3), 890–898. https://doi.org/10.3390/polym6030890.

[9] Lahaye, M.; Rochas, C. Chemical Structure and Physico-Chemical Properties of Agar. *Hydrobiologia* **1991**, 221 (1), 137–148. https://doi.org/10.1007/BF00028370.

[10] Zhang, Y.; Fu, X.; Duan, D.; Xu, J.; Gao, X. Preparation and Characterization of Agar, Agarose, and Agaropectin from the Red Alga Ahnfeltia Plicata. *J. Oceanol. Limnol.* **2019**, 37 (3), 815–824. https://doi.org/10.1007/s00343-019-8129-6.

[11] Marinho-Soriano, E.; Bourret, E. Effects of Season on the Yield and Quality of Agar from Gracilaria Species (Gracilariaceae, Rhodophyta). *Bioresour. Technol.* **2003**, 90 (3), 329–333. https://doi.org/10.1016/S0960-8524(03)00112-3.

[12] Givernaud, T.; El Gourji, A.; Mouradi-Givernaud, A.; Lemoine, Y.; Chiadmi, N. Seasonal Variations of Growth and Agar Composition of Gracilaria Multipartita Harvested along the Atlantic Coast of Morocco. *Hydrobiologia* **1999**, 398–399, 167–172. https://doi.org/10.1007/978-94-011-4449-0_19.

[13] Schuh, E.; Hofmann, S.; Stok, K.S.; Notbohm, H.; Müller, R.; Rotter, N. The Influence of Matrix Elasticity on Chondrocyte Behavior in 3D. *J. Tissue Eng. Regen. Med.* **2012**, 6 (10), e31–42. https://doi.org/10.1002/term.501.

[14] Van De Velde, F.; Knutsen, S.H.; Usov, A.I.; Rollema, H.S.; Cerezo, A.S. 1H and 13C High Resolution NMR Spectroscopy of Carrageenans: Application in Research and Industry. *Trends Food Sci. Technol.* **2002**, 13 (3), 73–92. https://doi.org/10.1016/S0924-2244(02)00066-3.

[15] Adharini, R.I.; Suyono, E.A.; Suadi; Jayanti, A.D.; Setyawan, A.R. A Comparison of Nutritional Values of Kappaphycus Alvarezii, Kappaphycus Striatum, and Kappaphycus Spinosum from the Farming Sites in Gorontalo Province, Sulawesi, Indonesia. *J. Appl. Phycol.* **2019**, 31 (1), 725–730. https://doi.org/10.1007/s10811-018-1540-0.

[16] Anderson, N.S.; Dolan, T.C.S.; Penman, A.; Rees, D.A.; Mueller, G.P.; Stancioff, D.J.; Stanley, N.F. Carrageenans. Part IV. Variations in the Structure and Gel Properties of κ-Carrageenan, and the Characterisation of Sulphate Esters by Infrared Spectroscopy. *J. Chem. Soc. C Org.*, **1968**, 602, 602–606. https://doi.org/10.1039/J39680000602.

[17] Freile-Pelegrín, Y.; Robledo, D. Carrageenan of Eucheuma Isiforme (Solieriaceae, Rhodophyta) from Yucatán, Mexico. II. Seasonal Variations in Carrageenan and Biochemical Characteristics. *Bot. Mar.* **2006**, 49 (1), 72–78. https://doi.org/10.1515/BOT.2006.009.

[18] Dawes, C.J.; Lawrence, J.M.; Cheney, D.P.; Mathieson, A.C. (Rhodophyta, Gigartinales). III. Seasonal Variation of Carrageenan, Total Carbohydrate, Protein, and Lipid. *Bull. Mar. Sci.* **1973**, 24 (2), 286–299.

[19] Zinoun, M.; Cosson, J.; Deslandes, E. Influence of Culture Conditions on Growth and Physicochemical Properties of Carrageenans in Gigartina Teedii (Rhodophyceae – Gigartinales). *Bot. Mar.* **1993**, 36, 131–136. https://doi.org/10.1515/botm.1998.41.1-6.299.

[20] Hilliou, L.; Larotonda, F.D.S.; Abreu, P.; Ramos, A.M.; Sereno, A.M.; Gonçalves, M.P. Effect of Extraction Parameters on the Chemical Structure and Gel Properties of κ/ι-Hybrid Carrageenans Obtained from Mastocarpus Stellatus. *Biomol. Eng.* **2006**, 23 (4), 201–208. https://doi.org/10.1016/j.bioeng.2006.04.003.

[21] Thành, T.T.T.; Yuguchi, Y.; Mimura, M.; Yasunaga, H.; Takano, R.; Urakawa, H.; Kajiwara, K. Molecular Characteristics and Gelling Properties of the Carrageenan Family, 1: Preparation of Novel Carrageenans and Their Dilute Solution Properties. *Macromol. Chem. Phys.* **2002**, 203 (1), 15–23. https://doi.org/10.1002/1521-3935(20020101)203:1<15::AID-MACP15>3.0. CO;2-1.

[22] Anderson, N.S.; Campbell, J.W.; Harding, M.M.; Rees, D.A.; Samuel, J.W.B. X-Ray Diffraction Studies of Polysaccharide Sulphates: Double Helix Models for κ- and ι-Carrageenans. *J. Mol. Biol.* **1969**, 45 (1), 86–99. https://doi.org/10.1016/0022-2836(69)90211-3.

[23] Arnott, S.; Scott, W.E.; Rees, D.A.; McNab, C.G.A. I-Carrageenan: Molecular Structure and Packing of Polysaccharide Double Helices in Oriented Fibres of Divalent Cation Salts. *J. Mol. Biol.* **1974**, 90 (2), 263–267. https://doi.org/10.1016/0022-2836(74)90371-4.

[24] Morris, E.R.; Rees, D.A.; Robinson, G. Cation-Specific Aggregation of Carrageenan Helices: Domain Model of Polymer Gel Structure. *J. Mol. Biol.* **1980**, 138 (2), 349–362. https://doi.org/ 10.1016/0022-2836(80)90291-0.

[25] Michel, A.S.; Mestdagh, M.M.; Axelos, M.A.V. Physico-Chemical Properties of Carrageenan Gels in Presence of Various Cations. *Int. J. Biol. Macromol.* **1997**, 195–200. https://doi.org/ 10.1016/S0141-8130(97)00061-5.

[26] Norton, I.T.; Goodall, D.M.; Morris, E.R.; Rees, D.A. Role of Cations in the Conformation of Iota and Kappa Carrageenan. *J. Chem. Soc. Faraday Trans. 1 Phys. Chem. Condens. Phases* **1983**, 79 (10), 2475–2488. https://doi.org/10.1039/F19837902475.

[27] Spera, R.; Gellan Gum for Tissue Engineering Applications: A Mini Review. *Biomed. J. Sci. Tech. Res.* **2018**, 7, 2. https://doi.org/10.26717/BJSTR.2018.07.001474.

[28] Kool, M.M.; Gruppen, H.; Sworn, G.; Schols, H.A. The Influence of the Six Constituent Xanthan Repeating Units on the Order–Disorder Transition of Xanthan. *Carbohydr. Polym.* **2014**, 104, 94–100. https://doi.org/10.1016/j.carbpol.2013.12.073.

[29] Petri, D.F.S.; Xanthan Gum: A Versatile Biopolymer for Biomedical and Technological Applications. *J. Appl. Polym. Sci.* **2015**, 132, 23. https://doi.org/10.1002/app.42035.

[30] Kumar, A.; Rao, K.M.; Han, S.S. Application of Xanthan Gum as Polysaccharide in Tissue Engineering: A Review. *Carbohydr. Polym.* **2018**, 180, 128–144. https://doi.org/10.1016/j. carbpol.2017.10.009.

[31] Severian, D. *Polysaccharides*; Dumitriu, S., Ed.; CRC Press, **2004**. https://doi.org/10.1201/ 9781420030822.

[32] Okamoto, Y.; Yano, R.; Miyatake, K.; Tomohiro, I.; Shigemasa, Y.; Minami. S. Effects of Chitin and Chitosan on Blood Coagulation. *Carbohydr. Polym.* **2003**, 53 (3), 337–342. https://doi.org/10.1016/S0144-8617(03)00076-6.

[33] Je, J.-Y.; Kim, S.-K. Antimicrobial Action of Novel Chitin Derivative. *Biochim. Biophys. Acta* **2006**, 1760 (1), 104–109. https://doi.org/10.1016/j.bbagen.2005.09.012.

[34] Zheng, H.; Zuo, B. Functional Silk Fibroin Hydrogels: Preparation, Propert es and Applications. *J. Mater. Chem. B* **2021**, 9 (5), 1238–1258. https://doi.org/10.1039/D0TB02099K.

5 Synthetic and natural extracellular matrix

5.1 What is the ECM?

The extracellular matrix (ECM) is defined as the non-cellular portion of tissue. It is made of an ensemble of macromolecules that are making the three-dimensional network that provides structural and biochemical support to the cells (Figure 5.1). The ECM assumes diverse functions ranging from mechanical support, cell adhesion, and intercellular communication. In the human body, the ECM composition varies to assume the diverse mechanical and biological properties of the various tissues. For instance, the ECM of the skin will be elastic whereas the ECM of bone will be mineralized to resist compression. The ECM is mainly composed of cell-secreted macromolecules such as polysaccharides and proteins that form the hydrogel materials that provides the 3D mechanical network required for the cells to form a tissue. In addition, growth factors (GFs) and chemokines secreted by the cells are diffusing through or bound to the ECM and assure intercellular communication. Cell-secreted matrix metalloproteinase (MMP) enzymes are capable to degrade the ECM.

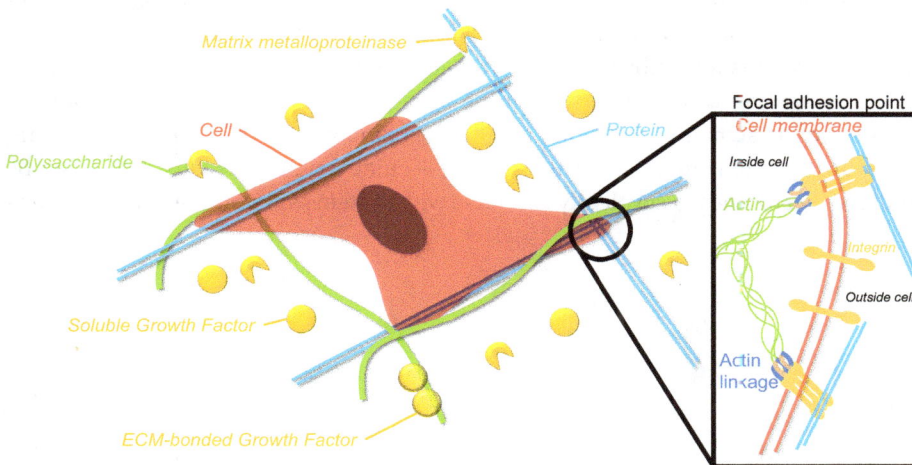

Figure 5.1: The cell in the extracellular matrix (ECM) is composed of macromolecules, growth factors, and degrading enzymes.

https://doi.org/10.1515/9781501515736-005

5.2 The composition of the ECM

The ECM is made of polymeric materials comprising polysaccharides and proteins. These biomacromolecules are secreted by cells and are assembled to form complex ECM structures with specific biological and mechanical properties (Table 5.1).

5.2.1 Glycosaminoglycans

In the ECM, glycosaminoglycans (GAGs) are anionic polysaccharides that can be subdivided into two categories: sulfated and non-sulfated. Heparin, a sulfated extracellular GAG first identified in the blood in 1919 [1], plays a major role as a blood thinner and during the wound healing process to inhibit blood coagulation. Following this discovery, a range of GAGs have been identified including hyaluronic acid (HA), which is the only non-sulfated polysaccharide found in mammalian tissues. HA is one of the major components of the ECM and plays a role in cell proliferation and migration. In particular, HA can bind to the cell receptor CD44, which is involved in cell adhesion and allows cell migration [2]. Like other GAGs found in the ECM such as chondroitin sulfate and heparin sulfate, HA exhibits specific viscoelastic properties that play a role in regulating the tissue mechanical properties such as high compressive modulus and lubrication of articulations [3]. The anionic charge of GAGs promotes binding with GFs, including the fibroblast growth factor family, allowing them to act as signaling reservoirs capable of immobilizing or releasing GFs as required [4, 5]. While already forming high molecular weight polymers, GAGs can be modified by enzymes to covalently assemble with proteins and form the proteoglycans

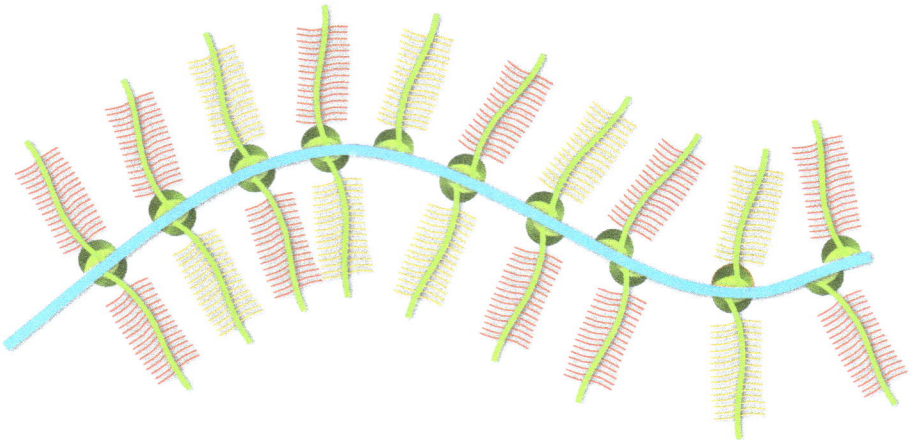

Figure 5.2: Glycosaminoglycan (GAG) formed by a hyaluronic acid backbone (blue), link to a core protein (green) on which keratan sulfate (orange) and chondroitin sulfate (red) can attach.

(PGs) super-family. In that assembly, the GAGs form branche perpendicular to a core protein and are often connected to it by a tetra-saccharide composed of a glucuronic acid (GlcA), two galactoses, and a xylose residue (Figure 5.2) [6]. However, due to the complex biosynthesis of GAGs and PGs that involve different enzymes, most of their role in the ECM has not yet been fully understood. Currently, only heparin and HA have found clinical and cosmetic applications and can be affordably sourced either as a recombinant molecule or directly from animals [7, 8].

5.2.2 Proteins

In ECM, proteins are secreted by cells and controlled by the activation of specific genes. The most abundant protein, collagen is found in different variations depending on the gene code utilized and post-transcriptional modification that occurs inside and outside of the cells. Collagen, like the protein laminin, forms a triple helix and can interact with other proteins such as nidogen which linked these two proteins together to form the basal lamina of cardiac and lung tissues [9]. The helical secondary structures of collagen and laminin produce resistant fibrillar materials that contribute to the mechanical stiffness of tissue. Similar to GAGs, proteins of the ECM can serve as signaling reservoirs by binding, and when required releasing GFs [4]. In addition, specific cell surface receptors can anchor to these proteins, including the binding of fibronectin and collagen to the integrin receptor [10–12], laminin to the 67 kDa laminin receptor (67 LR) [13], and elastin to the 67 kDa elastin receptor (67-ER) [14]. As required, cells can also secret enzymes, MMPs, to degrade ECM proteins. MMPs are released into the ECM and upon metal complexation, cleave specific amino acid sequences in the ECM proteins. For example, collagen types I, II, and III are cleaved by MMP-1 [15, 16]. These processes provide control over the maintenance of the ECM and its degradation in a process called homeostasis.

5.2.3 Proteoglycans, PGs

PGs are glycosylated proteins in which GAGs are typically covalently or physically bound to a protein core. These super-macromolecules can be very large with molecular weights over 1,000 kDa [17]. The specific biological functions of many PGs are still emerging, but interaction with GFs and cell receptors has already been identified [18]. However, the main function of PGs appears to be related to their physicochemical properties. For example, the high surface charge of PGs allows them to retain large amounts of water through osmotic effects, making them the material of choice for hydrogel-like tissues such as articular cartilage [19].

Table 5.1: Macromolecular composition of the extracellular matrix (ECM).

Name	Tissue location	Function	Structure	Mw (kDa)
Polysaccharide Heparin	Blood, wound	Anticoagulant		10–12
Heparan sulfate	Cartilage, mucosa	Compressive load Shock absorption Water retention Formation of proteoglycans Reservoir of soluble signals		10–70
Hyaluronic acid	Eye, cartilage			800
Chondroitin sulfate	Cartilage			300

				60
Dermatan sulfate	Cartilage	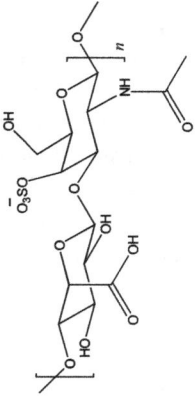		60
Keratan sulfate	Cornea, cartilage bone	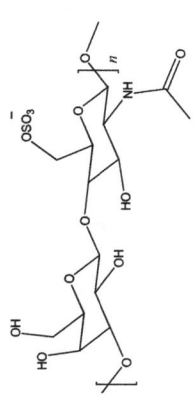		20

Protein				
Collagen	Skin, bone, scar tissue	Stretch strength of the tissue	Triple strand	290
Laminin	Basal lamina	Network organization	Triple strand	900
Nidogen	Basal lamina	Connect collagen and laminin	Single strand	150
Keratin	Hair, cornea, nails	Protect from exterior	Single strand	200
Elastin	Skin, arteries,	Elasticity of the tissue	Single strand	64*
Fibronectin	all	Cell adhesion	Double strand	250

*only the precursor: tropoelastin.

5.3 The tissue repartition of ECM components

To form the proper matrix that will support their function and fate, cells can synthesize these biomacromolecules in different proportions. The combination of GAGs, PGs, and proteins in different ratios allow for the formation of tissues with mechanical properties tailored to their task. For example, in articular cartilage tissue, the ability of PGs to retain large amounts of water upon compression is crucial to support the load bared by the joints [20]. In the skin, cells organize at the tissue interface to form a protective barrier in which a collagen type I matrix provides high resistance and collagen type III the flexibility [21, 22]. Similarly, the ECM of the vasculature is composed of resistant protein-based biomacromolecules, including collagen type IV and laminin. Connective tissues are hydrogel-like materials made up of fibrillar and hydrogel components that usually bind different tissues type-together [23]. Connective tissues can be easily degraded and remodeled by cells through the secretion of enzymes, such as MMPs [24].

Rearrangement of the ECM is a critical feature for the repair of injured tissue, however, the imbalance between its degradation and synthesis rates may be related to various diseases [25]. As an example, heart diseases such as myocardial infarction have been related with enhanced levels of tissue inhibitor of metalloproteinases – 1 thereby inhibiting collagen degradation leading to the formation of non-elastic tissue which deforms upon heart contraction [26].

By varying the composition of the ECM, different tissue-mechanical properties can be achieved. From the smoothness of soft brain tissues to the toughness of bones, the mechanical properties of tissues vary widely across the mammalian body. As proof of this difference, the elastic moduli, or Young's moduli of different tissues have been characterized *in vivo* or *in vitro* by using different techniques. It was measured that the range of modulus values varies over a factor of 10^9 between the smoothest and the toughest tissues (Table 5.2).

Table 5.2: Young's modulus of human tissues assessed by different techniques.

Tissue	E (MPa)	Method	ref
Bone	18×10^3	Compression (ex vivo)	[27]
Cartilage	1.2	Compression (ex vivo)	[28]
Skin	421×10^{-3}	Rheology (*in vivo*)	[29]
Muscle	18×10^{-3}	Magnetic resonance elastography (*in vivo*)	[30]
Brain	240×10^{-6}	Aspiration device (*in vivo*)	[31]

ECM biomacromolecules provide mechanical support to the tissue structures but more importantly, the ECM provides physical anchors for cell attachment. As a result of the diversity of the ECM composition, cells evolve in a system composed of variable chemical and biological stimuli. Through cell receptors, cells can sense and interact with different ECM biomacromolecules. Since tissues exhibit diverse mechanical properties, researchers have recognized the need to expose cells to environments with appropriate stiffness, deformation, or viscous properties to mimic the native environment. It is now well accepted that the ECM mechanical properties play a major role in the fate and function of the cells. It is well known that apart from soluble signaling, including GF or chemokines, mechanical properties of the cell-substrate can also direct cellular differentiation and movement (migration) [32]. There are several strategies to reproduce this complex environment for biofabrication of tissue models or implants.

5.4 Natural ECM for biofabrication

Given the complexity of the ECM, using naturally derived ECM-based biomaterials is a compelling strategy to provide an accurate cell culture substrate. One way to achieve this is to isolate *via* decellularization the useful biomaterials of the ECM. From allogeneic or xenogeneic natural tissues, cells are removed leaving behind the natural ECM composed of various biomacromolecules (Figure 5.3).

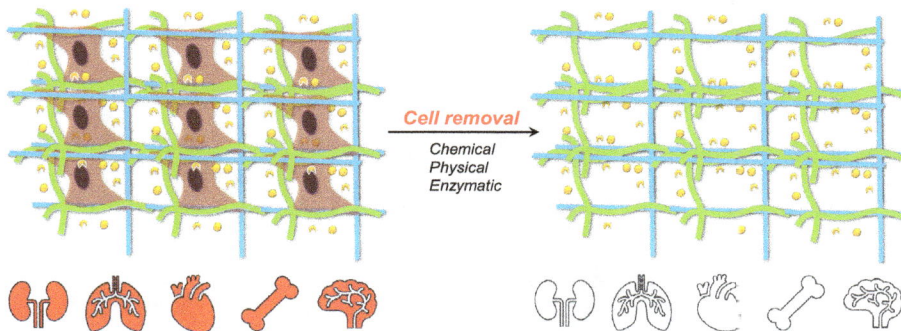

Figure 5.3: Cell removal from mammalian tissues to obtain DNA, RNA, and cell-free biomaterials. The decellularization can be carried out on full organ or tissue biopsy.

The ratio of each biomacromolecule varies depending on the type of tissue and the species from which it originates. The decellularized matrix can then be repopulated with the patients' cells *in vitro* or directly implanted into the host wherein the cells migrate and repopulate the empty matrix. Various types of decellularized organs are under development, among them, pancreas [33], liver [34], trachea [35], heart [36], cornea [37], or kidney [38] have been reported. Despite the apparent simplicity of this method, a complicated bioengineering process is required to remove the cells and all their components (e.g., DNA, RNA) without deteriorating the native ECM. To achieve this, chemically-based processes using a detergent like triton

X-100 or sodium dodecyl sulfate are used. Alternatively, enzymatic degradation with trypsin or nuclease can be used [39]. But these methodologies suffer from several drawbacks. In particular, the need for an organ donor, potential immune responses due to the use of cross-species biological materials caused by remaining genetic material, loss of mechanical properties, and difficulty in obtaining detailed characterization of the matrix materials.

5.5 Synthetic ECM for biofabrication

As an alternative to decellularization, a synthetic hydrogel that reproduces the function of the ECM: mechanical support, adhesion, and communication media can be used (Figure 5.4). Several polysaccharides forming hydrogels such as alginate or agarose have been studied and proteins such as gelatin have been brought to use as synthetic ECM. But naturally, most of the synthetic hydrogel-forming macromolecules do not reproduce cell adhesion or matrix degradation. Therefore, modification of these hydrogels is needed to accurately reproduce ECM function. For applications in biofabrications, these hydrogels must be processable, by 3D printing for example, and alginate and methacrylated gelatin are the two main molecules of choice.

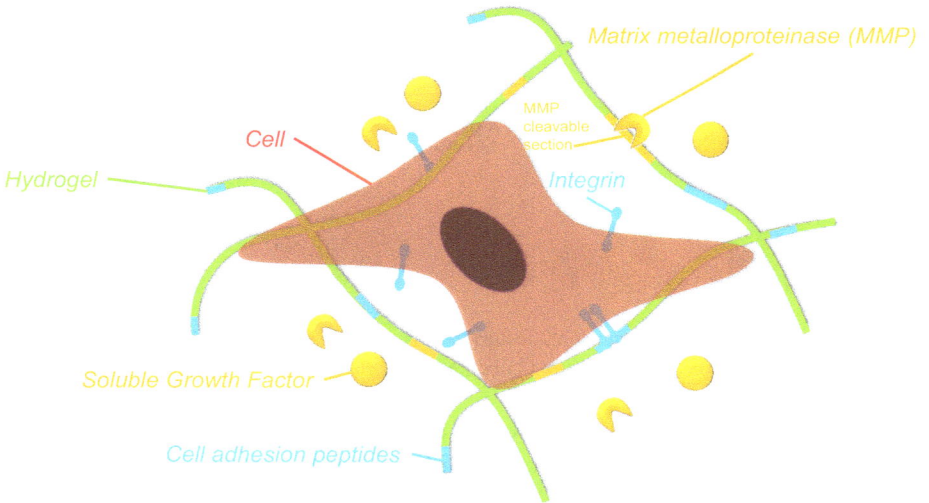

Figure 5.4: The cell in its microenvironment, showing the interactions of the macromolecules of the ECM with the cell and soluble growth factors.

However, inducing cell adhesion in alginate or MMP's induced cleavage, modification of the macromolecule is needed. To induce cell attachment different strategies can be utilized. Full-length ECM proteins when extracted from animals can be used for directing cell-based processes. But These ECM proteins are associated with certain disadvantages such as cost and potential immunogenic response. These disadvantages can be avoided by the application of short-cell recognition peptide motifs derived from macromolecules and these short peptides can be manufactured synthetically. The cell adhesion peptides (CAPs) are usually designed for a specific receptor molecule and permit to interrogate a specific adhesion pathways. Synthetic hydrogels have been modified with a variety of CAP reproducing macromolecules such as collagen, laminin, fibronectin, and elastin. These cell-binding peptide motifs mimic the signaling provided by natural ECM biomacromolecules and can direct the cell fate in a synthetic cell microenvironment. The peptide motif containing arginine (R), glycine (G), and aspartate (D) is the most prominent integrin-binding adhesion sequence and is found in fibronectin. Since RGD binds to a variety of integrins, RGD based peptides are generally applicable across most biomaterial applications. In addition to cell adhesion, cleavable peptides can be introduced in the hydrogel matrices. The crosslinking of hydrogel with the GGPQG↓IWGQK peptide cleavable by the MMP1 allows cells to remodel their synthetic environment. In turn, the introduction of a cleavable sequence upon cell generates a feedback loop where the mechanical properties of the matrices are softened much like the natural ECM. But designing this advanced system can quickly become expensive as longer peptide sequences are more expensive to synthesize . Furthermore, such advanced systems can add much complexity and the advantage of using simple systems based on synthetic hydrogen is lost. Thus, such complex systems compete with natural ECM in term of cost and model.

Table 5.3: Cell-adhesion peptide (CAP) motifs used in synthetic systems to mimic the ECM cell adhesion and to mimic ECM degradation. The arrow indicates the cleavage site.

	Macromolecule					Matrix degradation
	Collagen I	Laminin	Fibronectin	Elastin	Collagen IV	MMP1
Peptide motif	GFOGER	YIGSR	GRGDS	VAPG	GDR	GGPQG↓IWGQK
	DGEA	IKVAV	PHSRN		GRD	

Combining different CAPs with hydrogel-forming biomaterials of different stiffness permits to elaborate libraries of stiffness and adhesions to recreate several tissue environments. These materials form then the basis for several biofabrication processes depending on their compatibility with manufacturing techniques.

5.6 Quiz

1.   What is the most abundant macromolecule of the extracellular matrix (ECM)?
2.   What are the two main types of polymer composing the ECM?
3.   What are the main roles of the ECM?
4.   How are decellularized extracellular matrices made?
5.   What is a cell-adhesion peptide?
6.   What is the most used cell-adhesion peptide?
7.   What is homeostasis?
8.   What molecules can degrade the ECM?
9.   Which molecules can stop the degradation of the ECM?
10. Which molecules can be found in solution or bound to the ECM?

References

[1]   Wardrop, D.; Keeling, D. The Story of the Discovery of Heparin and Warfarin. *Br. J. Haematol.* **2008**, No. March, 757–763. doi:https://doi.org/10.1111/j.1365-2141.2008.07119.x.

[2]   Necas, J.; Bartosikova, L.; Brauner, P.; Kolar, J. Hyaluronic Acid (Hyaluronan): A Review. *Vet. Med. (Praha).* **2008**, 53 (8), 397–411.

[3]   Swann, D.A.; Radin, E.L.; Nazimiec, M.; Weisser, P.A.; Curran, N.; Lewinnekt, G. Role of Hyaluronic Acid in Joint Lubrication. *Ann. Rheum. Dis.* **1974**, 33, 318–326. doi:https://doi.org/10.1136/ard.33.4.318.

[4]   Schultz, G.S.; Wysocki, A. Interactions between Extracellular Matrix and Growth Factors in Wound Healing. *Wound Repair Regen.* **2009**, 17 (2), 153–162. doi:https://doi.org/10.1111/j.1524-475X.2009.00466.x.

[5]   Gandhi, N.S.; Ricardo, L. The Structure of Glycosaminoglycans and Their Interactions with Proteins. **2008**, 455–482. doi:https://doi.org/10.1111/j.1747-0285.2008.00741.x.

[6]   Sasisekharan, R.; Raman, R.; Prabhakar, V. Glycomics Approach to Structure-Function Relationships of Glycosaminoglycans. *Annu. Rev. Biomed. Eng.* **2006**, 8, 181–231. doi:https://doi.org/10.1146/annurev.bioeng.8.061505.095745.

[7]   Liu, L.; Liu, Y.; Li, J.; Du, G.; Chen, J. Microbial Production of Hyaluronic Acid: Current State, Challenges, and Perspectives. *Microb. Cell Fact.* **2011**, 10 (1), 99. doi:https://doi.org/10.1186/1475-2859-10-99.

[8]   Björk, I.; Lindahl, U. Mechanism of the Anticoagulant Action of Heparin. *Mol. Cell. Biochem.* **1982**, 48 (3), 161–182.

[9]   Ho, M.S.P.; Böse, K.; Mokkapati, S.; Nischt, R.; Smyth, N. Nidogens-Extracellular Matrix Linker Molecules. *Microsc. Res. Tech.* **2008**, 71 (5), 387–395. doi:https://doi.org/10.1002/jemt.20567.

[10] Larsen, M.; Artym, V.V.; Green, J.A.; Yamada, K.M. The Matrix Reorganized: Extracellular Matrix Remodeling and Integrin Signaling. *Curr. Opin. Cell Biol.* **2006**, 18 (5), 463–471. doi:https://doi.org/10.1016/j.ceb.2006.08.009.

[11] García, A.J.; Boettiger, D. Integrin-Fibronectin Interactions at the Cell-Material Interface: Initial Integrin Binding and Signaling. *Biomaterials.* **1999**, 20 (23–24), 2427–2433. doi:https://doi.org/10.1016/S0142-9612(99)00170-2.

[12] Barczyk, M.; Carracedo, S.; Gullberg, D. Integrins. *Cell Tissue Res.* **2010**, 339 (1), 269–280. doi:https://doi.org/10.1007/s00441-009-0834-6.

[13] Nelson, J.; McFerran, N.V.; Pivato, G.; Chambers, E.; Doherty, C.; Steele, D.; Timson, D.J. The 67 KDa Laminin Receptor: Structure, Function and Role in Disease. *Biosci. Rep.* **2008**, 28 (1), 33–48. doi:https://doi.org/10.1042/BSR20070004.

[14] Park, P.W.; Broekelmann, T.J.; Mecham, B.R.; Mecham, R.P. Characterizat on of the Elastin Binding Domain in the Cell-Surface 25-KDa Elastin-Binding Protein of Staphylococcus Aureus (EbpS). *J. Biol. Chem.* **1999**, 274 (5), 2845–2850.

[15] Verma, R.P.; Hansch, C. Matrix Metalloproteinases (Mmps): Chemical-Bio ogical Functions and (Q)SARs. *Bioorg. Med. Chem.* **2007**, 15 (6), 2223–2268. doi:https://doi.org/10.1016/j.bmc.2007.01.011.

[16] Williams, K.E.; Olsen, D.R. Matrix Metalloproteinase-1 Cleavage Site Recognition and Binding in Full-Length Human Type III Collagen. *Matrix Biol.* **2009**, 28 (6), 373–379. doi:https://doi.org/10.1016/j.matbio.2009.04.009.

[17] Iozzo, R.V.;. Matrix Proteoglycans: From Molecular Design to Cellular Function. *Annu. Rev. Biochem.* **1998**, 67, 609–652. doi:https://doi.org/10.1146/annurev.biochem.67.1.609.

[18] Forsten-Williams, K.; Chua, C.C.; Nugent, M.A. The Kinetics of FGF-2 Binding to Heparan Sulfate Proteoglycans and MAP Kinase Signaling. *J. Theor. Biol.* **2005**, 233 (4), 483–499. doi:https://doi.org/10.1016/j.jtbi.2004.10.020.

[19] Scott, J.E.;. Elasticity in Extracellular Matrix "Shape Modules" of Tendon, Cartilage, etc. A Sliding Proteoglycan-Filament Model. *J. Physiol.* **2003**, 553 (Pt 2), 335–343. doi:https://doi.org/10.1113/jphysiol.2003.050179.

[20] Knudson, C.B.; Knudson, W. Cartilage Proteoglycans. *Semin. Cell Dev. Biol.* **2001**, 12 (2), 69–78. doi:https://doi.org/10.1006/scdb.2000.0243.

[21] Mcgrath, J.A.; Eady, R.A.J.; Pope, F.M. Anatomy and Organization of Human Skin. *Rook's Textb. Dermatol.* **2010**, 45–128. doi:https://doi.org/10.1002/9781444317633.ch3.

[22] Cheng, W.; Yan-hua, R.; Fang-gang, N.; Guo-an, Z. The Content and Ratio of Type I and III Collagen in Skin Differ with Age and Injury. *African J. Biotechnol.* **2011**, 10 (13), 2524. doi:https://doi.org/10.5897/AJB10.1999.

[23] Culav, E.; Clark, C.; Merrilees, M. Connective Tissues: Matrix Composition and Its Relevance to Physical Therapy. *Phys. Ther.* **1999**, 79 (3).

[24] Cameron, R.;. Connective Tissues. *Br. Med. J.* **1962**, 2 (5320), 1664.

[25] Mott, J.D.; Werb, Z. Regulation of Matrix Biology by Matrix Metalloproteinases. *Curr. Opin. Cell Biol.* **2004**, 16 (5), 558–564. doi:https://doi.org/10.1016/j.ceb.2004.07.010.

[26] Moore, L.; Fan, D.; Basu, R.; Kandalam, V.; Kassiri, Z. Tissue Inhibitor of Metalloproteinases (Timps) in Heart Failure. *Heart Fail. Rev.* **2012**, 17 (4–5), 693–706. doi:https://doi.org/10.1007/s10741-011-9266-y.

[27] Cuppone, M.; Seedhom, B.B.; Berry, E.; Ostell, A.E. The Longitudinal Young's Modulus of Cortical Bone in the Midshaft of Human Femur and Its Correlation with CT Scanning Data. *Calcif. Tissue Int.* **2004**, 74 (3), 302–309. doi:https://doi.org/10.1007/s00223-002-2123-1.

[28] Jurvelin, J.S.; Buschmann, M.D.; Hunziker, E.B. Mechanical Anisotropy of the Human Knee Articular Cartilage in Compression. *Proc. Inst. Mech. Eng. H.* **2003**, 217 (3), 215–219. doi:https://doi.org/10.1243/095441103765212712.

[29] Agache, P.G.; Monneur, C.; Leveque, J.L.; Rigal, D.J.; de Rigal, J.; Rigal, D.J. Mechanical Properties and Young's Modulus of Human Skin in Vivo. *Arch. Dermatol. Res.* **1980**, 269, 221–232. doi:https://doi.org/10.1007/BF00406415.

[30] Uffmann, K.; Maderwald, S.; Ajaj, W.; Galban, C.G.; Mateiescu, S.; Quick, H.H.; Ladd, M.E. In Vivo Elasticity Measurements of Extremity Skeletal Muscle with MR Elastography. *NMR Biomed.* **2004**, 17 (4), 181–190. doi:https://doi.org/10.1002/nbm.887.

[31] Schiavone, P.; Chassat, F.; Boudou, T.; Promayon, E.; Valdivia, F.; Payan, Y. In Vivo Measurement of Human Brain Elasticity Using a Light Aspiration Device. *Med. Image Anal.* **2009**, 13 (4), 673–678. doi:https://doi.org/10.1016/j.media.2009.04.001.

[32] Engler, A.J.; Sen, S.; Sweeney, H.L.; Discher, D.E. Matrix Elasticity Directs Stem Cell Lineage Specification. *Cell.* **2006**, 126 (4), 677–689. doi:https://doi.org/10.1016/j.cell.2006.06.044.

[33] Salvatori, M.; Katari, R.; Patel, T.; Peloso, A.; Mugweru, J.; Owusu, K.; Orlando, G. Extracellular Matrix Scaffold Technology for Bioartificial Pancreas Engineering: State of the Art and Future Challenges. *J. Diabetes Sci. Technol.* **2014**, 8 (1), 159–169. doi:https://doi.org/10.1177/1932296813519558.

[34] Uygun, B.E.; Soto-Gutierrez, A.; Yagi, H.; Izamis, M.-L.; Guzzardi, M.A.; Shulman, C.; Milwid, J.; Kobayashi, N.; Tilles, A.; Berthiaume, F., et al. Organ Reengineering through Development of a Transplantable Recellularized Liver Graft Using Decellularized Liver Matrix. *Nat. Med.* **2010**, 16 (7), 814–820. doi:https://doi.org/10.1038/nm.2170.

[35] Weymann, A.; Patil, N.P.; Sabashnikov, A.; Korkmaz, S.; Li, S.; Soos, P.; Ishtok, R.; Chaimow, N.; Pätzold, I.; Czerny, N., et al. Perfusion-Decellularization of Porcine Lung and Trachea for Respiratory Bioengineering. *Artif. Organs.* **2015**, 39 (12), 1024–1032. doi:https://doi.org/10.1111/aor.12481.

[36] Weymann, A.; Loganathan, S.; Takahashi, H.; Schies, C.; Claus, B.; Hirschberg, K.; Soós, P.; Korkmaz, S.; Schmack, B.; Karck, M., et al. Development and Evaluation of a Perfusion Decellularization Porcine Heart Model–Generation of 3-Dimensional Myocardial Neoscaffolds. *Circ. J.* **2011**, 75 (4), 852–860. doi:https://doi.org/10.1253/circj.CJ-10-0717.

[37] Xiao, J.; Duan, H.; Liu, Z.; Wu, Z.; Lan, Y.; Zhang, W.; Li, C.; Chen, F.; Zhou, Q.; Wang, X., et al. Construction of the Recellularized Corneal Stroma Using Porous Acellular Corneal Scaffold. *Biomaterials.* **2011**, 32 (29), 6962–6971. doi:https://doi.org/10.1016/j.biomaterials.2011.05.084.

[38] Bonandrini, B.; Figliuzzi, M.; Papadimou, E.; Morigi, M.; Perico, N.; Casiraghi, F.; Dipl, C.; Sangalli, F.; Conti, S.; Benigni, A., et al. Recellularization of Well-Preserved Acellular Kidney Scaffold Using Embryonic Stem Cells. *Tissue Eng. Part A.* **2014**, 20 (9–10), 1486–1498. doi: https://doi.org/10.1089/ten.TEA.2013.0269.

[39] Zhang, X.; Chen, X.; Hong, H.; Hu, R.; Liu, J.; Liu, C. Decellularized Extracellular Matrix Scaffolds: Recent Trends and Emerging Strategies in Tissue Engineering. *Bioact. Mater.* **2022**, 10, 15–31. doi:https://doi.org/10.1016/j.bioactmat.2021.09.014.

6 Surface functionalization and topography for biofabrication

6.1 Surfaces for cell culture

Typical cell culture protocol involves the isolation of cells from living tissues. This requires to digest the extracellular matrix (ECM) with enzymes degrading collagen and other proteins. Once digested, the cells are free from the ECM and are collected in a container. While some cells, typically from the blood circulation, can be cultured in suspension, most of the cells need adhesion to survive. But now the cells free of ECM need to have a substrate to attach to. For a surface to provide adhesion, it must allow for adsorption of serum proteins. The mechanism of cell adhesion is a two-step process. First, proteins from the serum, supplied in the cell culture media, must deposit on the substrate's surface. This adsorbed protein cocktail can be diverse but it mainly is composed of albumin and fibronectin. When cell culture media is added into the container, the surface is brought into contact with these proteins which start to form a thin coated layer on the substrate. Simultaneously, the cells are falling by gravity at the bottom of the container and will recognize the adsorbed protein forming the coated layer (Figure 6.1). To achieve a proper coating, the tissue culture substrate must capture the protein and not alter their folding. If the protein folding is altered, it might not be recognized and thus not binding with cell receptors such as integrins which in turn will not trigger cell adhesion mechanism. However, Once attached, cells will spread and move on the surface.

Figure 6.1: Mechanism of cell adhesion on surfaces. First, the proteins contained in the serum form a thin layer on the substrate through adsorption. Cell meets the layer of proteins and upon recognition of cell-adhesion motifs of the protein will start to spread and adhere.

It has not always been as trivial as it is today to grow cells on surfaces. Indeed, pioneers of cell cultures had only glass containers to grow the precious cells that they had harvested. The glass containers due to their brittleness could break and had to be cleaned and sterilized after each experiment. The typical sterilization process in laboratories was to autoclave the glass containers. With the development of the plastic industry, the availability of inexpensive non-brittle containers that could be supplied sterile and ready to use was a breakthrough in increasing the safety, and the reproducibility of cell-culture protocols. Indeed, one-time-use containers provided a

https://doi.org/10.1515/9781501515736-006

sterile, free of virus, germs, and fungi container. This was made possible through several developments. In the early 1960s, it was first hypothesized that poly(styrene) (PS) surfaces treated with sulfuric acid could be transformed into a cell adhesive surface. However, sulfuric acid treatment can be difficult to bring to use in an industrial setup because the toxic liquid must be then removed from the container and washed away. In the 1970s, it was then demonstrated that plasma treatment of PS can replace the sulfuric acid treatment to transform the plastic surface into a cell culture substrate.

6.2 Surface plasma treatment

Plasma is the so-called fourth state of the matter. A gas brought under a vacuum at room temperature is ignited into plasma by inducing an electrical current. This high-energy state creates charged gas molecules that can modify the surface of PS. Using oxygen as the plasma gas precursor introduces oxygen atoms onto the PS surface. As a result, the PS surface becomes less hydrophobic, and the wettability of the surface increases (Figure 6.2). This means that the contact angle between a droplet of water and the PS surface reduces as the amount of oxygen introduced on the surface is increased. This increased wetting of the surface enable proteins to adsorb on the surface and to form the required thin coating of protein necessary for cell adhesion.

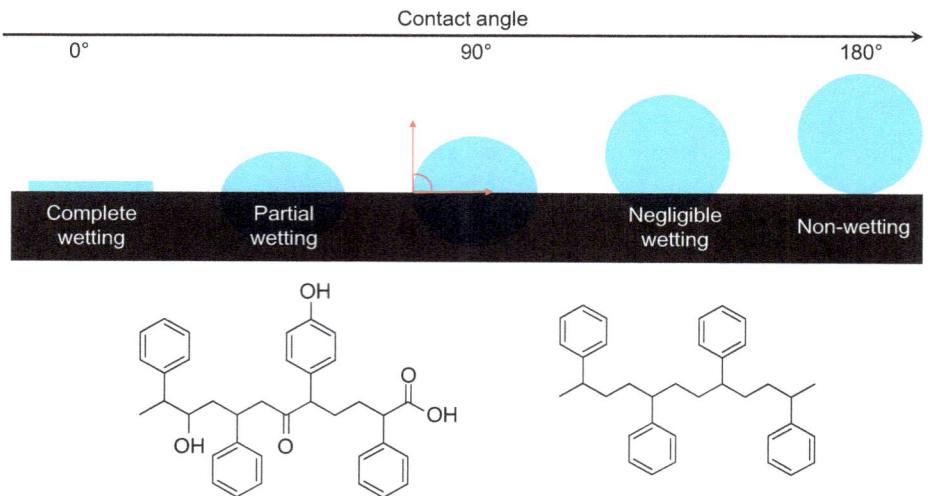

Figure 6.2: Contact angle and wettability. Air plasma treatment introduces oxygen-based functional groups on the poly(styrene) surface thus modifying its wetting capability from being negligible wetting to partial wetting.

To treat the surfaces with plasma, a dedicated setup is required (Figure 6.3). This consists of a sealed chamber into which the surface to be treated is inserted. The chamber is then brought under vacuum and different gas can be introduced into the chamber using a second inlet. Once the setup is ready, energy is provided to the system using a radiofrequency generator or current generator connected to the electrode. The stage on which the substrate is placed is grounded and the plasma can be ignited.

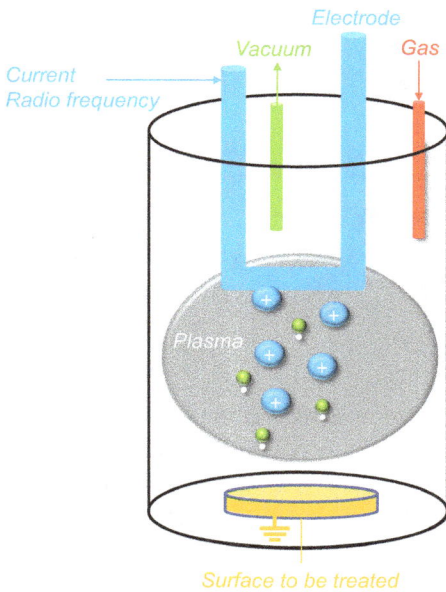

Electrode

Vacuum   Gas

Current
Radio frequency

Plasma

Surface to be treated

Figure 6.3: The plasma setup consists of an electrode that provides energy by current or radiofrequency into a sealed chamber. The substrate is placed into the chamber and gas can be introduced at a controlled flow rate using a vacuum. The ignition of the plasma leads to a surface modification.

The plasma once ignited gives a typical plasma glow that can vary from grey, blue for oxygen to a bright pink for argon. Using different gas for the plasma precursor will change the chemistry of the treated surface. For instance, plasma generated with a nitrogen high content gas will lead to more nitrogen atom being introduced on the exposed surface.

6.A Plasma

Increasing the energy from solid, liquid,gas and plasma the fourth state of matter. As a higher energy state than gas, plasma is an electrically conductive state. As such, long-range electrical and magnetic fields dominate the behavior of the ionized particles constituting the plasma.

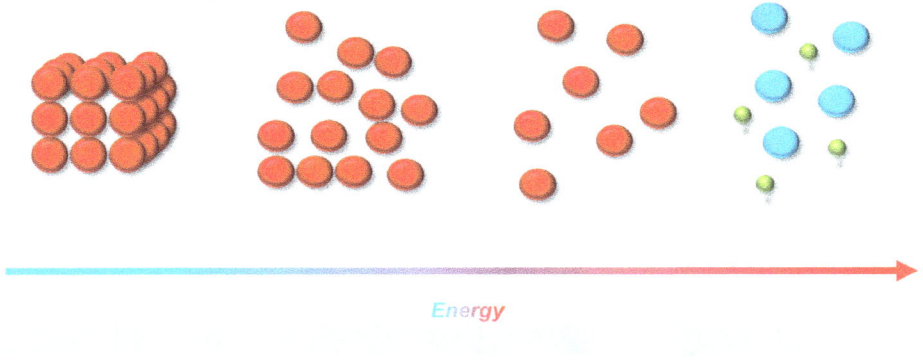

Energy

6.3 Plasma polymerization surface coating

Going further with the plasma treatment, one can introduce in the vacuum chamber a monomer that has a low boiling point, so that under vacuum the monomer is introduced as a gas in the sealed chamber. Monomers with a high boiling point can be heated to allow their vaporization. Once the chamber is filled with the monomer, the gas is ignited into a plasma. The high energy of the plasma will generate radicals and charged monomers. The radical species generated by the plasma polymerize as they would do in a typical radical or anionic polymerization reaction. The monomers start to polymerize and the molecular weight increases forming on the substrate's surface polymers chains. As the reaction continues, a thin polymer coating is formed on top of the substrate made of the polymerized monomers (Figure 6.4).

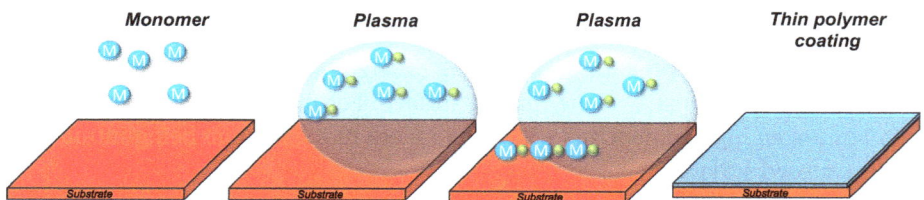

Figure 6.4: The plasma polymerization process. Where a volatile monomer is introduced into the sealed chamber plasma reactor. The monomer is ignited into plasma and it starts to polymerize. As it polymerizes, a thin polymer coating is created at the surface of the substrate.

Variation of the reaction time and monomer's flow allows for precise control of the polymer coating. Typical coating height range within 100s of a nanometer. The characterization of the surface chemistry is usually conducted with X-ray photoelectron spectroscopy (XPS) or Fourier-transformed infrared (FTIR). Several monomers have been used for the surface coating of the substrates for biofabrication purposes (Table 6.1). One of the main uses of plasma polymerization is to modify the hydrophobicity of the substrate to modulate the cell adhesion on the substrate. For instance, the incorporation of hydrophilic functional groups using ethanol and acrylic acid permit to increase the surface's wettability. Conversely, using perfluorooctane increases the hydrophobicity and gives a fluorinated coated like PTFE. Other modifications using functional groups such as glycidyl ether gives surfaces that can be reacted with amine functional groups. Allyl amine monomers are used when a surface that can react with carboxylic acid groups is needed. The main challenge in obtaining reactive functional groups with plasma polymerization is that the plasma process involves a high energy state of the matter. In this state, all kinds of active species are generated from ions to radicals, which can then react with the functional moieties rendering the final surface functionalization inert. It is thus, a matter of careful

Table 6.1: The monomers used in plasma polymerization with their associated surface coating chemistry.

Monomer	Chemical formula	Surface functionality
Acrylic acid		Carboxyl
Allyl alcohol		Hydroxyl
Ethanol		Hydroxyl
Allylamine		Amine
Allyl glycidyl ether		Epoxy
Glycidyl methacrylate		Epoxy
Alkyloxazoline		Oxazoline
Alkylamine		Amine
Propanal		Aldehyde
1,7 Octadiene		Alkyl
Perfluoro octane		Fluoro
Propanethiol		Thiol

optimization that is needed to not give too long reaction time and energy during the coating phase to not compromise the targeted reactive species [1]. For instance, glycidyl ether function is highly reactive and is thus polymerized by giving a pulse of energy to avoid the total reaction of the epoxy groups during the coating [2].

With a constantly growing library of monomers that can be used for plasma polymerization coating, a great range of applications are being developed [3]. Such modifications are easily scalable for industrial processes without the need to modify the properties of the bulk material such as the mechanical properties and resistance to constraints. These processes lead to surfaces that have antibiotic properties, can select cell adhesion, or even incorporate drugs or active molecules that can be later released once implanted or used *in vitro* [4]. These releasing surfaces are made by sandwiching the active ingredients between two layers of the plasma polymer coating. Then, the top layer wettability will regulate the release profile of the active ingredient. The higher the wettability of the surface, the quicker the ingredient is released. Complex surfaces can be achieved by using masks that can create a pattern of functional groups and gradients of functions [5]. One great advantage of plasma polymerization is that the functionalization of the surface occurs in the gas phase. This means that an intricate surface can be coated. This was demonstrated with a 3D printed porous scaffold and fibers mesh [6]. The gas will penetrate all the pores and void volumes between the fiber mesh. Upon ignition of the plasma, the polymerization occurs and the surfaces that would not be accessible with a solution can be reached. In comparison to the coating or functionalization of a surface with wet techniques, washing steps and rinsing of the substrate are not required, which has tremendous advantages for industrial processes as it greatly simplifies the functionalization method.

6.B X-ray photoelectron spectroscopy

To identify elements, present on a surface in a non-destructive manner, X-ray elemental and functional composition photoelectron spectroscopy (XPS) is one of the best methods. It is based on the photoelectric effect. For analytical purposes, it is an extremely powerful method as it gives information on the element but also to which another element they are bound to. Therefore, functional groups can be identified with XPS. The test substrate is placed in a vacuum and is irradiated with a beam of X-ray. The photon interacts with the matter ejecting electrons from the substrate and their kinetic energy is measured. From this information, spectra of binding energy are calculated, and information on the electronic shell and binding of an element is obtained. Electrons engaged in a covalent bond will have a different energy, and thus a precise map of the material composition can be obtained (Figure 6.5).

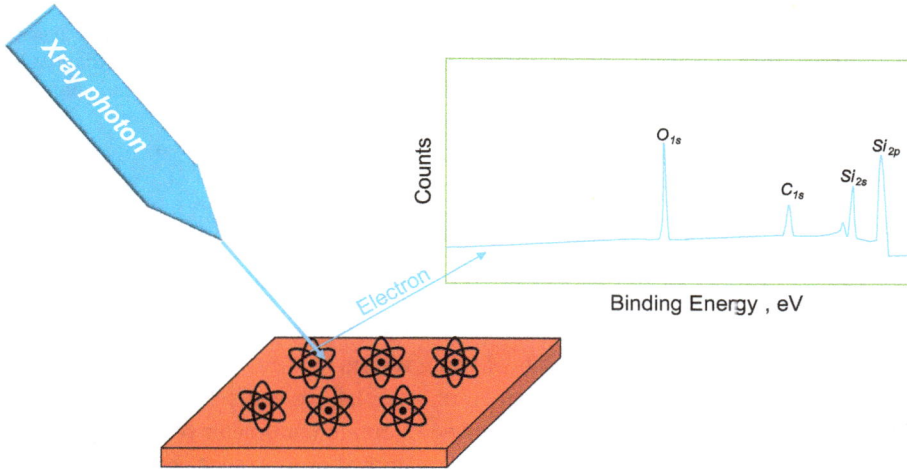

Figure 6.5: Principle of X-ray photoelectron spectroscopy. Photons are bonbarded on the surface to be analyzed. The energy of the electrons ejected from the surface will provide a binding energy profile. This results let precisely analyze the chemical composition of the surface.

6.C Fourier-transformed infrared spectroscopy

The infrared principle utilizes the capacity of molecules to absorb specific frequencies that correlate with their structure. In a molecule, the atoms vibrate. Two atoms of carbon bound together will vibrate differently than an atom of hydrogen bound to an atom of oxygen. Shinning an infrared light through the sample, if the light frequency is equal to the frequency of vibration of said two atoms of carbon bound together, the molecule will absorb the radiation. So, looking at the radiation passing through the samples will show a missing radiation corresponding to the vibration frequency of two carbon atoms bound together (Figure 6.6). Now, if instead of looking at one frequency at the time but all frequencies at the same time,

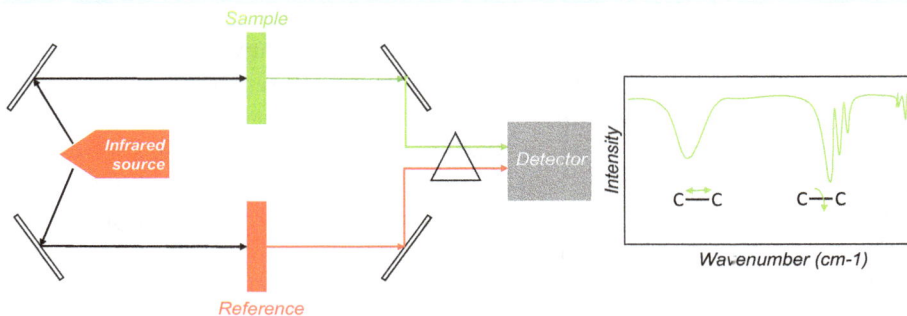

Figure 6.6: Principle of Fourier-transform infrared spectroscopy. This permits to investigate the covalent bonds.

an interferometer is obtained and is transformed by the Fourier function to obtain a frequency distribution. The signal obtained is an interferogram that contains all the frequencies that have been let through. The mathematical operation of the Fourier transforms permits decoupling the interferogram and obtaining the wavenumber spectra.

6.4 Photolithography to create shapes on surfaces

Originally developed for microprocessor manufacturing, the photolithography process has moved into the biomedical space to enable the fabrication of precisely shaped surfaces. The process consists of using wafers that are made of silicon with a layer of silicon oxide. Using a mask and a photoresist, we can selectively expose the silicon oxide layer to be etched and removed to create nanometer resolution precise surface topography. Wet etching is a material removal process that uses liquid chemicals or etchants to remove materials from a wafer. The specific patterns are defined by mask made of a photoresist that is deposited on the wafer. Materials that are not protected by this mask are etched away by liquid chemicals. These masks are deposited on the wafer in an earlier fabrication step called lithography. The whole process consists of four steps (Figure 6.7). First, a viscous resin that can be photocrosslinked is spin-coated on the silicon wafer. Then, a mask is applied above the surface of the coated wafer. The whole object is exposed to UV light. The typical photoresist used for such process is the SU-8. This is an epoxy-based negative photoresist: which means that the parts exposed to UV become crosslinked, while the unexposed part remains soluble. The SU-8 consists of eight epoxy groups that can crosslink through UV activation of photoacid: hexafluoro antimonic acid which in turn protonates the epoxides groups. Unmasked areas are activated by light and then polymerized during the baking step. The uncrosslinked photoresist (the part masked from the UV) is then washed made of. The photoresist protects the silicon wafer below it from the etching solution of hydrofluoric acid which remove the silicon from the wafer. Different types of etching can be conducted: isotropic (uniformly in all directions) or anisotropic etching (uniformity in a vertical direction). Some of the anisotropic wet-etching agents for silicon are potassium hydroxide (KOH), ethylenediamine pyrocatechol, or tetramethylammonium hydroxide. Etching a silicon wafer would result in a pyramid-shaped etch pit. The etched wall will be flat and with an angle. For isotropic wet etching, the most common agent is a mixture of hydrofluoric acid, nitric acid, and acetic acid. As the reaction takes place, the material is removed vertically and laterally at the same rate. Once etched, the photoresist is removed to reveal the pattern surface. This technique permits to create nanometer-scale structures and surface topography at high precision. However, dry etching, which followed the same workflow, is much more precise. Usually, dry etching is done using a plasma reactive gas such as fluorocarbons, oxygen, and chlorine with an addition of other inert gases like nitrogen or argon.

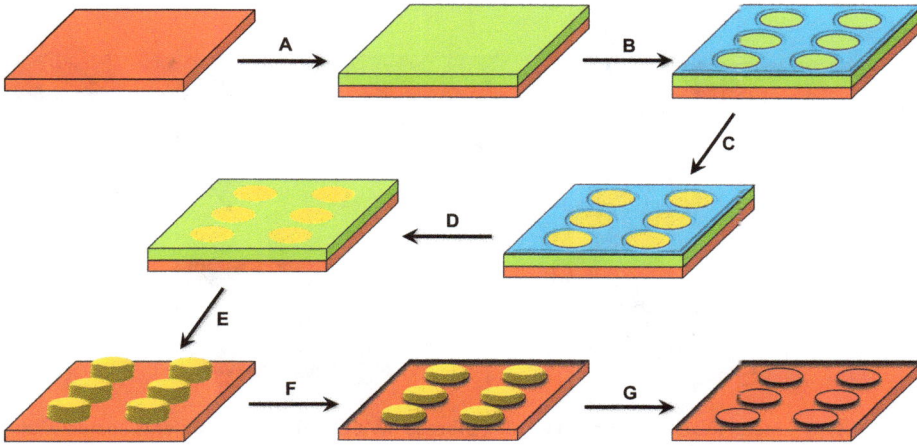

Figure 6.7: The photolithography process: (A) The silicon wafer is coated with a photoresist, (B) a mask is applied onto the surface, (C) the wafer is exposed to UV light (D) crosslinking only exposed area. (E) uncrosslinked photoresist is washed away exposing a (F) bare silicon wafer that can be etched. (G) Finally, the crosslinked photoresist is removed.

6.5 Nanopillar and nanoneedles

The photolithography etching method described above has one drawback, that is, the etching of the silicon wafer does not give a straight angle. Typical wet etching using SU-8 gives a pyramidal etched pattern with a 54.5°.

Figure 6.8: (left) Wet etching using a photoresist giving a pyramidal topography versus (right) metal-assisted wet etching giving a straight etching.

If we want to get straight etching, a more conductive layer is needed. This process is called metal-assisted chemical etching which can be obtained by adding a metal layer on top of the photoresist to coat the wafer with a metal such as silver or gold and masking the metal with PS beads (Figure 6.8). Several variation of these methods have been developed over the years using various metal coating and different masking techniques. The metal coating catalysts the etching of the silicon to obtain an aniso-tropic process that allows the fabrication of pillar. These Si pillars have many applica-tions in modern electronics ranging from sensors, batteries, and solar cells (Figure 6.9).

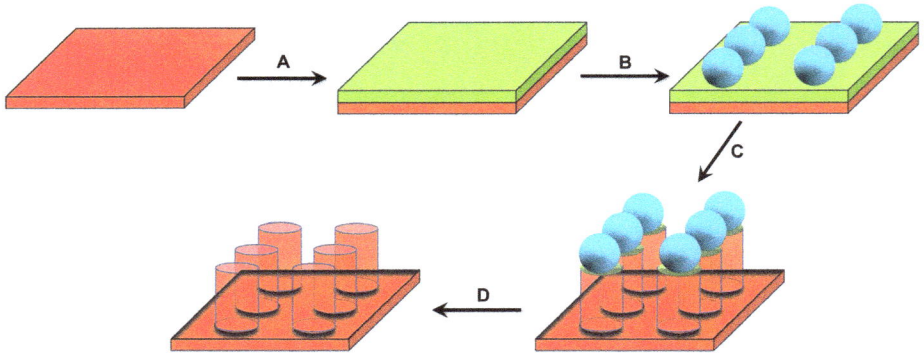

Figure 6.9: The metal-assisted chemical etching process (MACE). (A) A silicon wafer is coated with a metal layer, (B) an isolator such as poly(styrene) beads can be used to form the patterns, (C) chemical wet etching occurs in an isotropic manner giving pillar or wires, and (D) the PS beads are dissolved.

Beyond their applications in electronics, nanowires and nanopillars have been pro-posed for biomedical applications [7], for example, as a research tool to interrogate the traction force of cells. Cells culture on the pillar will deflect the pillar to a cer-tain angle depending on the traction force applied to the pillars. Measuring the de-flection angle permits the measurement of the cell traction forces (Figure 6.10) [8].

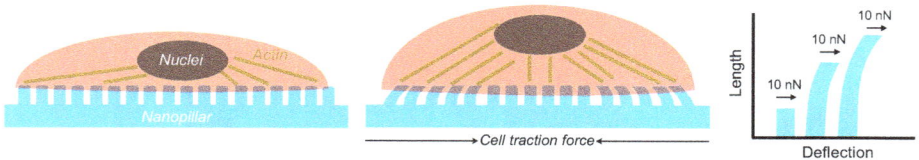

Figure 6.10: Silicon nanopillar used to measure the traction forces involved during cell adhesion and movement by measuring the deflection of the pillar.

Further development of the etching technique has led to the manufacturing of nanoneedles, that is, pillars that are pyramidally shaped and which can pierce the cell membrane for transmembrane delivery [9]. By creating an array of these nano-needles, drug delivery devices can be made. Intending to replace the often-painful needle injection, needle array permits to multiply the surface area to deliver the same amount of drug through the skin and reduce the pain. Different types of nee-dles have been proposed such as coated needles that leave the active ingredient in the skin once removed (Figure 6.11). Other design can come as a solid needle that pierces the skin to allow deposition of the drug. Further needle design consists of a porous or breakable needles that once implanted are left behind and degrades overtime to release their active pharmaceutical ingredient (API). More complex hollow needles act as a classical needle through which the API is pushed into the skin. Further applications of these nanoneedles array are for constant implantation and monitoring and sensing of physiological and biochemical markers [10].

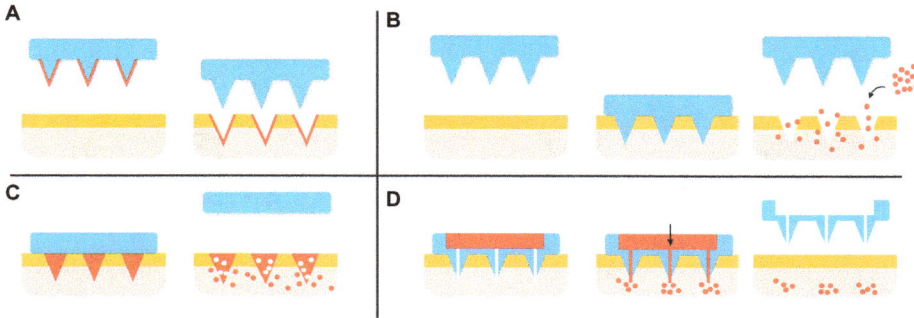

Figure 6.11: Different types of nanoneedles array technology proposed for drug delivery: (**A**) Coated microneedle, (**B**) solid removable nanoneedle, (**C**) dissolving microneedles, and (**D**) hollow microneedle.

6.6 Soft lithography

Compared to photolithography, soft lithography is a process that instead of using a photoresist to create a topology, is using an elastomeric stamp to create the surface topography. The most common materials to make the stamp is poly(dimethylsiloxane) (PDMS) due to its biocompatibility, viscosity, chemical versatility, and low surface tension. Soft lithography is a family of techniques for fabricating surface topology that is then used to sort cells, create synthetic organs on a small surface called organ-on-a-chip (OOC) [11, 12].

6.6.1 Replica molding

When talking about soft lithography the first method that comes to mind is the replica molding technique. This is by far the most common technique (Figure 6.12). Starting from a master mold that is made out of a silicon wafer manufactured through photolithography or other manufacturing technique, PDMS viscous solution is then poured on top of the master mold. The PDMS is cured (crosslinked) and peeled off the mold giving a negative replication of the master mold. The PDMS mold can then be replicated after being rendered non-adherent through silanization. This PDMS stamp can be used several times. The replica molding process is usually made with PDMS, but other soft materials have been used. Hydrogel made of gelatin methacrylate or agarose is also a good candidate for this technique which is particularly suited for making microwells.

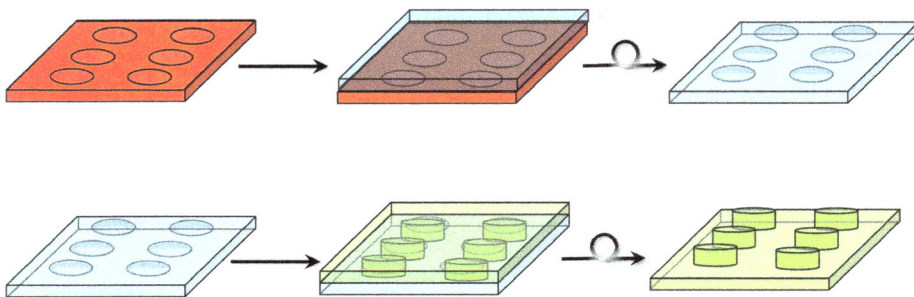

Figure 6.12: Replica molding technique. A master mold (red) is used to make a negative PDMS stamp (blue). The stamp can be used several times to stamp another PDMS or hydrogel (green).

6.6.2 Capillary molding

A technique particularly suited for the fabrication of small linear patterns is the capillary molding method (Figure 6.13). Using a PDMS mold with capillary size channels applied to a substrate, a polymer precursor is sucked into the channel through capillary force. This pre-polymer solution can be poly(ethylene glycol), PDMS, or hydrogels such as gelatin methacrylate and agarose. The polymer is cured and the polymeric mold is peeled off giving a structured polymer microstructure. This sort of patterning is interesting when cells need to be aligned on a substrate or guided through a linear pattern, for instance to replicate the architecture of blood vessels [13].

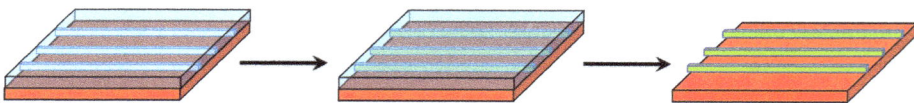

Figure 6.13: Capillary molding using a PDMS mold (blue): a pre-polymer solution (green) is infused into the channel of the mold. After curing, the mold is removed and the polymer stays on the substrate (red).

6.6.3 Microcontact printing

Printing of proteins or peptides motifs on a precise pattern on a surface can be made with a PDMS stamp (Figure 6.14). The stamp has a protrusion that has ink on the surface. The stamp is inverted and brought in contact with the substrate and the molecule of interest is either adsorb on the substrate or covalently bonded. Typical chemical reactions are with an amine of the biological molecules that can react with an epoxy-functionalized surface (for instance, plasma coated surfaces). Different shapes can be made, and these can be used to study how cell adhesion occurs on these coated surfaces [14]. Other application of this microcontact printing is particularly useful for high-throughput screening where a high density of environment is to be tested. The proteins on the stamp can be varied thus allowing a rapid fabrication of test surfaces for cell adhesion [15].

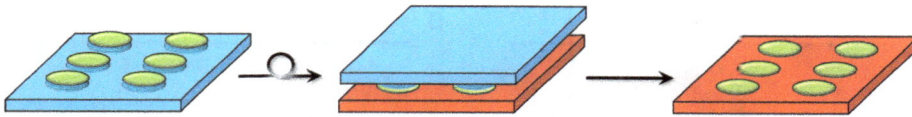

Figure 6.14: Microcontact printing the transfer support is covered with ink. Inverted and stamped on the substrate.

6.6.4 Microtransfer molding

A stamp can also be used to transfer a polymer on a surface (Figure 6.15). The polymer precursor solution is loaded on the microwell of the PDMS stamp. The stamp is inverted and brought into contact with the substrate. The polymer solution is cured and the stamp is peeled off leaving a pattern of cured polymer. A typical polymer that can be used is PDMS, but also hydrogels such as gelatin methacrylate, collagen, or agarose. Into the patterned hydrogel, a cell can be cultured within the confined space of the hydrogel. This permits a reduction of the size of the cultured sample and forces the cells to organize into a defined shape [16].

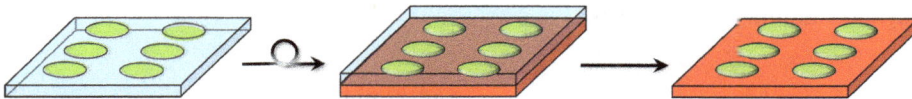

Figure 6.15: Micotransfer molding precursor polymer is loaded in the well of the transfer support. The support is inverted, the polymer is deposited on the substrate, cured, and the transfer peeled off the substrate.

6.7 Microfluidic devices

6.7.1 Laminar flow

With all the techniques available for surface functionalization and fabrication such as plasma functionalization, photolithography, silicon etching, and soft lithography, complex microfluidic devices can be designed and manufactured. One of the particularities of a microfluidic chip is the small channel diameter that is used to reproduce the blood flow. Using a smaller diameter channel, the fluid will be under high shear flow and will have a laminar flow meaning that the fluid will difficulty mix (Figure 6.16). The small diameter used for these OOCs permits to reduce the volume of liquid needed, thus reducing the cost of development by using a minimum amount of reagents and cells.

A

B

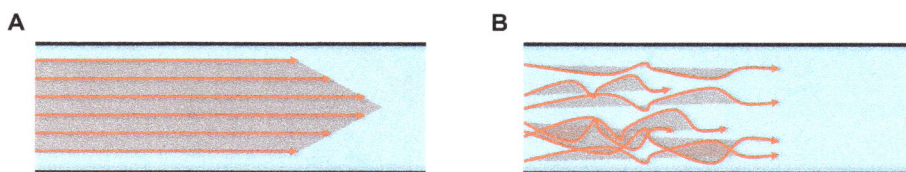

Figure 6.16: Schematic of (**A**) laminar flow in microfluidic channels as compared to (**B**) turbulent flow.

This laminar flow in small channels can be leveraged in chemistry to build small reactors and simulate industrial setup [17, 18]. When a fluid is flowing through a closed channel, two types of flows may occur depending on the velocity and viscosity of the fluid: laminar flow or turbulent flow. Turbulent flow is characterized by rough flow where mixing of the fluid can happen. Laminar flow is characterized by smooth flow where the mixing of components is mainly driven by diffusion, a slow process.

The Reynolds number quantifies the relationship between fluid viscosity and fluid velocity and allow for the calculation of a threshold between the turbulent and laminar flow. The Reynolds number, that is, the values where laminar flow occurs, will depend on the geometry channel and flow. In a microfluidic channel, the Reynolds number is defined as

$$\mathrm{Re} = \frac{\rho u D_\mathrm{H}}{\mu} = \frac{u D_\mathrm{H}}{v} = \frac{Q D_\mathrm{H}}{vA}$$

with D_H for the hydraulic diameter of the channel (m); Q for volumetric flow rate (m^3/s); A for the pipe's cross-sectional area (m^2); u for mean speed of the fluid (m/s); μ for dynamic viscosity of the fluid (Pa $\cdot$ s); v for kinematic viscosity of the fluid (m^2/s); and ρ for density of the fluid (kg/m^3).

6.7.2 Cell sorting

The laminar flow easily reached in microfluidic devices can be leveraged to create cell sorting devices with a small surface ratio. Different chip designs with either passive or active cell sorting capabilities have been proposed. Using the laminar flow, cells can be sorted out by size in the channel. An array of posts can also be used to sort cells by size by disrupting the flow of cells [19] (Figure 6.17).

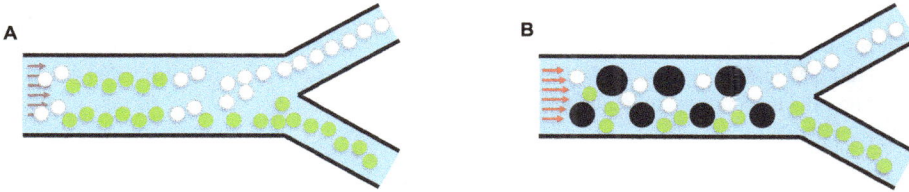

Figure 6.17: Microfluidic cell-sorting device capable of sorting the cell by size. The bigger cells (green) are separated from the smaller ones (white). (A) By laminar flow and (B) using post that disrupt the flow.

But not all cells can be sorted based on their size. When a cell population must be sorted based on their protein expression, more complex devices that rely on an active substrate are needed (Figure 6.18). Microdevices that can retain cells by presenting an antibody that recognizes cell surface markers are quite efficient in purifying a cell population. An alternative technique used a laser that can recognize a fluorescently tagged cell marker that then can trigger an active sorting mechanism to direct a type of cell toward a dedicated channel. Microbeads tagged with an antibody that can selectively bind to a cell population can be useful in creating a size gradient between two cell populations to sort them in a passive sorting device. Building on microbeads, antigenic and fluorescence technologies microfluidic cell sorting devices can become extremely efficient in sorting cells on a small footprint. The emergence of this technology has been seen as having a great potential in miniaturizing biomedical devices for diagnostic purposes but also to provide affordable tools for cell therapies currently in the development pipeline [20].

6.8 Organ-on-a chip

Most of the drugs discovered in the past were done by serendipity. Case in point, the discovery of penicillin was possible as Fleming left a culture plate on a window for several days. When he came back, he found that fungi had grown and inhibited the growth of bacteria. Many other drugs have been discovered from natural compounds. Today, pharmacy is a business and as such, it cannot rely on sheer luck. Therefore,

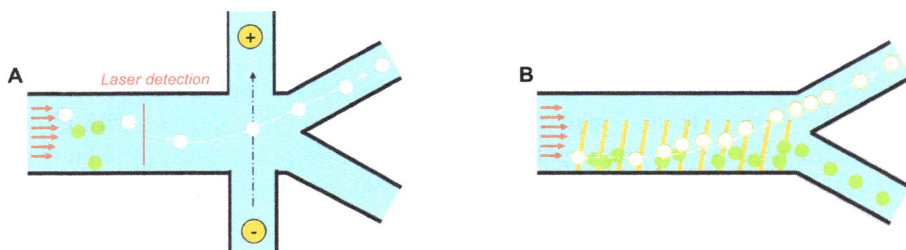

Figure 6.18: (**A**) Active cell sorting where a laser detects the fluorescent tag of a cell before actively sorting the cell through an electromagnetic field (if the cell is loaded with magnetic particle), or with a lateral microfluidic flow. (**B**) Sorting through antigen adhesion the yellow strips that guide the white cells toward the upper outlet.

methods and processes to foster the discovery of new compounds have been put in place to ensure the capability to bring new molecules to the market.

Currently, the drug discovery process starts by creating a library of molecules that you test on a cell line or cellular assay that reproduces the disease you want to target. The disease has been selected as part of an industrial decision with the marketing and medical department. You will screen the different molecules, refine the structure of the molecule using *in silico* computer methods. Once you have your molecule, you will test it *in vivo* using animal models that reproduce the targeted disease. Once your molecule passes these steps, you can start the human phase with the clinical trial. First assess the safety of the drug, then the efficacy, and finally the efficacy on different populations and monitor the side effects. But this development takes decades, but the patent on the molecule, which gives you exclusivity, expires after 20 years.

In addition to this cost, not all the molecules that are discovered make it to the market. Some molecules are not better than current treatment, some molecules are not safe leading to failure of the clinical trials. Indeed, dozens of clinical trials failed every year and sometimes led to fatality when the tested molecule is found not safe. While many molecules work on laboratory animals, many are not working in humans.

There is therefore a great interest in reducing the part of animal testing to get drug development on models that better reproduce human physiology. This would enable to earlier assess the potential of the molecules and avoid failure of clinical trials. On top of this industrial and scientific need, there is a great societal interest in removing the reliance on animals for the testing of active pharmaceutical ingredients (API). Decades ago, the cosmetic industry had successfully banned animal testing and replaced the safety testing of new molecules with tissue-engineered skin. But while cosmetics are contained to the skin, API are traveling into the bloodstream and can reach all organs. Therefore, models that reproduce human physiology and drug circulation *in vitro* are needed if we want to reduce or replace the reliance on animal models for the commercialization of safe and efficient drugs.

Not all drugs are equal. While some drugs are injected directly into the bloodstream (intravenously), some drugs are taken orally, and this makes a huge difference. Indeed, when you inject a drug into the bloodstream, it will directly travel to all the organs. An oral drug will have to pass through the digesting track and the liver before entering the bloodstream. As the drug goes through the liver, it might be treated by the liver enzymes and modify its chemical structure or the dose that is available in the blood circulation is drastically different than the amount taken. This field of study called pharmacokinetic is a big part of understanding how drugs work and generates key data to design the molecule delivery mode, the molecule dose, and interval of injection or ingestion. Conversely, the field of pharmacy focusing on how the drug works once ingested is called pharmacodynamic. One of the best methods to assess cokinetics is to use animal models. Although all animals have physiology as humans (blood temperature in mouse 36.6 °C, in rat 35.7 °C), it can help to extrapolate the pharmacokinetics in humans. Organ-on-a-chip (OOC) approach consist of using soft litography fabrication techniques to create model organs that uses human cells organised in a microfluidic chip. OOC, by combining different synthetic organs, is proposed to serve as a simulation of human pharmacokinetic *in vitro* (Figure 6.19) [21].

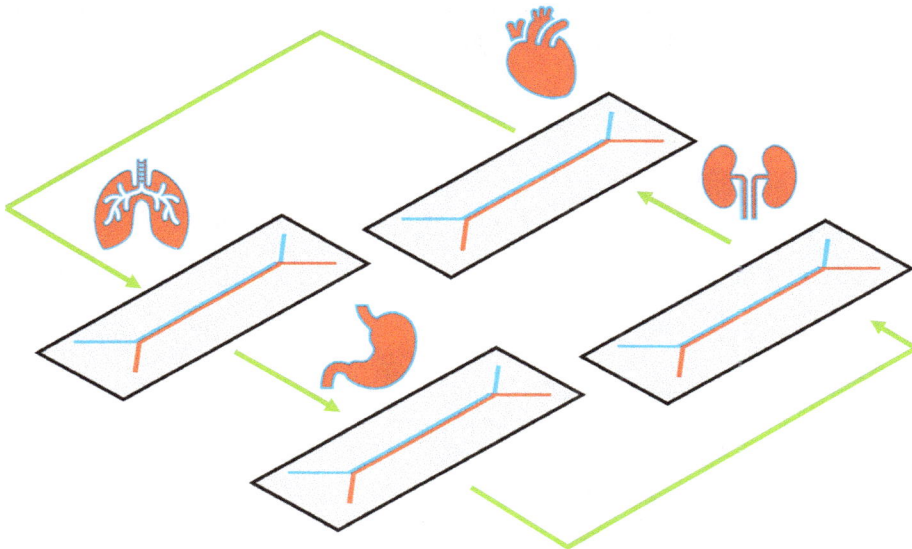

Figure 6.19: Serial assembly of microfluidic chips to simulate the human physiology and blood circulation between organs from the heart, lung, stomach, pancreas, and liver.

6.8.1 Simulated organs

When reproducing an organ synthetically *in vitro*, two strategies can be followed: reproducing the minimum required function or mimicking the design from nature. The soft lithography methods used to manufacture microfluidic chips while quite advanced suffer from many limitations that render challenging to perfectly mimic the natural architecture of human organs (Figure 6.20). Therefore, most efforts have been focusing on mimicking the key function of organs. Some organs have a tissue barrier: they protect one organ system from the other and thus serve as a selective exchange of molecules. For instance, the lung serves for the exchange of gas between the air and the bloodstream. Other organs have been mimicked on a chip. For instance, the blood–brain barrier (BBB) serves as a selective exchange hub which able to protect the brain from detrimental molecules or parasites that could have made it into the bloodstream. This can be reproduced by designing a chip with two compartments composed of an interface to mimic the BBB. Other organs, such as the intestine, can be similarly reproduced with a two-compartment system. The liver function can be mimicked by using a hydrogel containing the different types of liver cells. Mimicking the kidney function also requires a hydrogel to serve as an exchange media between arteries and veins. Each of these organs can be redesigned with a simpler approach compatible with microfluidic chip manufacturing methods [22, 23].

One of the key advantages of OOCs is the modularity and versatility of the platform. Organ modules can be brought inline or not depending on the experiment, modules can be upgraded as the technologies evolve. Finally, the microfluidic chip

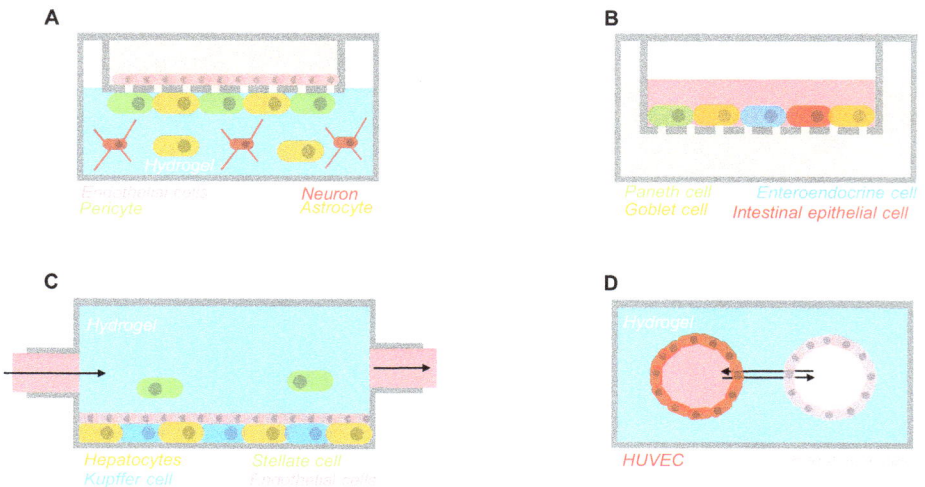

Figure 6.20: Schematic of the example of microfluidic cell models of (A) the blood-brain-barrier, (B) intestine, (C) liver, and (D) kidney.

setup can be scaled up and multiplied easily when maybe one day OOC can replicate the patient physiology *in vitro* and allow physicians to test a drug regiment before administrating it to their patients. With this potential, it is expected that OOC technology will gain much traction in the future and be a key player in our endeavor to replace animal testing for drug discovery.

6.9 Quiz

1. What is the contact angle of a non-wetting surface?
2. What are the two steps of cell adhesion on PS?
3. What is the most energetic state of the matter?
4. What is the Reynolds number useful for?
5. What is the difference between photolithography and soft lithography?
6. What is the method used to manufacture nanowires?
7. Which analytical techniques are best suited for the characterization of surface chemistry?
8. What are the applications of nanopillars?
9. Which societal and industrial needs does OOC technology promises to answer?
10. What is the difference between pharmacodynamic and pharmacokinetic?

References

[1]   Neděla, O.; Slepička, P.; Švorčík, V. Surface Modification of Polymer Substrates for Biomedical Applications. *Materials (Basel)*. **2017**, 10 (10), 1115. doi:https://doi.org/10.3390/ma10101115.

[2]   Macgregor, M.; Vasilev, K. Perspective on Plasma Polymers for Applied Biomaterials Nanoengineering and the Recent Rise of Oxazolines. *Materials (Basel)*. **2019**, 12 (1), 191. doi: https://doi.org/10.3390/ma12010191.

[3]   Bhatt, S.; Pulpytel, J.; Arefi-Khonsari, F. Low and Atmospheric Plasma Polymerisation of Nanocoatings for Bio-Applications. *Surf. Innov.* **2015**, 3 (2), 63–83. doi:https://doi.org/10.1680/sufi.14.00008.

[4]   Vasilev, K.; Ramiasa-macgregor, M. Nanoengineered Plasma Polymer Films for Biomedical Applications. *Adv. Mater. Lett.* **2018**, 9 (1), 42–52. doi:https://doi.org/10.5185/amlett.2018.1691.

[5]   Harding, F. J.; Clements, L. R.; Short, R. D.; Thissen, H.; Voelcker, N. H. Assessing Embryonic Stem Cell Response to Surface Chemistry Using Plasma Polymer Gradients. *Acta Biomater.* **2012**, 8 (5), 1739–1748. doi:https://doi.org/10.1016/j.actbio.2012.01.034.

[6]   Cools, P.; Mota, C.; Lorenzo-Moldero, I.; Ghobeira, R.; De Geyter, N.; Moroni, L.; Morent, R. Acrylic Acid Plasma Coated 3D Scaffolds for Cartilage Tissue Engineering Applications. *Sci. Rep.* **2018**, 8 (1), 3830. doi:https://doi.org/10.1038/s41598-018-22301-0

[7]   Mirbagheri, M.; Adibnia, V.; Hughes, B. R.; Waldman, S. D.; Banquy, X.; Hwang,
      D. K. Advanced Cell Culture Platforms: A Growing Quest for Emulating Natural Tissues. *Mater.*
      Horizons. **2019**, 6 (1), 45–71. doi:https://doi.org/10.1039/C8MH00803E.
[8]   Paulitschke, P.; Keber, F.; Lebedev, A.; Stephan, J.; Lorenz, H.; Hasselmann, S.; Heinrich,
      D.; Weig, E. M. Ultraflexible Nanowire Array for Label- and Distortion-Free Cellular Force
      Tracking. *Nano Lett.* **2019**, 19 (4), 2207–2214. doi:https://doi.org/10.1021/acs.
      nanolett.8b02568.
[9]   Chiappini, C.; De Rosa, E.; Martinez, J. O.; Liu, X.; Steele, J.; Stevens, M. M.; Tasciotti,
      E. Biodegradable Silicon Nanoneedles Delivering Nucleic Acids Intracellularly Induce
      Localized in Vivo Neovascularization. *Nat. Mater.* **2015**, 14 (5), 532–539. doi:https://doi.org/
      10.1038/nmat4249.
[10] Rzhevskiy, A. S.; Singh, T. R. R.; Donnelly, R. F.; Anissimov, Y. G. Microneedles as the
      Technique of Drug Delivery Enhancement in Diverse Organs and Tissues. *J. Control. Release.*
      2018, 270, 184–202. doi:https://doi.org/10.1016/j.jconrel.2017.11.048.
[11] Soffe, R.; Altenhuber, N.; Bernach, M.; Remus-Emsermann, M. N. P.; Nock, V. Comparison of
      Replica Leaf Surface Materials for Phyllosphere Microbiology. *PLoS One.* **2019**, 14 (6),
      e0218102. doi:https://doi.org/10.1371/journal.pone.0218102.
[12] Weibel, D. B.; DiLuzio, W. R.; Whitesides, G. M. Microfabrication Meets Microbiology. *Nat.*
      Rev. Microbiol. **2007**, 5 (3), 209–218. doi:https://doi.org/10.1038/nrmicro1616.
[13] Anna-Kristina Marel; Nils Podewitz; Zorn, M.; Rädler, J. O.; Elgeti, J. Alignment of Cell Division
      Axes in Directed Epithelial Cell Migration. *New J. Phys.***2014**, 16 (11), 115005. doi:https://doi.
      org/10.1088/1367-2630/16/11/115005.
[14] Théry, M. Micropatterning as a Tool to Decipher Cell Morphogenesis and Functions. *J. Cell Sci.*
      2010, 123 (24), 4201–4213. doi:https://doi.org/10.1242/jcs.075150.
[15] Belkaid, W.; Thostrup, P.; Yam, P. T.; Juzwik, C. A.; Ruthazer, E. S.; Dhaunchak, A. S.; Colman,
      D. R. Cellular Response to Micropatterned Growth Promoting and Inhibitory Substrates. *BMC*
      Biotechnol. **2013**, 13 (1), 86. doi:https://doi.org/10.1186/1472-6750-13-86.
[16] Sodunke, T.; Turner, K.; Caldwell, S.; Mcbride, K.; Reginato, M.; Noh, H. Micropatterns of
      Matrigel for Three-Dimensional Epithelial Cultures. *Biomaterials.* **2007**, 28 (27), 4006–4016.
      doi:https://doi.org/10.1016/j.biomaterials.2007.05.021.
[17] Luty-Błocho, M.; Wojnicki, M.; Grzonka, J.; Kurzydłowski, K. J. The Synthesis of Stable
      Platinum Nanoparticles in the Microreactor. *Arch. Metall. Mater.* **2014**, 59 (2), 509–512. doi:
      https://doi.org/10.2478/amm-2014-0084.
[18] Neumaier, J. M.; Madani, A.; Klein, T.; Ziegler, T. Low-Budget 3D-Printed Equipment for
      Continuous Flow Reactions. *Beilstein J. Org. Chem.* **2019**, 15, 558–566. doi:https://doi.org/
      10.3762/bjoc.15.50.
[19] Wyatt Shields IV, C.; Reyes, C. D.; López, G. P. Microfluidic Cell Sorting: A Review of the
      Advances in the Separation of Cells from Debulking to Rare Cell Isolation. *Lab Chip.* **2015**, 15
      (5), 1230–1249. doi:https://doi.org/10.1039/C4LC01246A.
[20] Bose, S.; Singh, R.; Hanewich-Hollatz, M.; Shen, C.; Lee, C.-H.; Dorfman, D. M.; Karp,
      J. M.; Karnik, R. Affinity Flow Fractionation of Cells via Transient Interactions with
      Asymmetric Molecular Patterns. *Sci. Rep.***2013**, 3 (1), 2329. doi:https://doi.org/10.1038/
      srep02329.
[21] Zhang, B.; Korolj, A.; Lai, B. F. L.; Radisic, M. Advances in Organ-on-a-Chip Engineering. *Nat.*
      Rev. Mater. **2018**, 3 (8), 257–278. doi:https://doi.org/10.1038/s41578-018-0034-7.

[22] Sosa-Hernández, J. E.; Villalba-Rodríguez, A. M.; Romero-Castillo, K. D.; Aguilar-Aguila-Isaías, M. A.; García-Reyes, I. E.; Hernández-Antonio, A.; Ahmed, I.; Sharma, A.; Parra-Saldívar, R.; Iqbal, H. M. N. Organs-on-A-Chip Module: A Review from the Development and Applications Perspective. *Micromachines*. **2018**, *9* (10), 536. doi:https://doi.org/10.3390/mi9100536.

[23] Zhang, Y. S.; Aleman, J.; Shin, S. R.; Kilic, T.; Kim, D.; Mousavi Shaegh, S. A.; Massa, S.; Riahi, R.; Chae, S.; Hu, N. et al. Multisensor-Integrated Organs-on-Chips Platform for Automated and Continual in Situ Monitoring of Organoid Behaviors. *Proc. Natl. Acad. Sci.* **2017**, 114 (12), E2293–E2302. doi:https://doi.org/10.1073/pnas.1612906114.

7 Porous scaffold

7.1 What is porosity?

In material science, porosity measures the void fraction, the empty spaces in a material. It defined the fraction of the volume that is empty over the total volume of the material as a percentage. Several techniques are available to calculate the void volume in materials. Choosing the proper method depends on the type of porosity and materials. Once the empty space of materials is measured, the porosity can be calculated as follows:

$$\varnothing = \frac{V_{\text{empty space}}}{V_{\text{total}}}$$

In nature, many materials are porous, that is, they contain empty space. A prime example is rocks such as sandstone or limestone made of pores that allow them to filter water. In mammalian, many tissues are made of a porous structure. A prime example is the bone inorganic matrix that has a porous architecture. But not all pores architectures are built equals. We must distinguish two types of porosity: the closed pores and the open pores architecture (Figure 7.1). The pores are not connected to each other in the close pores architecture, while most pores are connected in the open-pore architecture. This means that nutrients, cells, and gas can pass from one side of the porous structure to the other in the open pore architectures, while in the closed-pores architecture, the two sides of the porous architecture are not connected. Within these two architectures, pores can have different shapes. Pores can be square, rounded, or have an elongated structure, and this has to do with the way there are made.

A   **B**

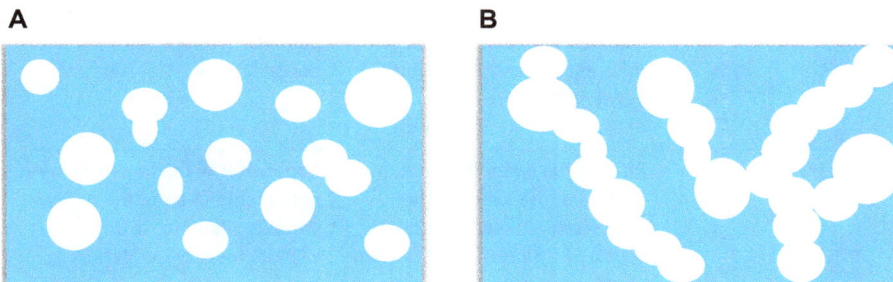

Figure 7.1: The two different pore architecture. (A) Closed-cell pores and (B) open pores.

There are several techniques to generate pores and create porosity in a scaffold or materials, and in this chapter, these different approaches are presented [1].

https://doi.org/10.1515/9781501515736-007

7.A Measuring porosity

Once a porous scaffold is obtained, several techniques can be used to measure the porosity of the scaffolds. Not all the techniques are compatible with all samples (Table 7.1). This will depend on the type of pores (closed or open pores), the size of the pores (micro- or mesopores), and the scaffold materials. Indeed, it can be extremely challenging to measure the porosity of scaffolds made of soft materials that can deform under stress. Therefore, the measuring technique must be carefully chosen to get an accurate measurement.

The pore definition from the International Union of Pure and Applied Chemistry (IUPAC) defines porous materials as being microporous when the pore diameter is less than 2 nm, mesoporous when the pore diameter is comprised between 2 and 50 nm, and macroporous when pore diameter is greater than 50 nm. In the case of porous scaffolds used in biofabrication usually, the pore size is greatly higher. Indeed, the porogen-made scaffolds have pores usually in hundreds of nanometer to micrometer scale. This great range of pore size must be taken into consideration when choosing the measuring technique.

Table 7.1: Techniques used for the measurement of porosity.

Technique	Principle	Pore type
Brunauer–Emmett–Teller (BET) theory	Gas (nitrogen) is adsorbed on the pore surface forming several layers. The pressure required to equilibrate the system provides information on the surface area.	Nanometer: micro- and mesopores according to IUPAC. The materials must resist the pressure.
Mercury intrusion	Mercury is forced into the pores under pressure. Then, the applied pressure lets us calculate the surface area and pore size.	Open-pore architecture between 250 μm to 3 nm. Specialized for hard materials needs to withstand the high pressure
Atomic force microscopy	The surface of the sample is scanned with a cantilever. The difference in surface height between pores reveals a 2D map of the surface topography.	Nanometer to micrometer open-pore architecture. Only surface pores
Scanning electron microscopy	An electron beam is shot at the surface of the sample. The backscattered electrons are captured by a receptor and give an image of the sample surface.	Nanometer to micrometer open-pore architecture of the surface porosity. Conductive and vacuum resistant materials
X-ray tomography	X-rays are used to scan the inner architecture of the sample. Each scan is computed to create digitally a 3D image of the object.	Provide a 3D image of the sample and the porosity. Depending on the resolution of the instrument micrometer, open or closed pores

Table 7.1 (continued)

Technique	Principle	Pore type
Nuclear magnetic resonance (NMR)	A medium such as water is used to probe the pore size. The capability of the media to interact with the movement of the probing media is measured.	Suitable for hydrogels with pores in the hundreds of a nanometer to tens of a micrometer.

7.2 Porogen-induced pores

One of the methods to create pores is to use a porogen. We can define porogen as a mass of particles of a specified shape and size used to make pores in molded structure, and the porogen is removed after the structure has set. These structures can, in turn, be used for different applications, from insulation foam to tissue engineering. Depending on the type of porogen, one can design and manufacture a variety of porous scaffolds.

7.2.1 Foam-based porous scaffolds

A scaffold that is constituted of small bubbles is called foam. Typically, foams in physical chemistry are defined as a colloids system constituted of gas particles dispersed in a continuous liquid medium. By extension, the term foam is applied to materials that are constituted of a cellular architecture, such as a sponge. One typical example of foam in polymer science is polyurethane foam, which combines the two monomers to make the polycondensation reaction (Figure 7.2). Upon reaction of the monomers with each other, carbon dioxide is generated. The gas bubble is generated and will then travel throughout the material, expanding the polymer matrix to create porous materials, a foam. Because we use a gas medium as porogen, the gas wants to escape in need of more space. Therefore, the gas bubbles will travel throughout the material to create an interconnected system of pores [2] (Figure 7.3).

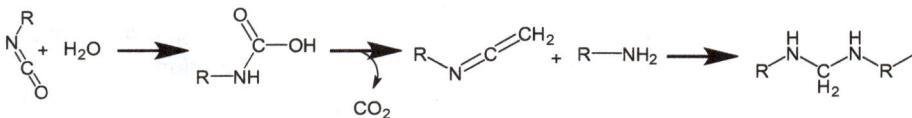

Figure 7.2: Chemical reaction of polyurethane.

In tissue engineering, open-cell materials are highly interesting as they permit the creation of materials that can reproduce natural tissue architecture. It allows nutrients, gas, and cell to travel throughout the materials. But creating sponges using gas that can be harmful to cells must be done *in vitro*, and only when the materials are cured and neutralized then cells can be added. Beyond polyurethane, other materials can be made as foam. The key here is to have a crosslinking reaction that generates a gas. If the gas is toxic, one must neutralize it and ensure that there are no more remaining toxic chemicals. But the major drawback of these foams is that it is challenging to control the size of the pores, the pore organization, and all the pores will have a rounded shape.

Figure 7.3: Scheme of the foaming process of polyurethane monomer A and B are mixed together in a container. The reaction occurs generating gas that travel in the polymer matrix creating the porosity.

7.2.2 Crystal-induced porous scaffold

To control the pore size of a polymer matrix in a precise manner, we can introduce a porogen of pre-determined size. Salt of NaCl or sugar that have been ground and sieved to obtain a controlled crystal size are mixed in a melted polymer matrix such as PLA or PCL. The melted mix can then be molded in a particular shape and allowed to set at a temperature below the melting temperature. The porogen, which is made of a water-soluble crystal, can be leached out of the matrix by simply immersing the object in deionized water and refreshing the water regularly to be consistently below the saturation concentration of the salt or sugar, in our example. After a few rinses, a porous scaffold is obtained. Sodium chloride (360 g/L) and sucrose (2,000 g /L) are good candidates as they have a high saturation concentration which permits rapidly leaching them out of the polymer matrix [3, 4] (Figure 7.4).

With this technique, we can control the pore size, but not the organization of the pores. In the case of a crystal sample having a broad size distribution, the physical effect of convection can be leveraged to create a gradient of particle size in the scaffold. In a sample made of different particle sizes, the bigger particles will rise to the top while the smaller ones will stay at the bottom. This phenomenon, also called the muesli effect, can be used to organize the porogen particle into a size gradient (7C).

Polymer melt

Figure 7.4: From left to right: Crystals are organized in a pattern, then the melted polymer matrix is added to the crystals, the material is set, and the porogen is removed.

When using the crystal technique, the second important consideration is that salt and sugar do not crystalize into a round shape. Since the shape of the crystal will be translated to the polymer matrix, it will govern the resulting pore shape. In our example, sodium chloride forms a face-centered cubic lattice, and sucrose forms a hexagonal prism. Using an extensive ball or roll milling can help change the shape of the crystals, but this can become a tedious process.

Similarly, to the melt process casting, the polymer can also be dissolved in a solvent that does not solubilize the porogen. The porogen is added to the polymer solution. The solvent is evaporated, and the bulk material is then immersed in the porogen's solvent. Depending on the packing of the porogen, an open- or closed-pore architecture can be obtained. If the porogen are pack in a manner that they are touching with each other, then a open-pore architecture is obtained. However, to obtain a closed-pore architecture, the vent used to remove the porogen needs to travel into the polymer matrix. Many small molecules and active pharmaceutical compounds can be crystallized. As crystals, they can be used as a porogen and in-corporated into a polymer matrix. Once processed into an implant, the porogen will be slowly leached out of the polymer matrix after implantation.

7.B Convection effect

When opening a bag of mixed nuts, have you ever wondered why the pecan nuts and other larger nuts are always on the top of the bag? Why at the end of the bag do you get all the smaller pistachios? This is due to the physical effect called convection or also called the Brazil-ian nuts effect or muesli effect. It states that when particles of various sizes are shaken in a container, the bigger particles will float at the surface (Figure 7.5). This principle is the key to avalanche safety inflatable backpacks. These safety devices have a balloon that can inflate when needed. This balloon expands when a skier is trapped in a snow stream and increase the skier size that can make him bigger than the snow aggregates and thus will help him float at the surface of the avalanche due to the convection effect. Using the convection effect when making scaffolds with particles of different sizes, will bring the bigger particles at the top and smaller particles at the bottom Thus, one can use the convection effect to create a gradient of particle size upon shaking container of porogen of various size. Once the particles are orga-nized in a gradient, the polymer melt or solution can be added following one of the porogen-based fabrication techniques presented in this chapter.

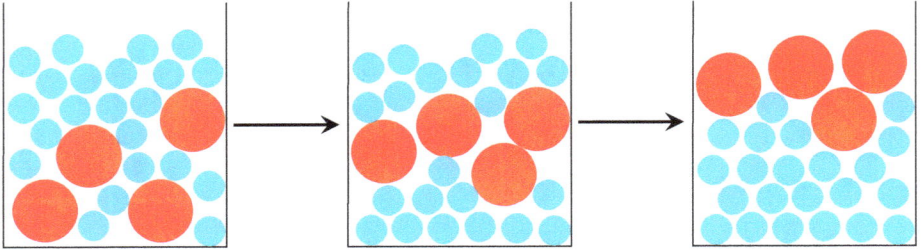

Figure 7.5: The convection effect: in a population of mixed-size particles, when shook the bigger particles float at the top of the mix.

7.2.3 Porogenic solvent

Because a porogen is a pore-generating phase, this can also come as a liquid. Solvents that are inert and can solubilize a polymer and have a boiling point that is lower than the melting point of the polymer matrix can be used to create pores. Typically, a polymer melt is mixed with a solvent. Once processed together, a phase separation must be induced to create a porous structure. Once the phase separation is obtained, the solvent can be removed by evaporation [5] (Figure 7.6).

Figure 7.6: From left to right: Solvent of the polymer is mixed with a polymer melt, then a homogenous solution is obtained, next a phase separation is generated by cooling down the matrix, and the solvent is evaporated, leaving a porous scaffold.

Different solvents can be used to create various type of pores. For instance, the use of methanol in a PLA melt leads to higher pore diameter as if a mix of ethanol and methanol is used or pure ethanol. This has to do with the solvent interaction with the polymer. A poor solvent of the polymer will give a larger pore than a good solvent because the phase separation will be more pronounced [6].

7.2.4 Water as a porogen

While few polymers are soluble in water, the case of water as a porogen medium is quite interesting. Indeed, water can be processed into a solid shape and sublime, that is, transition from the solid phase to the gas phase without becoming liquid. Present in major quantities in hydrogels, such as collagen, it can form highly porous sponges.

By depositing water droplets on a surface of known hydrophobicity, controllable water droplet size and shape can be obtained. The droplet can then be frozen. The resulting ice particles can be dispersed into a collagen solution. The ice crystals can then be recrystallized to create a connection between them leading to a open-pore architecture. Finally, the water is removed by sublimation leaving a porous scaffold (Figure 7.7). One can create precise and complex architecture by playing around with water droplet size, shape, and organization. [7, 8]

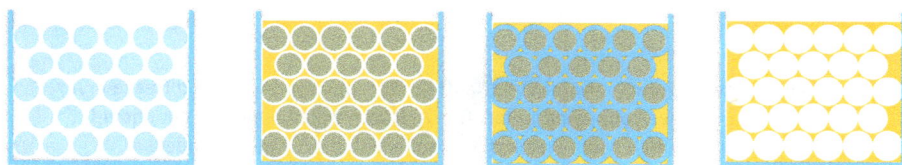

Figure 7.7: From left to right: Ice droplets are formed by freezing water droplet. These ice particles are then mixed with the polymer, and then connected with each other through a second freezing parameter. Finally, the water is sublimated, leaving a porous matrix.

In the case of hydrogels, removing the water by freeze-drying reveals the internal organization of the polymer chains. It is then possible to observe how the polymer organizes into the water media. However, one must acknowledge that the freezing process can modify the natural organization that the fibers have in a liquid phase. Nevertheless, agarose polymer chains organized into helical structure leave rounded pores when freeze-dried. Methylcellulose polymer chains that organize into a β-sheet leave a laminar porous scaffold once freeze-dried [9].

7.C Phase diagram of water

Water can be found in three different states on earth: solid (ice), liquid, and gas. On the phase diagram, we can see the energy required to shift between each phase. At atmospheric pressure, water becomes gas at 100 °C and solid at 0 °C (Figure 7.8). But on this diagram, we can also observe a triple point below which the ice can directly transform into a gas. This is reached at low pressure, and it is called sublimation. This particularity of water is used to efficiently remove water using a freeze dryer. In a freeze-dryer, solid water (frozen) is sublimated under a high vacuum typically below 0.1 mbar. The water gas, steam, is collected in a condenser usually operating at – 20 °C to – 50 °C to trap the gas as pure water (Figure 7.8). This technique is used to create dehydrated samples and is commonly used in the food and pharmaceutical industry. The removal of water increases the shelf life of pharmaceutical preparation. Indeed, the removal of water prevents the hydrolysis of active pharmaceutical ingredients or degradation of nutrients or vitamins in food samples.

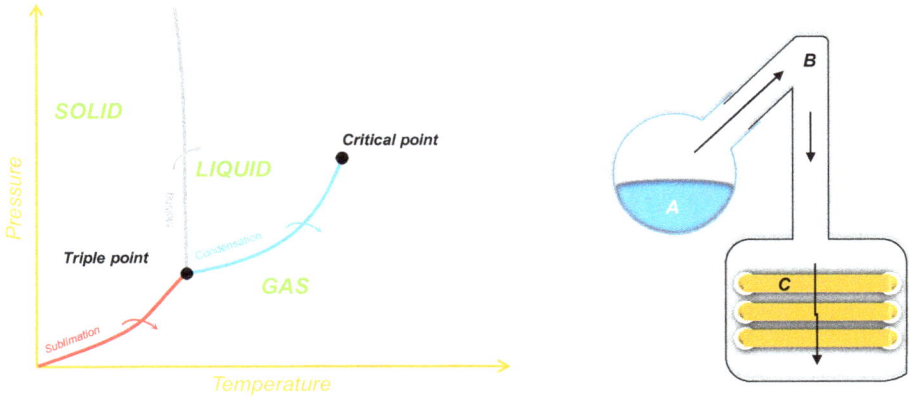

Figure 7.8: **(left)** Phase diagram for water and **(right)** scheme of the principle of a freeze-dryer utilized to dehydrate samples containing water. The (A) frozen water sample is put under (B) vacuum and the water sublimates into steam that is collected in a (C) water condenser.

7.2.5 Supercritical fluids

Polymer expansion technique uses CO_2 that can be infiltrated in the polymer matrix as a gas. CO_2 is adsorbed into the polymer matrix, and then the gas expands to create bubbles. However, the temperature at which the process occurs must be above the T_g of the polymer so as to be able to deform the matrix without inducing cracks in the polymer phase. But the addition of a gas in the polymer matrix impacts the physical properties of the polymer chains and can result in a change of T_g (Figure 7.9). Therefore, this process is highly dependent on the gas parameter and type of polymer used. Like water, carbon dioxide can be brought above its critical point under mild conditions. Above a certain temperature and pressure, the liquid and gas phases are not separated anymore, and the fluid can penetrate porous structures much like a gas. Using supercritical CO_2 permits to conduct the expansion process at a higher temperature than if it were in the gas phase. The parameter used in this process permits the creation of either closed-pore architecture or open-cell pore architecture. The technique of polymer expansion is derived from poly(styrene) foam which is a well-known consumer product for isolation materials. Applied to biomaterials, the expansion with supercritical CO_2 was adapted to PLA to create a porous biodegradable implant [12].

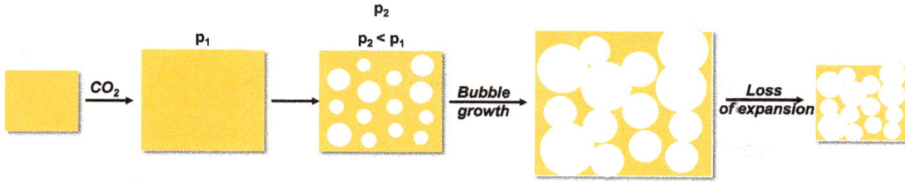

Figure 7.9: The expansion process of the polymer matrix. The polymer matrix swells upon adsorption of carbon dioxide. A decrease in the pressure lets the gas expand in the polymer matrix. Upon release to atmospheric conditions, the porous polymer matrix loses its expanded size but remains porous.

7.2.6 Block copolymer phase separation

In the late 1960s, Meier proposed using block copolymers to induce microphase separation in the polymer bulk. The microphase separation consists of a chain segregation in block copolymers in the bulk state and in concentrated solution. The structures generated by the packing of domains on the macrolattice are termed mesomorphic structures. Using Flory–Huggins interaction parameters of the polymer blends, one can predict the shape of the phase separation of the block copolymers. Microphase can be obtained depending on the ratio of each block and its miscibility to each other (Figure 7.10) [13]. If you have one block that can be leached out without affecting the other block, you can create a predictable microporous architecture. This can be used with block copolymers made of poly(butadiene) and poly(dimethylsiloxane) or PLA with poly(styrene). To reinforce each phase, one can cure or crosslink the phase that needs to be kept from the etching of the other block [14, 15].

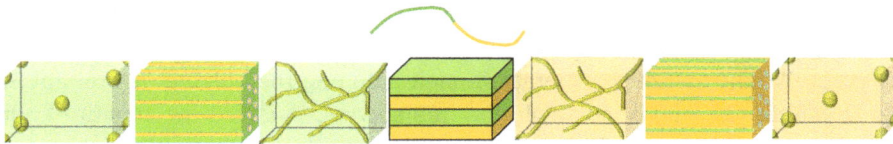

Figure 7.10: The phase separation transition observed in block copolymers can be calculated according to Meier's theory.

7.3 Microfiber-based scaffolds

If we are to reproduce the natural organization of tissues, we need to mimic the different types of natural porosity. For example, we have seen above that bone, and inorganic phases have a porous architecture that can be reproduced by using porogen. However, connective tissue, muscle, or tendon have a fibrous organization.

These tissues' organization is based on fibers made of extracellular matrix macro-molecules such as collagen proteins or chondroitin sulfate polysaccharides. These polymers will then form a fibrillar linear organization.

For centuries, we have made fibers that are used for making textile. These techniques have been refined over the years to become today's textile industry. Using three main processes: wet spinning, melt spinning, and dry spinning, fibers can be produced out of biomaterials, and these fibers can be brought to use for biofabrication applications (Figure 7.11).

7.3.1 Dry spinning

In the dry spinning process, the polymer is dissolved in a solvent. The polymer solution composed of the polymer and the solvent is extruded through a spinneret. The spinneret is an extrusion nozzle through which the polymer solution passes through. The size of the nozzle will determine the size of the fiber. The polymer chains will be forced to organize into the linear trend through the fiber extrusion. After the spinneret, the polymer solution must be dry to remove the solvent. The dry or semi-dry fibers are collected into a series of wheels called a draw twister to align the polymer chains further. This drawing tends to increase the crystallinity of the polymer and increases the strength of the fibers. At the end of this process, the fiber is collected on a spool.

7.3.2 Melt spinning

It is not favorable for all polymers to be processed through a dry spinning process in a solution. Thermoplastic polymer, if they have a low melting temperature might be preferable in terms of production cost and lack of solvent to be melt spun. The process is like dry spinning, whereas in melt spinning the polymer is brought to flow by melting it and not as a solution in a solvent. This has the benefit of not using solvents that can be potential ecologically and healthy harmful. The melted polymer that goes through the spinneret is then evaporated by air, drawn, and spun on a spool.

7.3.3 Wet spinning

In some cases, polymers can only be dissolved in a low-volatile solvent such as DMF or DMSO. These solvents cannot be evaporated in a dry spinning process. But, because they are water soluble, one can extract them in an aqueous bath. The dissolved polymer is passed through a spinneret that is immersed in a water bath. The fibers are then pulled into a draw twister for further drying of the water and then collected on a spool. The water bath can also be composed of calcium chloride and be used to crosslinked alginate polysaccharides [16, 17].

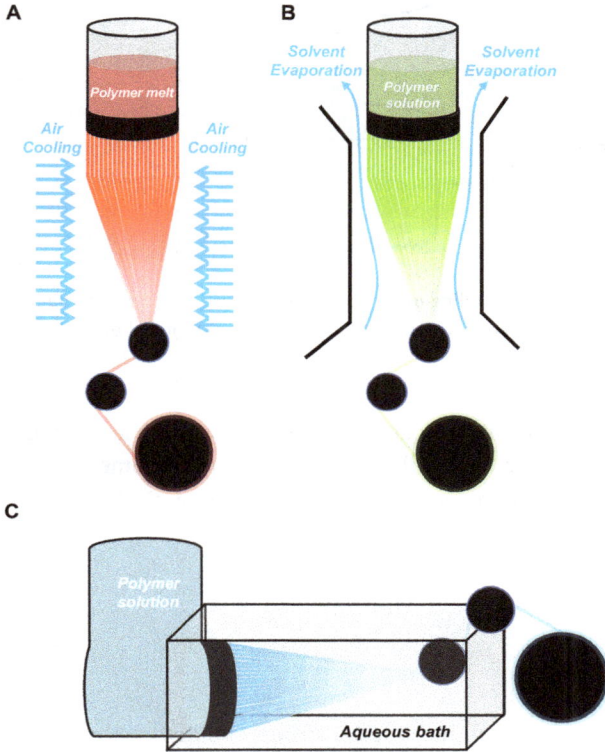

Figure 7.11: The fiber manufacturing process: (**A**) Melt spinning, (**B**) dry spinning, and (**C**) wet spinning.

7.3.4 Fiber bonding

Once fibers are obtained on a spinneret, they need to be bound to make a scaffold. One technique used for making micrometer scale fiber-based scaffold is to use the fiber bonding technique. Starting with individual fibers of a polymer A, the non-bonded fiber structure is immersed into a solution of a polymer B. The solvent containing the polymer B is then evaporated and we obtained a block made of polymer B with the fibers of the polymer A embedded in the polymer B matrix. Like this, the fibers made of polymer A are fixed. We can then apply a heat treatment to seal the fibers of polymer A together. Then, the polymer B can be dissolved using a non-solvent of polymer A and a fiber mesh is obtained (Figure 7.12).

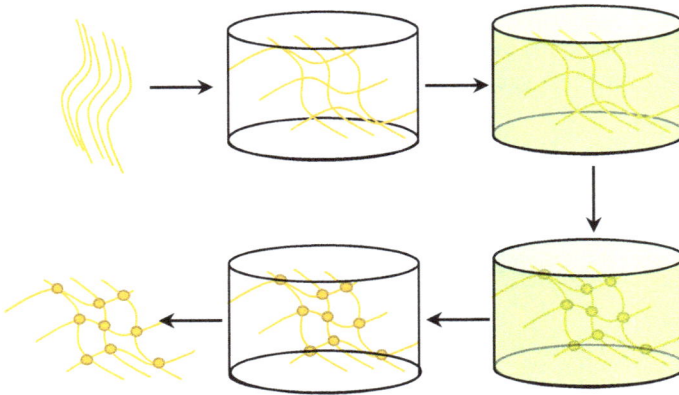

Figure 7.12: Schematic of the fiber technique bonding using a polymer matrix (green) to immobilize the fibers (yellow) and bond them through thermal treatment and subsequent removal of the green polymer.

While this technique is quite old, its efficacy has been proven for large fibers diameters. It is, however, challenging to handle micrometer- and nanometer-scale fibers and to precisely control their organization. Therefore, other techniques have been developed with the aim to gain control over the organization of the fiber in the scaffold.

7.3.5 Hollow fibers

One of the most sought-after architecture in biofabrication is hollow channels. Indeed, many tissues are composed of hollow channels such as blood vessels or lymphatic channels. Therefore, if we want to create functional tissues that can survive post-implantation in the patient, these tissues need to be connected to the host vasculature and lymphatic systems. As such, making hollow fibers has become an interesting way of engineering channels that could mimic the geometry of blood vessels. To make hollow fibers, a coaxial extrusion technique was developed using alginate hydrogel. Alginate forms a hydrogel upon crosslinking with calcium chloride. In a coaxial nozzle, a solution of calcium chloride is extruded in the middle of the alginate hydrogel precursor solution. The calcium chloride solution will induce crosslinking of the alginate and leave a hollow channel. The fiber is then extruded in a bath of calcium chloride to gel the outer shell of the fiber (Figure 7.13). Combined with a 3D bioprinter, these hollow fibers can be processed into 3D scaffolds with aligned hollow fibers that mimic the fibrillar organization observed in many tissues. Furthermore, the hollow architecture provides a channel to infuse endothelial cells that could be used to make blood vessel-like architecture [18–20].

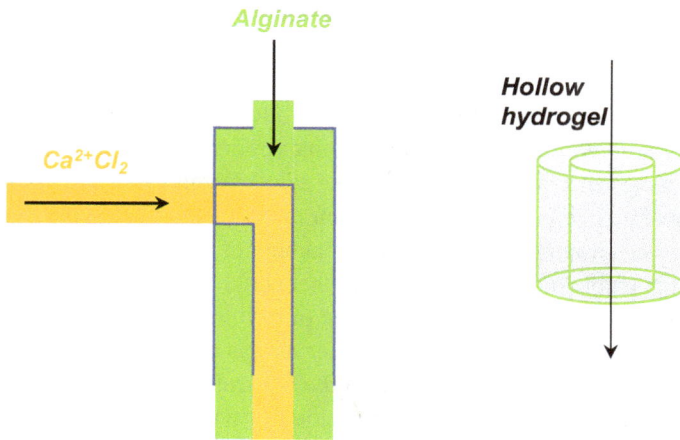

Figure 7.13: Schematic of (**A**) the coaxial extrusion setup and (**B**) resulting in hollow hydrogel fibers obtained with this setup.

7.4 Nanofibers fabrication

While these techniques allow making fibers with a diameter at the micrometer scale, the fiber diameter is limited by the extrusion nozzle diameter. To overcome this limitation a technique that could generate fibers that are smaller than the extrusion nozzle is needed. Therefore, the electrospinning process has been developed to overcome the limitation of the conventional fiber fabrication techniques.

7.4.1 Electrospinning

The electrospinning technique uses a high-voltage low-current electrical force to draw fibers out of an extrusion nozzle. The electric force permits elongating the fibers and getting fibers that are smaller than the diameter of the extrusion nozzle and obtaining polymer fibers at the nanometer scale. This technique is the result of centuries of observations and scientific advances. In the seventeenth century, it was observed that a water droplet can be deformed by an electric current generated by a rubber amber through electrostatic interaction. This observation shows that you can elongate a spherical droplet into a conical droplet. Later in the nineteenth century, Plateau and Rayleigh studied the formation of the water droplet. By letting water through a faucet, Plateau found experimentally that a vertically falling stream of water will break up into drops. Rayleigh showed theoretically that falling water with a circular cross-section should break up into drops. This is due to tiny perturbations in the stream. These perturbations exist initially in minuscule amplitudes, as time progresses, the perturbations will grow and pinches the stream into drops. Zeleny later studied the behavior of water under electric discharge. He observed that depending on the voltage, a droplet positioned at the opening of the meniscus was delivered differently as the electric potential increased. During the beginning of the twentieth century, following the work of Rayleigh and Zeleny, Taylor studied the behaviors of a droplet under electric current to understand storms and observed that to get jetting of the solution you need to be in a particular condition where a cone is generated. Variation of the potential will disrupt the cone and will give either droplet of multijetting. In parallel to the development of the industrialization of the textile industry, small fibers that could reproduce the silk's fibers properties were sought after. The electrospinning of cellulose acetate solution gave nanometer-scale fibers that could be used as a filter for gas masks. It is now an industrial process that can produce nanometer fibers at the ton scale.

The current electrospinning setup works using a loaded syringe. The solution of polymer is extruded at a constant flow rate using a syringe pump. At the tip of the syringe, a blunted needle is positioned, and a high potential is applied between the needle and the target. As the polymer solution reaches the needle, a Taylor's cone is formed, and the polymer solution jets onto the target. As the polymer solution travel, the solvent is evaporated, and the polymer reaches the target. Once onto the target, the polymer is further dried to remove the remaining solvent. The polymer can be jetted out of the needle and is not affected by gravity. Therefore, the setup of the solution electrospinning is usually horizontal [21] (Figure 7.14).

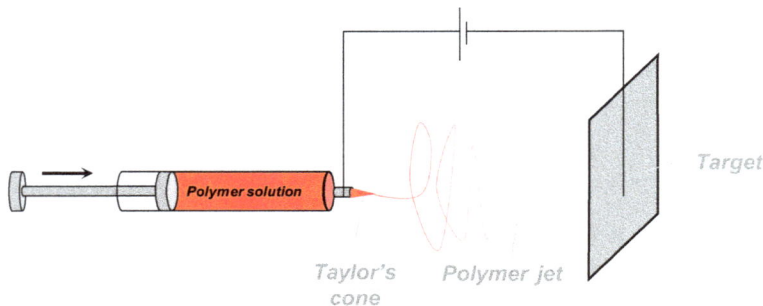

Figure 7.14: Scheme of the electrospinning setup showing the syringe, the polymer, and the target.

Following the first demonstrations on polysaccharides, research groups demonstrated that virtually every polymer solution can be electrospun. If the polymer cannot be solubilized in a volatile solvent, then melt electrospinning can be conducted, similarly to the melt spinning described above. However, because solutions of melted polymers are highly viscous, it requires a vertical setup and two current generators: one positive on the needle and one negative on the target to be able to push the polymer melt.

7.4.2 Advanced electrospinning

The extensive study of the electrospinning process has let scientists and engineers develop more advanced processes. The use of thermoplastic polymer has enabled to remove solvent and thus streamlined the process. Complex polymer solutions such as emulsion have been electrospun, and this enables the creation of complex fibers made of two polymers organized into a core–shell structure. Coaxial electrospinning like coaxial wet spinning now lets us makes fibers with nanometer diameter with a hollow core to be fabricated.

But still, the major limitation of electrospinning is the lack of control over the polymer fibers' organization. It is not possible to precisely organize the polymer fibers onto the target using conventional setups as only non-wave fibers are obtained. Different setups have been proposed to try to organize the fibers. Changing the flat smooth target with a metal grid lets the fiber concentrate cn the grid. Other systems used a cylinder as the collector plate. This setup gives the possibility to align the fibers along the rotation axis of the rotating collector (Figure 7.15). Combining the rotating collector with a grid setup permits to get grooves in the obtained tube. But still, a predictable deposition of the fiber and controlling the position of the fibers on the target needs to be achieved [22].

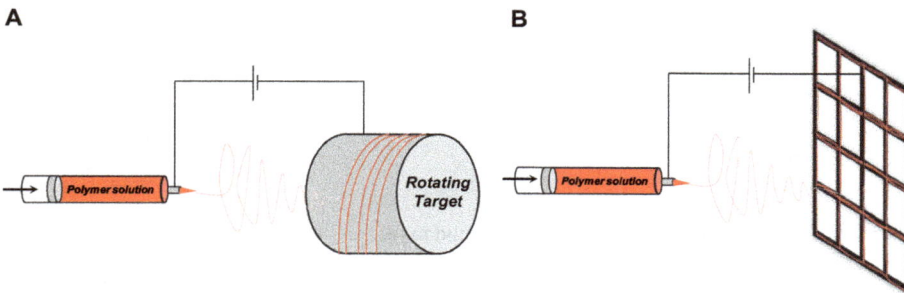

Figure 7.15: Electrospinning attempts at controlling the fiber organization: (**A**) on a rotating collector and (**B**) Electrospinning on a metal grid.

7.4.3 Melt electrowriting

The advances made in electrospinning-based fabrication processes and notably the use of polymer melt drastically streamlined the process. The lack of solvent made the technique simpler to apply. With the development of 3D printing technologies, extrusion setups that can be moved along three axes during extrusion have become widely accessible. Combining 3D melt printing with electrospinning in a process called melt electrowetting could help get a precise fiber organization. Conversely to extrusion-based 3D printing in electrospinning, the extrusion nozzle needs to be at a certain distance from the collecting plate to avoid an electric shortcut of the setup. When extruding a viscous liquid during movement, a drag of the fluid can be observed. Depending on the displacement speed of the extrusion head, further disruption can be observed, and the extruded viscous liquid can start to oscillate. Therefore, to get a precise deposition, in addition to the right electrospinning parameter (distance from the target, extrusion speed, and voltage) the optimal displacement speed needs to be found. This speed is called critical translational speed (CTS). Below the CTS, the polymer melt is buckling, and only above the CTS, a linear deposition can be observed (Figure 7.16). This phenomenon can be reproduced at home. If you deposit ketchup on a hot dog bun, you will squeeze the viscous ketchup while moving your bottle above the bun. Only if you move the ketchup bottle at a speed above the CTS, you will get a linear deposition below the ketchup will buckle and the deposition will not be linear. [23–25]

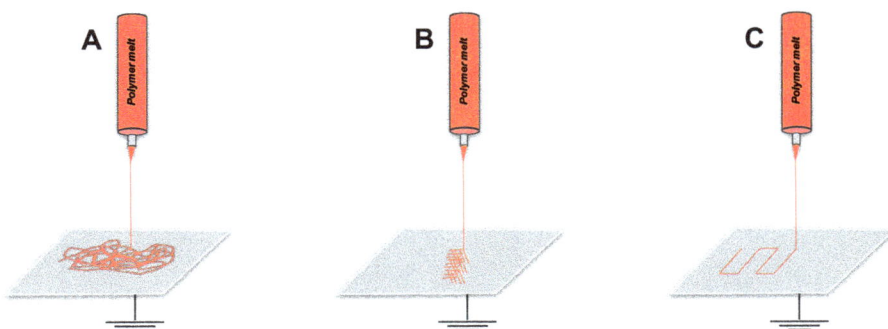

Figure 7.16: (**A**) Electrospinning deposition compared to (**B**) electrostatic writing below CTS and (**C**) above CTS.

The deposition of thermoplastic polymer with melt electrowetting permits to get high precision of the deposition of nanometer-scale fibers. Further development of this technique allows the user to choose the fiber size and deposit several layers of the fibers.

7.5 Cell organization on and in porous scaffolds

We have seen different techniques to create a porous scaffold. We have started this chapter by drawing the link between natural tissues that exhibit porosity and the need to reproduce this architecture to mimic this tissue *in vitro*. We have been able to fabricate porous scaffolds for several decades now, and a body of knowledge has been created by studying the interaction of these porous scaffolds with different types of cells.

7.5.1 Porous scaffold and the immune response

While presenting the different techniques to create a porous scaffold, one recurring criterion to assess the quality of the method was the ability or ease to control the pore size with a given technique. Why is this so important to be able to control the pore size? It is because the size of the pores in an implant can modulate the immune response. Like in *in-vitro* culture, an implanted material will interact with its surrounding environment once implanted in tissue. Proteins will start to adsorb on its surface and fold in a specific manner. The way the proteins fold and which type of protein can fold on the implanted materials will trigger a macrophage attack on the implant. The macrophage will then decide, depending on the type of protein they encounter, if they need to recruit fibroblast to create a collagen capsule around the implant to isolate this foreign host from the rest of the system. The material used to make the implant can modulate the immune response. But not only that, the pore size can also modulate the immune response to an implant. Depending on the size of the pore size, different types of macrophage will be interacting with the porous scaffold. An implant that has no pores will in most cases trigger the formation of a collagen capsule that will isolate the implant from the rest of the body. Pores that are about 30 μm will restrict the activation of macrophages and their infiltration into the pores while permitting the infiltration of blood vessels. Bigger pores (about 160 μm), however, will allow the infiltration of blood vessels and macrophages. Once infiltrated inside the pores, an isolation capsule will be formed. Therefore, having control over the pore size permits modulating the immune response of the implant [26, 27] (Figure 7.17).

Similarly, having interconnected pores will be a critical feature of the implant. If the goal is to have a blood vessel supply at the site of implantation, one would need to have pores that allow the infiltration of a blood vessel within the whole scaffold.

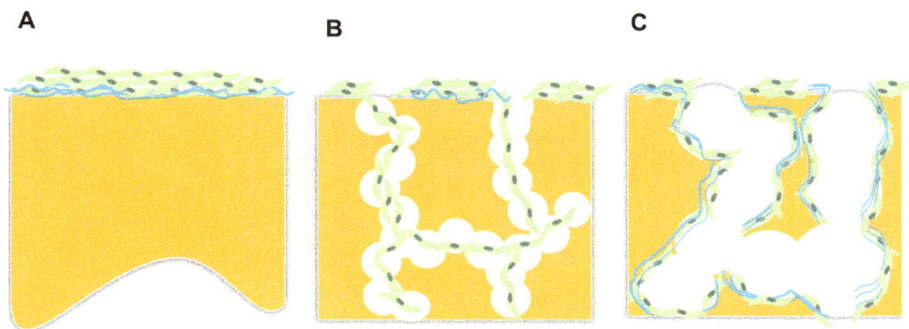

Figure 7.17: The impact of pore and pore size on cell infiltration and immune response. (A) On an implant with no pores, cells form a collagen capsule (in blue) to isolate the implant. (B) In small pores, some cells can penetrate the pore and no capsule is formed whereas (C) with an implant with broad pores a collagen capsule is generated.

For this to happen, the pores need to be interconnected throughout the entire scaffold. To have infiltration of the blood vessels, the pores need to be big enough to permit the infiltration of the blood vessel. Alternatively, the scaffold's pore needs to be degradable or elastic enough to permit the deformation of the pores for the infiltration of blood vessels [28]. Therefore, depending on the application of the implant, one must carefully design the pores' architecture and material composition. For instance, a drug-eluting implant might not require having access to the host vasculature, but a scaffold carrying functional cell such as an artificial pancreas system populated with insulin-secreting β-cell will need to have access to the host vasculature. Therefore, it is not required to have a precise pore definition for all applications.

7.5.2 The impact of fiber size

Like porogen-generated scaffolds, fiber-based scaffolds induced different cell behavior depending on the fiber size. The cell response to a fiber size can be used to define the required fiber-based scaffold for a specific application. Scaffolds with fiber diameters wider than cells are usually used to align cells in the fibers' direction. Whereas cells cultured on the scaffold with randomly organized nanofibers will tend to use the fiber as an attachment and spread between the fibers (Figure 7.18). But nanosized fiber can also be used to direct the organization of cells on the substrate. Using a rotatory target, electrospun fibers mat where all the fibers are aligned can be used to align neuron or muscle cells in one direction [17]. Within the two scales of fibers (micro- and nanofibers), subtle variation in the fiber size can also trigger different cell behavior in terms of gene expression and protein secretion. But these tremendously vary from cell type to cell type and need to be studied on a cell-type case basis.

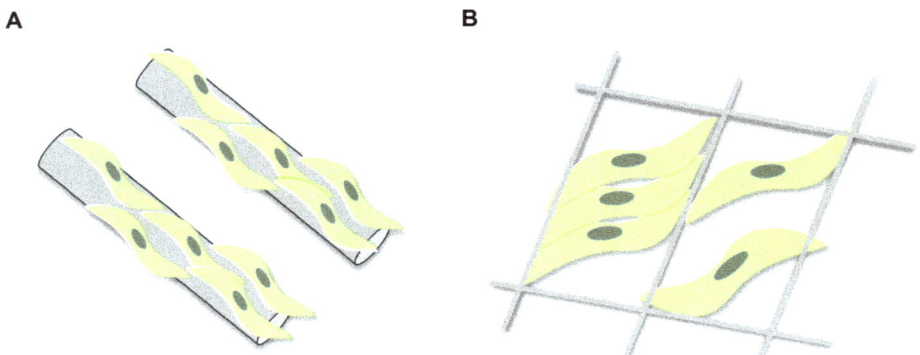

Figure 7.18: Typical organization of cells on (**A**) fibers with a diameter greater than the cell size (microfibers) and (**B**) fibers with diameters smaller than cells (nanofibers).

7.6 Quiz

1. What are the three main spinning methods?
2. Which criteria must be fulfilled to have optimal deposition in melt electrowriting?
3. What is the name of the transition of water from solid to gas?
4. What are the two types of porous networks?
5. How are polymers expanded? Example with expanded poly(styrene).
6. Which geometry is generated at the tip of the extrusion nozzle during electrospinning?
7. What type of extrusion needle is needed to fabricate hydrogel hollow fibers?
8. If I shake a recipient containing particles of different sizes, which particle size will be on top?
9. What is the definition of porosity?
10. What techniques are best suited to measure microporosity?

References

[1] Thompson, B.R.; Horozov, T.S.; Stoyanov, S.D.; Paunov, V.N. Hierarchically Structured Composites and Porous Materials from Soft Templates: Fabrication and Applications. *J. Mater. Chem. A.* **2019**, 7 (14), 8030–8049. doi:https://doi.org/10.1039/C8TA09750J.

[2] Ozcelik, B.; Blencowe, A.; Palmer, J.; Ladewig, K.; Stevens, G.W.; Abberton, K.M.; Morrison, W.A.; Qiao, G.G. Highly Porous and Mechanically Robust Polyester Poly(Ethylene Glycol) Sponges as Implantable Scaffolds. *Acta Biomater.* **2014**, 10 (6), 2769–2780. doi:https://doi.org/10.1016/j.actbio.2014.02.019.

[3] Yin, H.-M.; Qian, J.; Zhang, J.; Lin, Z.-F.; Li, J.-S.; Xu, J.-Z.; Li, Z.-M. Engineering Porous Poly (Lactic Acid) Scaffolds with High Mechanical Performance via a Solid State Extrusion/Porogen Leaching Approach. *Polymers (Basel).* **2016**, 8 (6), 213. doi:https://doi.org/10.3390/polym8060213.

[4] Forget, A.; Rojas, D.; Waibel, M.; Pencko, D.; Gunenthiran, S.; Ninan, N.; Loudovaris, T.; Drogemuller, C.; Coates, P.T.; Voelcker, N.H., et al. Facile Preparation of Tissue Engineering Scaffolds with Pore Size Gradients Using the Muesli Effect and Their Application to Cell Spheroid Encapsulation. *J. Biomed. Mater. Res. Part B Appl. Biomater.* **2020**, No. January, jbm.b.34581. doi:https://doi.org/10.1002/jbm.b.34581.

[5] Hermawan, B.A.; Mutakin; Hasanah, A.N. Role of Porogenic Solvent Type on the Performance of a Monolithic Imprinted Column. *Chem. Pap.* **2021**, 75 (4), 1301–1311. doi:https://doi.org/10.1007/s11696-020-01399-5.

[6] Yu, S.; Ng, F.L.; Ma, K.C.C.; Mon, A.A.; Ng, F.L.; Ng, Y.Y. Effect of Porogenic Solvent on the Porous Properties of Polymer Monoliths. *J. Appl. Polym. Sci.* **2013**, 127 (4), 2641–2647. doi: https://doi.org/10.1002/app.37514.

[7] Zhang, Q.; Lu, H.; Kawazoe, N.; Chen, G. Preparation of Collagen Porous Scaffolds with a Gradient Pore Size Structure Using Ice Particulates. *Mater. Lett.* **2013**, 107, 280–283. doi: https://doi.org/10.1016/j.matlet.2013.05.070.

[8] Zhang, Q.; Lu, H.; Kawazoe, N.; Chen, G. Preparation of Collagen Scaffolds with Controlled Pore Structures and Improved Mechanical Property for Cartilage Tissue Engineering. *J. Bioact. Compat. Polym.* **2013**, 28 (5), 426–438. doi:https://doi.org/10.1177/0883911513494620.

[9] Li, M.; Deng, H.; Peng, H.; Wang, Q. Functional Nanoparticles in Targeting Glioma Diagnosis and Therapies. *J. Nanosci. Nanotechnol.* **2014**, 14 (1), 415–432. doi:https://doi.org/10.1166/jnn.2014.8757.

[10] Cabezas, L.I.; Gracia, I.; García, M.T.; de Lucas, A.; Rodríguez, J.F. Production of Biodegradable Porous Scaffolds Impregnated with 5-Fluorouracil in Supercritical CO2. *J. Supercrit. Fluids.* **2013**, 80, 1–8. doi:https://doi.org/10.1016/j.supflu.2013.03.030.

[11] Chen, C.-X.; Liu, -Q.-Q.; Xin, X.; Guan, Y.-X.; Yao, S.-J. Pore Formation of Poly(ε-Caprolactone) Scaffolds with Melting Point Reduction in Supercritical CO 2 Foaming. *J. Supercrit. Fluids.* **2016**, 117, 279–288. doi:https://doi.org/10.1016/j.supflu.2016.07.006.

[12] Di Maio, E.; Kiran, E. Foaming of Polymers with Supercritical Fluids and Perspectives on the Current Knowledge Gaps and Challenges. *J. Supercrit. Fluids.* 2018, 134, 157–166. doi: https://doi.org/10.1016/j.supflu.2017.11.013.

[13] Li, L.; Molin, S.; Yang, L.; Ndoni, S. Sodium Dodecyl Sulfate (SDS)-Loaded Nanoporous Polymer as Anti-Biofilm Surface Coating Material. *Int. J. Mol. Sci.* **2013**, 14 (2), 3050–3064. doi:https://doi.org/10.3390/ijms14023050.

[14] Seo, M.; Hillmyer, M.A. Reticulated Nanoporous Polymers by Controlled Polymerization-Induced Microphase Separation. *Science (80-.).* **2012**, 336 (6087), 1422–1425. doi:https://doi.org/10.1126/science.1221383.

[15] Vidil, T.; Hampu, N.; Hillmyer, M.A. Nanoporous Thermosets with Percolating Pores from Block Polymers Chemically Fixed above the Order–Disorder Transition. *ACS Cent. Sci.* **2017**, 3 (10), 1114–1120. doi:https://doi.org/10.1021/acscentsci.7b00358.

[16] Tuzlakoglu, K.; Reis, R.L. Formation of Bone-like Apatite Layer on Chitosan Fiber Mesh Scaffolds by a Biomimetic Spraying Process. *J. Mater. Sci. Mater. Med.* **2007**, 18 (7), 1279–1286. doi:https://doi.org/10.1007/s10856-006-0063-4.

[17] Yang, Y.; Sun, J.; Liu, X.; Guo, Z.; He, Y.; Wei, D.; Zhong, M.; Guo, L.; Fan, H.; Zhang, X. Wet-Spinning Fabrication of Shear-Patterned Alginate Hydrogel Microfibers and the Guidance of Cell Alignment. *Regen. Biomater.* **2017**, 4 (5), 299–307. doi:https://doi.org/10.1093/rb/rbx017.

[18] Sarker, M.D.; Naghieh, S.; Sharma, N.K.; Chen, X. 3D Biofabrication of Vascular Networks for Tissue Regeneration: A Report on Recent Advances. *J. Pharm. Anal.* **2018**, 8 (5), 277–296. doi: https://doi.org/10.1016/j.jpha.2018.08.005.

[19] Lee, K.H.; Shin, S.J.; Park, Y.; Lee, S.-H. Synthesis of Cell-Laden Alginate Hollow Fibers Using Microfluidic Chips and Microvascularized Tissue-Engineering Applications. *Small.* **2009**, 5 (11), 1264–1268. doi:https://doi.org/10.1002/smll.200801667.

[20] Gao, Q.; He, Y.; Fu, J.Z.; Liu, A.; Ma, L. Coaxial Nozzle-Assisted 3D Bioprinting with Built-in Microchannels for Nutrients Delivery. *Biomaterials.* **2015**, 61. doi:https://doi.org/10.1016/j.biomaterials.2015.05.031.

[21] Zahedi, P.; Rezaeian, I.; Ranaei-Siadat, S.-O.; Jafari, S.-H.; Supaphol, P. A Review on Wound Dressings with an Emphasis on Electrospun Nanofibrous Polymeric Bandages. *Polym. Adv. Technol.* 2009, No. December **2009**, n/a-n/a. doi:https://doi.org/10.1002/pat.1625.

[22] Costa, P.F.; Vaquette, C.; Zhang, Q.; Reis, R.L.; Ivanovski, S.; Hutmacher, D.W. Advanced Tissue Engineering Scaffold Design for Regeneration of the Complex Hierarchical Periodontal Structure. *J. Clin. Periodontol.* **2014**, 41 (3), 283–294. doi:https://doi.org/10.1111/jcpe.12214.

[23] Huang, Z.M.; Zhang, Y.Z.; Kotaki, M.; Ramakrishna, S. A Review on Polymer Nanofibers by Electrospinning and Their Applications in Nanocomposites. *Compos. Sci. Technol.* **2003**, 63 (15), 2223–2253. doi:https://doi.org/10.1016/S0266-3538(03)00178-7.

[24] Dalton, P.D.; Grafahrend, D.; Klinkhammer, K.; Klee, D.; Möller, M. Electrospinning of Polymer Melts: Phenomenological Observations. *Polymer (Guildf).* **2007**, 48 (23), 6823–6833. doi: https://doi.org/10.1016/j.polymer.2007.09.037.

[25] Brown, T.D.; Dalton, P.D.; Hutmacher, D.W. Direct Writing by Way of Melt Electrospinning. *Adv. Mater.* **2011**. doi:https://doi.org/10.1002/adma.201103482.

[26] Ratner, B.D.;. A Pore Way to Heal and Regenerate: Twenty-first Century Thinking on Biocompatibility. *Regen. Biomater.* **2016**, 3 (2), 107–110. doi:https://doi.org/10.1093/rb/rbw006.

[27] Ussman, E.R.I.C.M.S.; Alpin, M.I.C.H.; Uster, J.E.M.; Oon, R.A.T.M.; Atner, B.U.D.R. Porous Implants Modulate Healing and Induce Shifts in Local Macrophage Polarization in the Foreign Body Reaction. **2013**. doi:https://doi.org/10.1007/s10439-013-0933-0.

[28] Madden, L.R.; Mortisen, D.J.; Sussman, E.M.; Dupras, S.K.; Fugate, J.A.; Cuy, J.L.; Hauch, K.D.; Laflamme, M.A.; Murry, C.E.; Ratner, B.D. Proangiogenic Scaffolds as Functional Templates for Cardiac Tissue Engineering. *Proc. Natl. Acad. Sci. U. S. A.* **2010**, 107 (34), 15211–15216. doi:https://doi.org/10.1073/pnas.1006442107.

8 Design strategies in biofabrication

8.1 What is design, and what is not design?

A definition of design looks at the discipline as the human power to conceive, plan, and realize products that serve human beings to accomplish any individual or collective purpose. A design is not only an object, but also a process. One can be designing a process or a way of manufacturing or achieving a task. In the more global sense, designing initiates change in a man-made thing. Therefore, the verb or the action of designing refers to the process of originating and developing a plan for a product structure, system, or component. Applied to tissue engineering, these are tissues that are man-made and not any more natural tissues. Therefore, they will undergo a design process as a human being conceives them. In developing biofabricated objects, one must have to have a plan on how to achieve this new structure. The discipline of design is at the intersection of three main fields: humanities, science, and art, as defined by Bruce Archer [1] (Figure 8.1). To evolve in the field of design, one must have scientific knowledge and know how to make an object.

Furthermore, one must have some humanities background to predict how people interact with the designed object. While design involves some artistic touch, it must not be confused with actual art. Indeed, art is made to make a statement; art makes you reflect, makes you wonder and think about something, while design works. Design is made to be functional. The difference between design and art is that design combine aesthetics and function while art is here to make you feel something, to make a statement. Going further, another difference must be clarified. It is between science and design. When sciences are concerned with how things are, design is concerned with how things are supposed to be. So, design is finding an answer to what the optimal shape of an object is. One can say that design combines aesthetics with function. Looking at two different objects with the same function, two different cars, for instance. A sports car might have a nice aesthetic but not the same purpose as a familial monospace car. Therefore, there is a different design for one object, one function, and the aesthetic of the object will affect our attitude toward the product. Aesthetic influences the pleasure we feel from the product and our long-term attitude about the product. As such, it can help us better interact with the designed object.

As an example, in the field of biofabrication, some of the 3D bioprinters currently on the market have received design awards. The bioprinter BioX manufactured by Cellink won the Red Dot Award in 2018. The bioprinter Rastrum manufactured by Inventia Life Science received the Good Design Award for the year 2020. We must now ask ourselves, what is the primary driver to design a shape for an object? The leading design theory coined by Louis Sullivan is that form follows function. That translates into the concept that the shape must be functional, and that the object's function

https://doi.org/10.1515/9781501515736-008

cannot be impaired by the form and shape of the object no matter how the aesthetic must be. If the aesthetic is the primary driver, then we leave the field of design, and the object becomes art to make a statement.

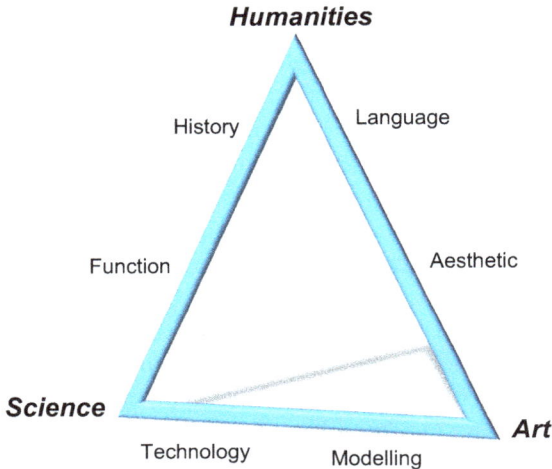

Figure 8.1: Design is at the intersection of many fields of research and performance. Combining humanities, science, and art in the design approach let designers find the best aesthetics for a functional object to predict how humans will interact with their design.

8.2 The design processes

How do we move from a design idea to an object? You will enter a discovery process following an initial idea, then build your design case, develop the idea, and test your design idea before launching your product. To assess the quality of a design, one can use different tools. One can create a matrix with selection criteria such as ease of use, performance, or aesthetics to evaluate a design. These criteria can be weighted depending on which priority or which customer your product is addressing. You will assess your design against the selected criteria using your evaluation matrix. You can give a score for each of the criteria and check if the design matches the criteria set at the beginning of the designing process. But to create this object and set of criteria, there are different approaches that will provide a conceptual framework around the design process and lead to a particular outcome. A typical design framework is called the user-centered design approach. This means that the design focuses on the end-user's needs, wants, and limitations, and these criteria will be used to define and improve the object during the design process. Therefore, you will have to identify who is the end-user and his wants and needs. In this design framework, you need to be having a great knowledge about the end-user and then test the design against the user's needs. Alternatively, you can have an issue-based design. This design framework will solve a current problem or issue through a design. It will convert a problem or limitation into something functional and aesthetic.

Another design framework consists of using a design theory to create your object. For instance, you want to create a webpage or an advertisement. There are theories that can be leveraged to help identify the best color to convey a specific message, how to use letter alignment to facilitate the reading, or how to place an image to capture the customer's attention. In architecture, you also have design theories or design schools that have emerged across the centuries. One famous school of design was the Bauhaus movement which redefined a method to design buildings. As a designer, one needs to predict a future state based on current information. An architect needs to predict how its clients will live and move within the house he is designing based on the current situation their clients are. One must be able to predict how the design of the house might also change the way his clients are currently perceiving space versus once they will be in the new house.

8.3 The design approach in biofabrication

How is design related to biofabrication? In biofabrication, we are trying to design synthetic tissue models to obtain living objects that are here to restore a function, regenerate, or reproduce natural tissue function. One must predict how the cells will react when exposed to a scaffold, how the human body will react to the shape of implants. We need to be aware of these concerns and limitations when designing artificial tissue for *in vitro* or *in vivo* applications. Biofabrication aims at engineering living tissue using industrial manufacturing approaches. These require four steps: design, material selection, manufacturing, testing, and evaluation. As we have previously discussed, designing an object can be achieved by following different approaches: design theory, user-centered, or issue-based approach. Designing an object following each of these approaches leads to objects with different forms, shapes, and performances while having the same function. Taking a biofabrication design challenge: designing vasculature, let see what the different design frameworks are, and what the resulting outcome could be.

8.3.1 Biomimicry

Biomimicry is defined as the design and production of materials, structures, and systems that are modeled on biological entities and processes. Therefore, in a biomimicry design approach, the blood vessels would be shaped following the shape of natural tissue. In this approach, the shape of the object will be acquired from the medical image of an adult vascular tissue [2, 3]. If we consider a blood vessel as an example, angioplasty can be used as a source of the design. The resulting image can then be processed into a computer aided design (CAD) file which can be used for manufacturing objects using biofabrication methods such as 3D printing (Figure 8.2). The synthetic reproduced adult tissue can then be compared to the original by comparing the blood flow and pressure. This is currently one of the main approaches followed to engineer synthetic tissue and artificial organs. However, this requires the precise organization of single cells into a hydrogel to form a three-dimensional structure, leading to an engineering race toward precision and speed of cell deposition. This approach is based on the hypothesis that the complexity of the mature functional tissue can be reproduced by copying the observed final architecture. But this reverse engineering approach omits the stepwise developmental processes that have led to the mature functional tissue. In an anthropomorphic analogy, we could say that tissues created by copying the final outcome lack the necessary experience for becoming mature.

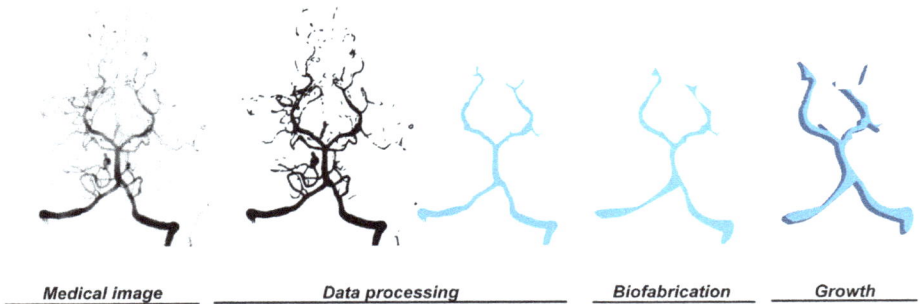

| Medical image | Data processing | Biofabrication | Growth |

Figure 8.2: Conceptual steps followed to create tissue models with a biomimicry approach. This consists in producing natural tissue from a medical image processed into a file that can be used for digital manufacturing process.

8.3.2 User-centered

Conversely, in a user-centered approach, one tries to answer the needs of the user. Therefore, one must understand the needs and limitations of the user. If we want to design a blood vessel network to supply oxygen and nutrient to pancreatic islets, then the pancreatic islets spheroids could be considered as a user. The amount of nutrients and oxygen required by the cells can be obtained from physiological studies, and pancreatic islets must be in a 150 μm vicinity of a blood vessel to not undergo hypoxia (Figure 8.3). We can generate a design where the spheroids are placed in a tissue with a blood vessel network designed to pass no further than 150 μm from each pancreatic islet. Using this approach, a design of blood vessel is obtained that answers to the needs of the user, here the pancreatic islets. [4]

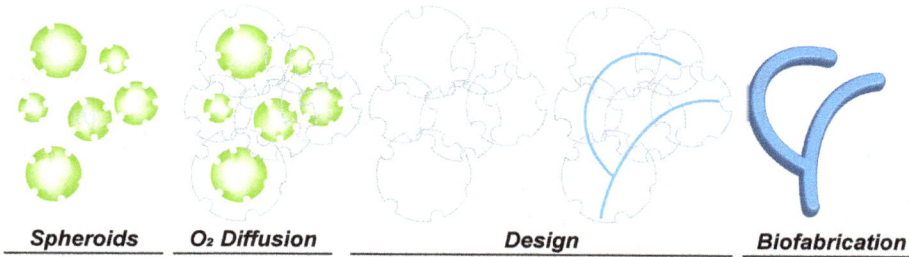

Figure 8.3: Conceptual steps followed to create tissue models with a user-centered approach, where the pancreatic islet (in green) is the user needing oxygen supply from blood vessels that are not further than 150 µm.

8.3.3 Design theory

Alternatively, a theory-based approach can leverage seminal tissue engineering results. In the case of designing a blood vessel network, it has been shown that the pore size of a scaffold can be used to control the infiltration of blood vessels [10]. Using this principle, one could design a scaffold with various pore sizes to control the shape of a blood vessel infiltrating a scaffold. This approach could also be done *in silico*, by creating an artificial porous matrix in a CAD software and letting a blood vessel finds the most efficient path in the virtual matrix. The blood vessel design can then be extracted, processed, and bioprinted [5] (Figure 8.4).

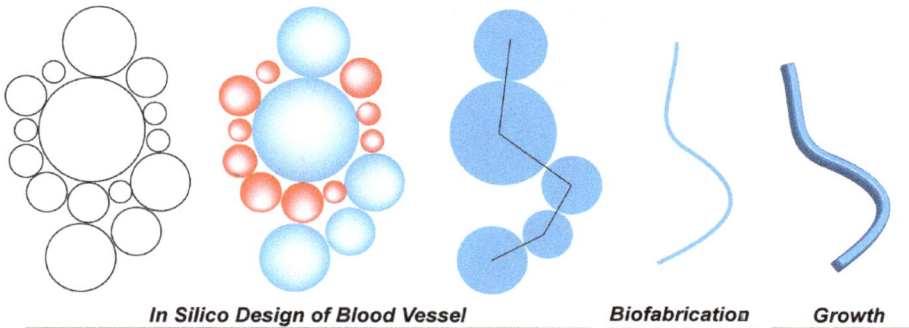

Figure 8.4: Conceptual steps followed to create tissue models with a design theory approach using *in-silico* design of a blood vessel, where a porous scaffold is used to guide the growth and path of a blood vessel.

8.3.4 Morphogenic approach

Incidentally with the establishment of robust gene sequencing instrumentation, most attention has been brought to the role of gene expression in regulating the self-organization of cells into functional organs, so-called morphogenesis. While an open-loop view where genes alone control morphogenesis has been proposed, increasing evidence for a closed-loop is collected where genes determine cell properties to execute morphogenesis based on the physicochemical properties of their environment, inducing new interactions and regulation of genes. In this closed-loop model, materials, which through physical and chemical interactions with cells, could be used to influence cell motility. During development and in a cancerous tumor, cells will undergo the epithelial-mesenchymal transition through which they will become motile. Once capable of movement, cells will move following a gradient of soluble signals, called *chemotaxis*; gradient of mechanical properties, *mechanotaxis*; and gradient of immobilized signal, named *haptotaxis*. Reproducing guided signals that induce a coordinated cell movement like what is observed during morphogenesis would allow for guiding cells through a natural developmental phase toward their final engineered adult functional tissue. Therefore, an alternative to mimicking the final organization of the blood vessel network like in the biomimicry approach and avoiding the necessity of having the technological advances to organize single cells rapidly and precisely would be to design a cellular environment that directs the self-organization of single cells into such mature tissue. In this way, one would have to create signaling patterns that will direct cell assemblies toward their self-organization into blood vessels. One could draw in a synthetic growth environment such as an hydrogel, a signaling gradient that will guide endothelial cells to organize into a blood vessel network (Figure 8.5). [8]

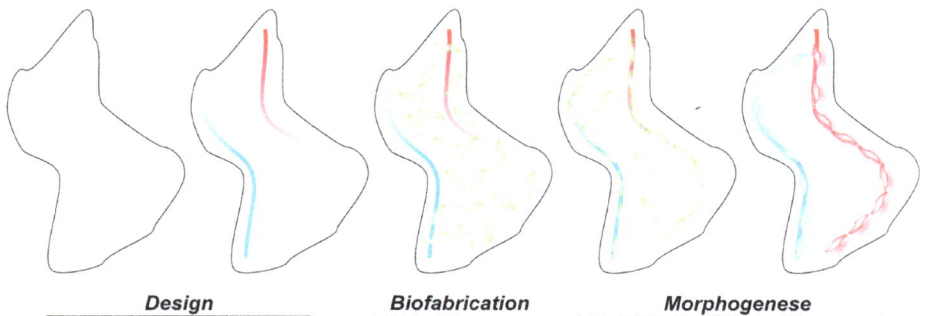

| Design | Biofabrication | Morphogenese |

Figure 8.5: Conceptual steps followed to create tissue models with a morphogenic approach. Here, a gradient of signals is coded into the cell culture milieu that will guide single cells to organize into a blood vessel network.

8.4 The tools for designing objects

Before the computer area, the engineering and designing of an object as complex as an airplane was made on paper, and each engineer was designing the parts by drawing them on paper in 2D following a strict nomenclature. This approach was manpower intensive and cost a lot of time. Then came the first computer-aided (CAD) design software which comprises a set of methods and tools to assist the product designers in creating a geometrical representation of the artifacts they are designing. This software can help with the dimensioning and tolerancing of the parts and permits the exchange of parts and part assemblies between teams and organizations, for instance, between conception and manufacturing or between conception and maintenance.

8.4.1 Computer-assisted design

Today, one can have a whole airplane with all its parts represented on a computer program. But to arrive at this stage, many improvements were made over the past decades. At the beginning, engineers were trained to draw on paper, and so the first CAD software was developed to be used with an interactive pen. Over the years, alongside the gain in computer power, programs have become more complex, and now it is also possible to conduct mechanical and aerodynamic tests on a virtual parts in a computer program. Initially, the development of the CAD programs was mainly driven by aviation companies such as Dassault and Boeing that wanted to improve the manufacturing process. These programs are based on vectorial drawing. So, there is a major difference between CAD programs and your common drawing software. Indeed, when

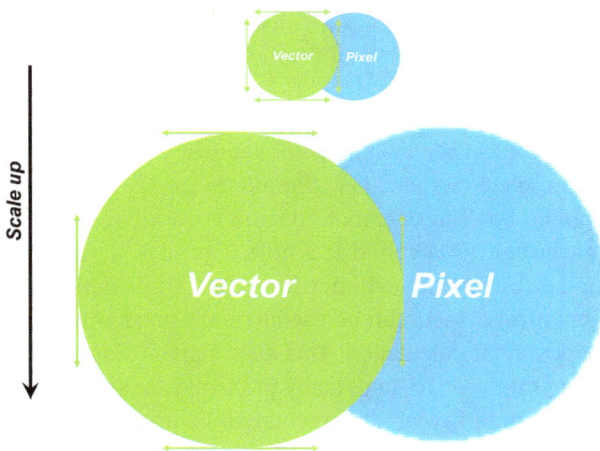

Figure 8.6: The difference in picture quality upon scaling up a vectorial drawing versus a pixel drawing. While the vector drawing do not loose resolution upon resizing, the pixel image becomes blurry.

we see a picture on a screen, we are looking at a map of pixels lighted by a diode of a different color. If we scale up this object, we will reach a limitation: the resolution of the screen used to display the object and so the resolution of the digital file will be visible, and soon the square pixels will be visible on the screen (Figure 8.6). Conversely, if we represent our object as a vector when we scale up the object, the computer has all the information to recalculate a bigger object using the given vectors. By doing so, we can scale up the object without loosing resolution. The other advantage of vector drawing is that the designer can zooms in and out of the virtual object on its screen and rotates the object to be able to inspect it from all angles. Something that was not possible previously when the object was drawn on paper.

8.4.2 Industrial drawing approach

Because all CAD programs represents objects as a vector, this necessitates a different drawing approach. The operator will not directly draw a line or a 3D object such as a cube. Instead, one must start with a point. Adding a direction will allow drawing a line. We have now an object with one dimension. One must add another vector (point, direction, and length) to add another dimension. We obtain a square plan from a line, and we have an object with two dimensions. The third dimension is obtained by adding another vector to obtain a cube (Figure 8.7).

Figure 8.7: In vector design approach, a point and a direction gives a line (1D), then adding a second vector gives a plan (2D) adding another vector through extrusion gives a volume (3D) arrow represent the vector.

Beyond the resolution, vector drawing has the advantage of managing mathematical objects (vectors) to which we can apply a mathematical formula to modify the object. The drawing process can incorporate Boolean operation to subtract, add, or combine objects. For instance, to create a cube with a cylindrical hole (Figure 8.8). One can create a cube and a cylinder and substrate the cylinder from the cube. Finally, the obtained objects that are mathematical objects can be used to make mechanical calculations such as strength or aerodynamic calculation. This allows optimizing the object before the manufacturing of a prototype. To date, many programs are available to design objects. Some are available online free of charge and can be used to create a simple shape that can then be used to create a scaffold, 3D printed objects, or even bioprinted objects. There are also many online libraries to exchange medical images, which can then be used for 3D bioprinting following a biomimicry approach.

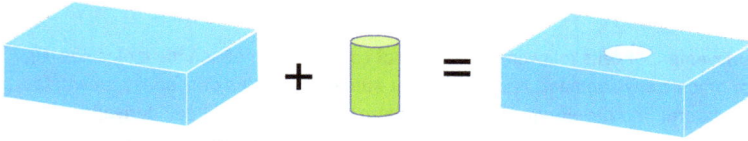

Figure 8.8: In CAD software, Boolean operations can be used to make complex objects through addition or subtraction of volumes.

8.4.3 Parametric design approach

Designs on a computer are expressed and treated as mathematical objects. We can create a parametric equation that is defined by a set of functions that expresses each variable of the equation as a function of another variable. This is particularly used for designing a building to take into consideration sun reflection and in designing mechanical parts to reduce the weight without affecting the mechanical resistance of the part. For example, designing a porous scaffold can be created *in silico* by providing the computer with a set of parameters as designing rules. As an example here, we designed a porous scaffold *in silico* using a parametric design software. Compacted beads have a set of points at their intersection that can be linked into a wireframe. The wireframe is then transformed into a volume giving a porous scaffold. Asking the computer to rearrange the beads in a random manner will give a different porous scaffold. Such scaffold is created without the intervention of the designer apart from the initial set of parameters. (Figure 8.9).

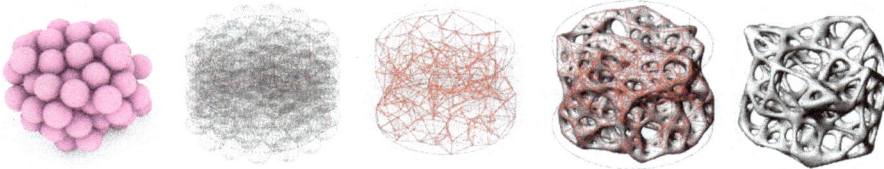

Figure 8.9: Porous scaffold created by parametric design. Beads are packed, and the intersection of these beads form a dense wireframe that is converted into a volume create a porous scaffold [6].

With all the tools available to create designs, today designers have many choices to choose from when they want to create an object. In this chapter, several ideas have been proposed on how to begin a design reflection. It was highlighted that the starting point is critical in dictating what the object will look like. Always driven by the function, a design can have different aspects depending on the point of view the designer begins with, driven by the end-user, solving a problem, or following a theory. It is the responsibility of the biofabricator to ask himself the right questions and test his hypothesis using evaluation criteria to obtain the best design for its functional synthetic tissue.

8.A Chemotaxis

Chemotaxis is the movement of cells following a gradient of concentration of soluble molecules (Figure 8.10). During embryonic development, one of the most studied examples of spatially growth factors is the zebrafish lateral line development guided by a gradient of stromal cell-derived factor 1 (SDF1) chemokine. In adult tissue, gradients of growth factors have been shown to guide the sprouting of new capillaries. This process is directed by a gradient of vascular endothelial growth factor A (VEGFA) that will be followed by tip cells leading the growth of the new blood vessel. [7]

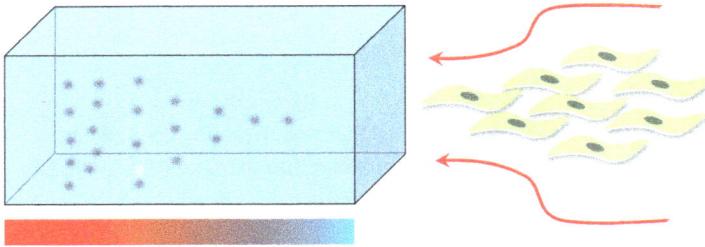

Figure 8.10: Chemotaxis: cells attracted by an increasing concentration of soluble molecules. These molecules are either trapped in a reservoir of the extracellular matrix or generated by a different cell population.

8.B Mechanotaxis

Independent of the presence of gradients of soluble signals, studies on embryonic and adult tissues have revealed the role of anisotropic mechanical properties in directing cell motility, the so-called mechanotaxis. Within this class of phenomena, it is to distinguish between *tensiotaxis* and *durotaxis* consisting in substrate tension migration and substrate stiffness migration, respectively [8, 9].

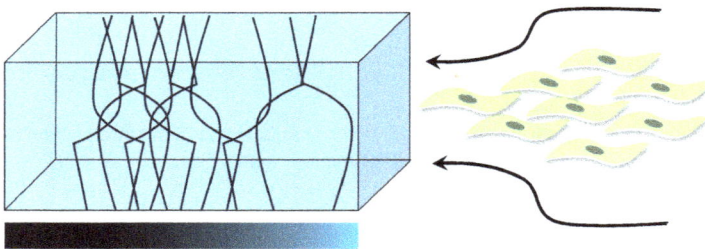

Figure 8.11: Mechanotaxis: cell movement guided by a gradient of mechanical properties generated by a gradient of material or crosslinker density.

8.C Haptotaxis

Haptotaxis is the directional motility of cells following a gradient of cellular adhesion. In this form of cell guidance, cell movement is directed by a gradient of immobilized signals on the extracellular matrix. During embryonic development, the ECM is changing to guide and support morphogenesis. For instance, in the avian embryo, it was discovered through texture analysis that the fibronectin network is organized in an anisotropic manner and that this organization is evolving through the different developmental stages [10, 11]. Thus creating a gradient of cell adhesion signalling.

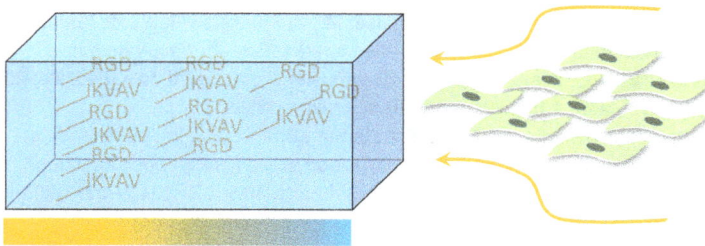

Figure 8.12: Haptotaxis: cell movement guided by the anisotropic organization of immobilized signals on the extracellular matrix, here represented by the cell adhesion peptides RGD and IKVAV.

8.5 Quiz

1. What is design?
2. What is the difference between design and art?
3. What is the relationship between function and form?
4. Which discipline intersects to fuel the design discipline?
5. Gives four different design approaches.
6. In biofabrication, what is the biomimicry approach?
7. What principles of developmental biology are used to create a morphogenetic design approach?
8. What is CAD software?
9. What is the difference between a vectorial and a pixel image?
10. What mathematical operator can we use in CAD to create an object?

References

[1] Archer, B. Design as a Discipline. *Des. Stud.* **1979**, 1 (1), 17–20. doi:https://doi.org/10.1016/0142-694X(79)90023-1.

[2] Organogenesis News and Views Biomimicry in Biomedical Research. **2012**, No. December, 101–102.

[3] Chia, S.L.; Tay, C.Y.; Setyawati, M.I.; Leong, D.T. Biomimicry 3D Gastrointestinal Spheroid Platform for the Assessment of Toxicity and Infl Ammatory Effects of Zinc Oxide Nanoparticles. **2015**, 6, 702–712. doi:https://doi.org/10.1002/smll.201401915.

[4] Hilderink, J., Spijker, S., Carlotti, F., Lange, L., Engelse, M., van Blitterswijk, C., de Koning, E., Karperien, M. and van Apeldoorn, A. (2015), Controlled aggregation of primary human pancreatic islet cells leads to glucose-responsive pseudoislets comparable to native islets. J. Cell. Mol. Med., 19: 1836–1846. https://doi.org/10.1111/jcmm.12555

[5] Sussman, E.M.; Halpin, M.C.; Muster, J.; Moon, R.T.; Ratner, B.D. Porous Implants Modulate Healing and Induce Shifts in Local Macrophage Polarization in the Foreign Body Reaction. *Ann. Biomed. Eng.* **2014**, 42 (7), 1508–1516. doi:https://doi.org/10.1007/s10439-013-0933-0.

[6] Forget, A.; Derme, T.; Mitterberger, D.; Heiny, M.; Sweeney, C.; Mudili, L.; Dargaville, T.R.; Shastri, V.P. Architecture-inspired Paradigm for 3D Bioprinting of Vessel-like Structures Using Extrudable Carboxylated Agarose Hydrogels. *Emergent Mater.* **2019**, 2 (2), 233–243. doi:https://doi.org/10.1007/s42247-019-00045-5.

[7] Ambrosi, D.; Bussolino, F.; Preziosi, L. A Review of Vasculogenesis Models. *J. Theor. Med.* **2005**, 6 (1), 1–19. doi:https://doi.org/10.1080/1027366042000327098.

[8] Li, S.; Huang, N.F.; Hsu, S. Mechanotransduction in Endothelial Cell Migration. *J. Cell. Biochem.* **2005**, 1110–1126. doi:https://doi.org/10.1002/jcb.20614.

[9] Hadjipanayi, E.; Mudera, V.; Brown, R.A. Guiding Cell Migration in 3D: A Collagen Matrix with Graded Directional Stiffness. *Cell Motil. Cytoskeleton.* **2009**, 66 (3), 121–128. doi:https://doi.org/10.1002/cm.20331.

[10] Lühmann, T.; Hall, H. Cell Guidance by 3D-Gradients in Hydrogel Matrices: Importance for Biomedical Applications. *Materials (Basel).* **2009**, 2 (3), 1058–1083. doi:https://doi.org/10.3390/ma2031058.

[11] Rhoads, D.S.; Guan, J.L. Analysis of Directional Cell Migration on Defined FN Gradients: Role of Intracellular Signaling Molecules. *Exp. Cell Res.* **2007**, 313 (18), 3859–3867. doi:https://doi.org/10.1016/j.yexcr.2007.06.005.

9 Bioprinting

9.1 Manufacturing

The usual techniques used for manufacturing parts consist in removing materials; for instance, machining parts using computer numerical control (CNC) machines. In this process, you start with a metal or woodblock, and you will manufacture the object by removing the metal, or wood (Figure 9.1). The second highly used industrial manufacturing method is molding. It is brought to use for metal and plastic. In the plastic industry, melt extrusion is one of the favorite manufacturing techniques. Pellets of thermoplastic are heated up in an extruder. The melted plastic is pushed through a dye by an infinite spin screw, and then the plastic is molded or extruded as a filament.

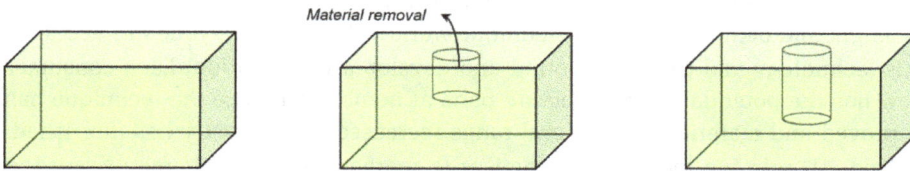

Figure 9.1: Material removal approach starting with a block, the materials that are not needed is removed and wasted.

Conversely, additive manufacturing (AM) does not rely on removing materials but on adding the material needed to create the part. Only the material needed to make the part is utilised. In terms of materials use it is similar to the extrusion molding approach. Indeed, materials such as a thermoplastic is added layer after layer to make the final desired object (Figure 9.2). However, AM parts need to be designed with a different concept and approach. removing versus adding materials makes a significant difference. Some objects can only be manufactured with a material removal approach and others only with the AM approach. In general, the AM techniques permit to manufacture more complex parts than the machining and molding approach.

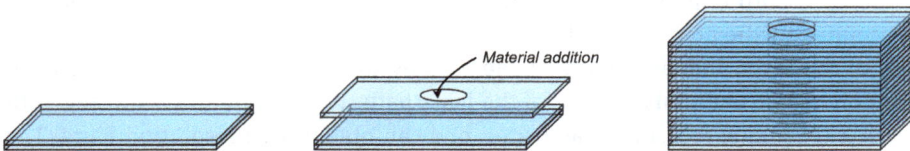

Figure 9.2: Additive manufacturing approach where only the needed material is utilized to make the part.

https://doi.org/10.1515/9781501515736-009

9.2 3D printing

In early 1984, 3D bioprinting emerged with the first patent from Charles Hull, who then launched 3D Systems with the stereolithography (SLA) process. Two years later, the first patent for the fuse deposition modeling process consisted of a thermoplastic extruder placed on an x,y,z-axis. The SLA process works with a resin that is UV activated for crosslinking, giving a thermoset material, while fuse deposition modelling (FDM) is reserved for thermoplastic. In early 1990, the first industrial 3D printer was released. At the beginning of the technique, 3D printing was mainly reserved for industrial processing and reserved for prototyping. As the method developed, the cost was optimized, and processes fastened, it slowly became a possible means for manufacturing functional parts. In 2009, as the initial patents on 3D printing expired, MakerBot proposed a 3D printer for the consumer market. Made of assembled wooden panels, the MakerBot 3D printer brought 3D printing into the limelight and democratized this technique worldwide. Now, everyone has access to this technology and can play around and develop an application. Each consumer now has the potential to manufacture parts at home. Ever since the technique has improved and materials quality and range increased, the industry has flourished. Indeed, 3D printing technology promises to reorganize the whole manufacturing and supply chain industry by relocating the manufacturing closer to the end-user or even at the end-user site. Thus, reducing the manufacturing time, shipping, and reliance on external manufacturing suppliers. With the now availability of cost-effective 3D printers, new applications in the biomedical field emerged. Over the past decades, different manufacturing 3D printing techniques have been brought to the market. Each with their advantage and drawbacks, which ultimately gave the digital manufacturing world a plethora of techniques to play with.

9.2.1 Stereolithography

A resin that can be crosslinked by UV light is brought into a bath that can move along the z-axis (height). Then, a laser is shined into the matrix to crosslink the material . Only the resin lightened by the laser is crosslinked. The part of the resin not in the laser path remain liquid and uncrosslinked. The resolution of the object is determined by the precision of the laser and the precision of the crosslinking chemistry of the resin. Indeed, radicals generated by the laser can travel into the resin and polymerize materials outside of the laser path. But by carefully choosing with the SLA technique crosslinker activator, one can obtain high-resolution parts. The part received is a thermoset object which cannot be melted anymore and cannot be reused.

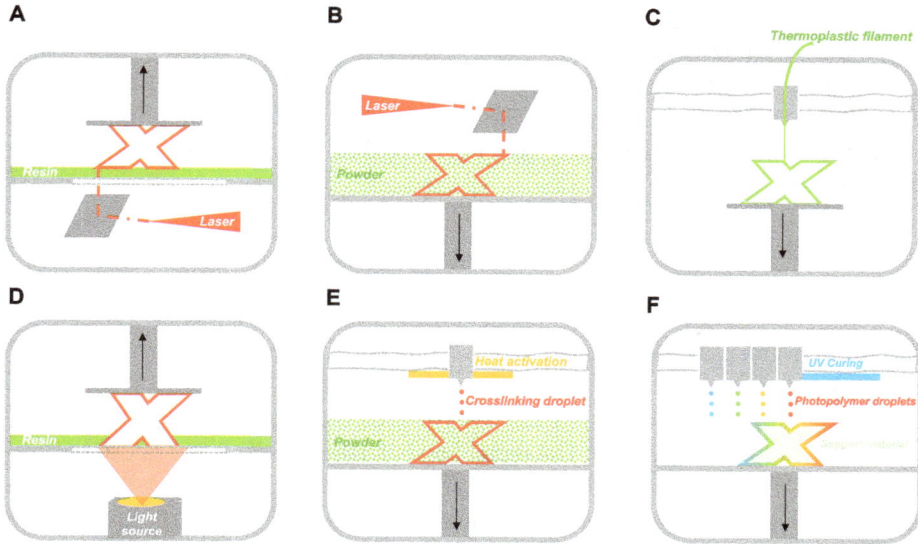

Figure 9.3: The three main additive manufacturing techniques: (**A**) stereolithography (SLA), (**B**) selective laser sintering (SLS), (**C**) fused deposition modeling (FDM), (**D**) digital light processing (DLP), (**E**) multijet fusion (MJF), and (**F**) polyjet technology.

9.2.2 Selective laser sintering

The sintering process consists of compacting materials by heat without melting. A powder of polymer materials (or metal) is placed on a stage in the selective laser sintering (SLS) printer. A laser will then heat the surface of the powder to sinter (compact the powder) to form a solid object. Because the material does not have to be melted to be sintered, high melting temperature materials can be processed by this technique. For polymer materials, polyamide and PEEK are usually processed by SLS due to their semi-crystalline properties and high melting point (268 °C and 343 °C). During this process, the heat permits the polymer molecules in the powder particle to travel from one particle to another one, thus fusing the particles together.

9.2.3 Fused deposition modeling

Using thermoplastic materials, fuse deposition modeling (FDM) enables to create plastic parts from biodegradable polymers such as PLA or PCL. First, the material is provided as a filament rolled on a spool. Then, the thermoplastic polymer filament is guided into an extruder that heats the filament to melt the polymer. The melted polymer is pushed out of the extrusion nozzle (dye) by the incoming filament continuously feeding the extruder. In this technique, the nozzle diameter is the limiting factor for the resolution. In contrast to the laser-activated 3D printing technique, the resolution achieved with fused deposition modeling (FDM) is lower due to the mechanical limitation of the extrusion nozzle. The smaller the nozzle is, the more force is required to push the materials out of it. Therefore, this technique is usually reserved for a polymer with a low melting temperature.

9.2.4 Digital light processing

Digital light processing (DLP) is similar to SLA as it is using a photopolymer that crosslink into a resin upon light activation. But instead of using a laser to trace a layer, digital light processing (DLP) uses a projected image that allows cross-linking an entire layer at once. The technique is faster than SLA and is used to create intricate designs, but the size of the printed object is limited by the light source and thus reserved for small objects. In addition, because it uses light instead of a laser, the resolution of the parts is often lower. However, the latest improvement of the light source quality and resolution has permitted to increase the quality of the DLP-produced parts.

9.2.5 Multijet fusion

Usually reserved for nylon 12, multijet fusion uses a bed containing the polymer powder. Then, a liquid crosslinking agents are added; these do not react directly. The powder and the crosslinking agents are then heated, fusing the activated nylon particles. Once the build is cooled, the unreacted powder can be removed and reused for other manufacturing. The printed parts are then post-processed using water or air blasting to remove unreacted polymer powder. One of the limitations of this technology is the grey colour of the obtained objects and the few materials range available, although now different types of polyamide and polyurethane are becoming available which is expected to extend the range of applications for this technique.

9.2.6 Polyjet

Parts that need to be colored are usually printed with polyjet technology. This method uses droplets of liquid polymer, which can be colored. The polymer droplets are deposited on a bedplate and are directly cured by UV once they are deposited on the bed. Because it uses several inkjet printer heads, parts can be printed in a broad range of colors and materials properties.

9.3 3D printing for medical applications

It has always been of high interest for surgeons to obtain information on their patient's anatomy and condition before surgery. Typically, this is achieved by medical imaging obtained by X-ray computed tomography or magnetic resonance imaging (MRI) which directly give a digital 3D representation of the internal organs. Although virtual reality glasses have permitted to help the surgeon to have a better feeling about the organs in 3D, nothing can beat a real object. This is how 3D printing made its way to the hospital by giving the possibility to reproducing an organ outside the body. In preparation for complex surgeries, the surgeon can transfer the digital images obtained by CT or MRI into a 3D printed plastic replicate of the organ made using one of the techniques described above (Figure 9.4). Using the 3D printed model permits visualization and drawing a strategy for the surgery before getting to the patient. Some of these digital images obtained by CT or MRI into plastic models can be used for training purposes as well. The doctor can rehearse the required movement before the operation using the exact model of the patient's pathology. These replicated plastic organs have also been helpful for doctors to explain complex pathologies to a patient and to reduce anxiety by showing and explaining the planned procedure.

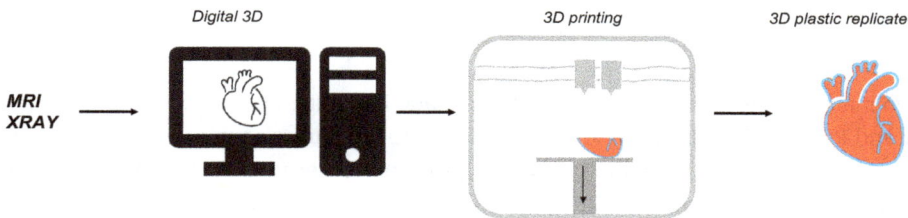

Figure 9.4: Workflow of 3D printing of plastic replicated organs for educational and surgery preparation purposes.

In parallel to this educational and training application of 3D printing, the potential of 3D printing became apparent for the manufacturing of patient-specific implants. For example, patients suffering from bone injuries often require a metal implant, which needs to be custom-made to fit their anatomy and pathology. Highly skilled and precise mechanics can machine such implants. The introduction of digital manufacturing enables to optimize the process of designing the implant from imaging file (MRI, X-ray) into a manufactured part that can directly be fitted to the patient.

9.3.1 Metal printing

Typical bone implants are manufactured out of metal alloys. With the emergence of metal 3D printing using laser and electron beam sintering, it is now possible to create patient-specific metallic implants with these technologies. Following a typical digital manufacturing workflow, metal implants can be custom-made for patients.

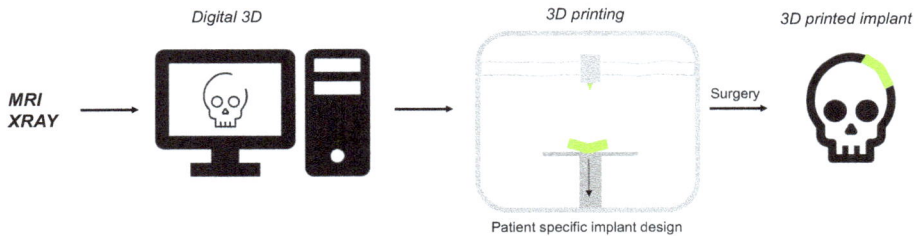

Figure 9.5: Workflow for the digital manufacturing of patient-specific implants. Anatomical information is obtained from medical imaging. Next, the implant is designed *in silico*, 3D printed, and then implanted.

9.3.2 Plastic printing

Two materials are currently used for plastic-made implants: polymer of the PEEK family and PCL. While PEEK polymers are extremely tough and can replace metal implants in some applications, they are challenging to process because of their high melting temperature. Conversely, PCL has a low melting temperature and is biodegradable. Therefore, PCL is mainly used for applications where osteoregeneration is needed, meaning the implant degrades as the bone fracture heals. The ongoing development of PCL implants aims to generate drug-eluting implants to support tissue regeneration. Due to their low melting temperature, it is possible to add drugs to the polymer melt without damaging the active pharmaceutical ingredient. Furthermore, by optimizing the workflow of implant manufacturing, a reduction of manufacturing cost is expected, and better integration of the scaffold with the patient anatomy.

Dentistry is the leading consumer of medical implants. Because implants are not positioned in a critical part of the body (lung, heart, liver, or kidney), and the implant is not inside the patient, it can thus be easily removed in case of failure. Also, many dental prosthetics are not kept by the patient and are removed at night. Therefore, the time in contact with the body is restricted. Because of these, the regulation for materials used in dental applications are less stringent than for other implants. In addition, the UV-activated materials that are used for 3D printing use similar chemistry than the resin used for dental repair. So, there is a long-standing relationship between dentistry and polymer implant which has led to a great experience with these materials and an excellent knowledge of the interaction of these polymers with the body. With the aim of replacing complex dental prosthetic manufacturing, reducing the cost of manufacturing and revolutionising the prosthetic workflow manufacturing, dental 3D printing is slowly gaining market share and adopter.

9.4 3D bioprinting

Cells have been cultured on a planar substrate such as a petri dish for over a century. But this type of culture does not support cells to organize into a natural structure. For instance, endothelial cells cultured on petri dishes have a random organization, while in the liver, they organize into luminal shapes to form blood vessels. If we want to reproduce the function of natural tissues *in vitro*, then the cells must resume their biological organization *in vitro*. Indeed, endothelial cells have a principal role in forming blood vessels that carry blood. However, endothelial cells cannot achieve this function when the culture is in 2D on a petri dish. Therefore, culture models that can force endothelial cells to recapitulate this function are needed.

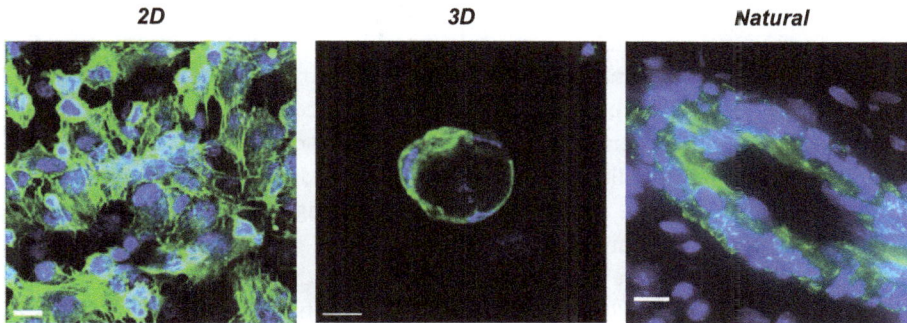

Figure 9.6: Endothelial cells cultured on a petri dish (**2D**) compared to the same cells in a hydrogel (**3D**) and a mouse kidney (**natural**). Adapted from Labor und More Issue L&M 2, 2014.

This is where 3D cell cultures come. As tissue engineering emerged in the late 1990s rapidly, hydrogel-forming materials were used to culture cells in 3D, which provides cells with the necessary volume to resume their natural function. But as the engineered tissues evolved, it quickly reached a limitation as most of these tissues were made through tedious manual methods. When the 3D printing technologies were democratized, tissue engineers saw a potential tool to process living biological materials reproducibly and in an automated manner. Replacing bones and teeth with plastic or metal materials only replaces and mimics a small percent of the body tissue and does not include living cells. So, hydrogels used in tissue engineering were adapted into a 3D bioprinting process where the cells and the hydrogel could be processed into a tri-dimensional object.

9.4.1 The materials of bioprinting

Leveraging the advances made on 3D cell culture, a range of bioprinting compatible hydrogel materials are now available. Pluronic, a synthetic block copolymer, is often used as reference material for testing bioprinting parameters. Because of its low critical saturation temperature, it exhibits a gel to solution transition at about 4 °C [1]. This is particularly interesting for using Pluronic as support materials that can be later removed by lowering the temperature. Other materials are extracted from animals: collagen can form a hydrogel by crosslinking the polymer chain into a triple-helix. A high concentration of collagen can be extruded and to form a support for cell proliferation [2, 3]. Gelatin modified with methacrylate groups permits to obtain an extrudable hydrogel that is then crosslinked under UV light [4]. Other hydrogels are extracted from marine seaweed: agarose [5] and alginate polysaccharides [6, 7]. Agaroses undergo a thermal solution to gel transition at about 37 °C and a gel to solution transition at 90 °C. This gives bioinks that are thermally stable. Alginate forms a gel when mixed with calcium chloride solution. Finally, nanocellulose is extracted from plants and forms a shear-thinning colloidal suspension [8]. Some hydrogel materials are used alone or in a blend, combination of different hydrogel forming materials permits to achieve the optimal mechanical properties for a specific tissue and a particular bioprinting method. Loading cells in hydrogel materials was inherited from the advances made in tissue engineering. However, not all hydrogel materials are compatible with every type of cell and every bioprinting method. Loading cells into a hydrogel in a formulation that is processable by a bioprinter was named *bioink*. The formulation has to be optimized for each bioprinting method using an established repertoire of biomaterials [9] (Table 9.1).

Table 9.1: The principal hydrogels used in 3D bioprinting.

Name	Type	Crosslinking
Pluronic	Synthetic	Physical (LCST)
Gelatin methacrylate	Natural (animal)	UV crosslinking
	Chemically modified	
Collagen	Animal	Physical (triple helix)
Agarose	Natural (algae)	Physical (UCST)
Alginate	Natural (algae)	Physical complex
Nanocellulose	Natural (plants)	Colloidal suspension

9.4.2 The bioprinting methods

The bioprinting technology was pioneered by modifying an inkjet printer and shown that inkjet technology could be used to print living cells in a hydrogel matrix. Now, advances made by the 3D printing industry are adapted to bioprinting, and the number of methods is constantly increasing (Figure 9.7). But processing a living material is quite challenging as the method needs to be compatible with the precious and fragile cells [10].

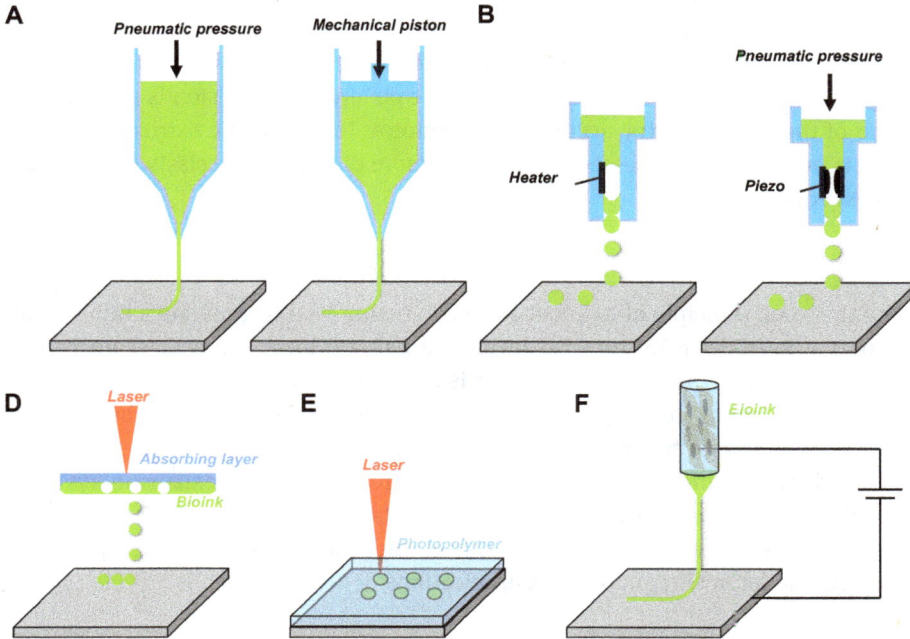

Figure 9.7: The main bioprinting methods are (**A**) extrusion printing, (**B**) inkjet printing, (**D**) laser-assisted droplet, (**E**) stereolithography, and (**F**) electrohydrodynamic printing.

Inkjet/drop-on-demand

The inkjet and drop-on-demand bioprinting consist of extruding materials as a droplet on a substrate. But these methods differ in the way they generate droplet and their size. Inkjet technology is directly derived from the printing industry. A droplet is extruded by generating pressure in a loaded channel. The one method to generate pressure is by heating the bioink, which dilates the materials and creates a pressure point. In this technique, the droplet is continuously ejected by the printer. Alternatively, the bioink container can be pressurized, and a piezo element or a valve controls the opening of the cartridge, providing control over the deposition of the droplet, giving a drop-on-demand process [11–13].

Extrusion

With the democratization of fuse deposition modeling 3D printers, now access to reliable, cost-effective extrusion platforms was granted. It was then possible to change the filament extruder of a commercially available 3D FDM printer by a syringe pump to push a bioink out of a cartridge. The extrusion process relies on a syringe loaded with the bioink (hydrogel and cell) that is pushed out of the nozzle through pneumatic pressure or by a mechanical piston. While both techniques are currently available, mechanical extrusion offers several advantages such as precision as a motor rotates a screw that pushes the piston. The precision is given by the thread of the screw and the accuracy of the motor. In pneumatic extrusion the materials is pushed by compressed air, which is more difficult to precisely control with low cost equipments.

Laser-assisted

Using the same principle of to inkjet extrusion (using heat to generate a bubble), an alternative to inkjet printing was developed using a laser. On a laser absorbing layer, a bioink layer is coated. The laser is directed on the absorbing layer, which will store the energy creating a heated area that generates a hydrogel bubble that falls to the substrate by gravity. Conversely to the extrusion process, the inkjet, drop-on-demand, and laser-assisted printing are non-contact extrusion systems. This allows minimizing the potential cross-contamination between the extrusion nozzle and the substrate [14, 15], and this is important for future development of the technique for *in vitro* bioprinting applications.

Stereolithography

Directly translated from the 3D printing industry, SLA methods have been applied to 3D bioprinting. Using a laser that excites a photopolymer such as gelatin methacrylate, or alginate methacrylate, the hydrogel is crosslinked using non-cytotoxic radical activators. Using the methacrylate chemistry, iterations of this technique are pursued to apply DLP to 3D bioprinting. The DLP technique permits to gain speed and resolution to create complex shapes made of living materials [16, 17].

Electrohydrodynamic

The development of advanced electrospinning technologies, has made it possible to apply this method to bioinks. Although the methods use high voltage, because the distance between the nozzle and collector plate is low, the current can be kept minimal to have no impact on the cell viability. By changing the current, viscosity of the bioink, and the distance between the nozzle and plate, the bioink can be deposited as a droplet or as a continuous filament. Bioink formulations have been proposed based on gelatin methacrylate and alginate. While this technique is quite promising in terms of resolution, it still suffers limitations on the height of the object and size of the printed living materials [18, 19].

Needle array

Spheroids dispersed in a hydrogel can be printed using extrusion-based techniques, but this requires the use of biomaterials [20]. Yet, we have seen that spheroids are particularly useful for biomaterials-free manufacturing aproach as they can themselves generate the needed biomaterials to maintain their mechanical structure. The needle array manufacturing technique or Kazan permits a biomaterials-free 3D bioprinting approach. The spheroids are skewed on a needle array that is patterned in the desired shape. Layers upon layers, the spheroids are positioned on the needles. The spheroids are then incubated, and fused with each other. Once the tissue is matured, the needles are removed, and a final 3D shape is obtained. This method can be automated, and different shapes can be obtained by changing the array pattern. This approach enables to grow a tissue mass rapidly [21] (Figure 9.8).

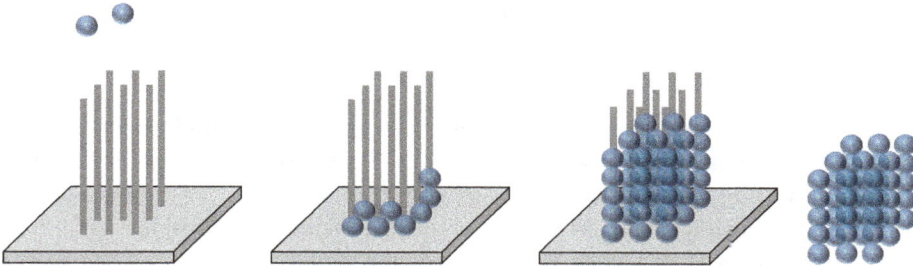

Figure 9.8: A needle array is used to organize spheroids into the desired shape. Then, the needles are removed and the spheroids fuse as the cells grow and generate an extrace lular matrix.

Freeform reversible embedding of suspended hydrogels

The complexity of objects manufactured by extrusion 3D bioprinting is limited by gravity. If printed on a planar substrate, an overhang cannot be created without support to avoid the object to collapse on itself. To overcome this limitation, the freeform reversible embedding of suspended hydrogels (FRESH) technique permits to print directly into a hydrogel that behaves like a Bingham plastic, at low stress, it is a solid, and at high stress, it is a liquid. This means that when the extrusion nozzle moves within the FRESH bath, it is under stress and the hydrogel bath behaves like a liquid. Once the bioink is deposited, there is no longer stress applied to the hydrogel bath, so it acts like a solid and can support the bioink mass. Usually, the FRESH bath is made of gelatin.

So that once the bioink is deposited, the gelatin bath can be removed by heating the bath thus melting the gelatin. To obtain a Bigham plastic with gelatin, it must be processed into a particulate that forms an assembly of small gelatin beads. This method has been applied to different bioinks and is also used for silicon printing in a poly (acrylic acid) microparticulate bath. This approach can be used to scale up the bioprinting process as the impact of gravity is removed and thus the risk for the printed structure of collapsing [22].

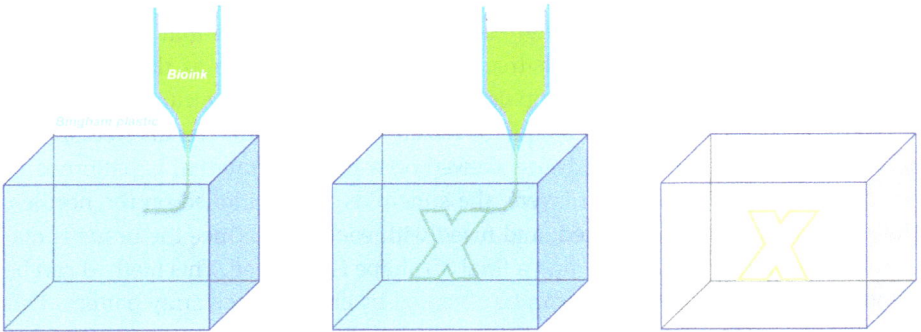

Figure 9.9: FRESH process: the hydrogel-based bioink is deposited in a support media (blue) that maintain the ink in place, after the printing, the media is removed.

Sacrificial hydrogel

Hollow geometries cannot be printed when using an extrusion bioprinting approach because adding material above a hole is not possible. To overcome this limitation, in FDM 3D printing support materials are added. Similarly, in extrusion 3D bioprinting, a sacrificial support hydrogel can be used. Channels can be manufactured using a bioprinter with two printer heads, one for the bioink, one for the sacrificial hydrogel. The sacrificial hydrogel is used to fill the channel. Then, the sacrificial hydrogel is removed, leaving a hollow tube. The typical sacrificial hydrogel used for this sort of bioprinting is Pluronic that has an LCST. So, the part can be printed at room temperature, and when incubated at 4 °C, the Pluronic becomes liquid and can be washed out of the channel. Once the hollow channel is obtained, it can be used to simulate blood vessels [23] (Figure 9.10).

9.5 *In vivo* 3D bioprinting

With the development of medical robots and the introduction of surgical robot arms that a surgeon (for now) controls, the potential of *in vivo* 3D bioprinting is slowly emerging. Surgical robotic arms are now proposing a solution for remote surgery and more precise keyhole procedure. This paves the way toward the deposition of materials with regenerative or reparative potential during the surgery. To achieve this, 3D bioprinters are now

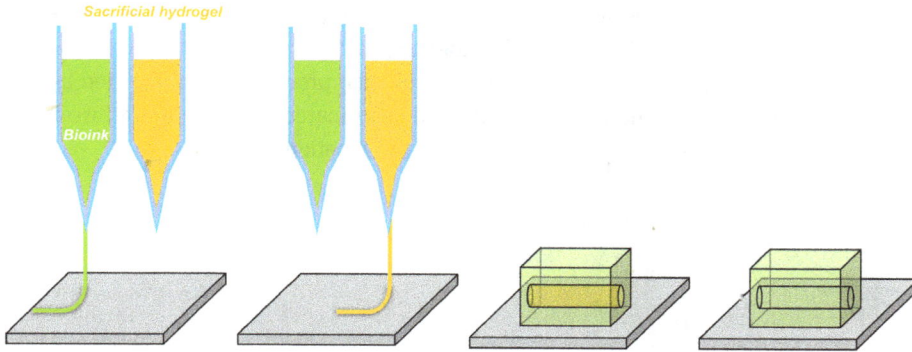

Figure 9.10: Sacrificial hydrogel (yellow) is printed alternatively with the bioink (green). Once the multimaterials object is printed, the sacrificial hydrogel is removed by activating an external stimulus, lowering the temperature below 4 °C for Pluronic.

transitioning from a 3-axis geometry to a 6-axis arm architecture (Figure 9.11). As a result, more freedom of movement permits access to difficult areas and perfectly follows anatomical features.

The current proposal for *in-vivo* bioprinting consist of a workflow that first incorporates a scanning step to acquire the geometry of the area to be treated and the size of the pathology. For instance, to treat a bone fracture, scanning of the bone and its fracture is first needed. Then, the robotic arm changes the tool and uses an extrusion nozzle to deposit a bioink onto the fracture. The bioink, once put in place, is expected

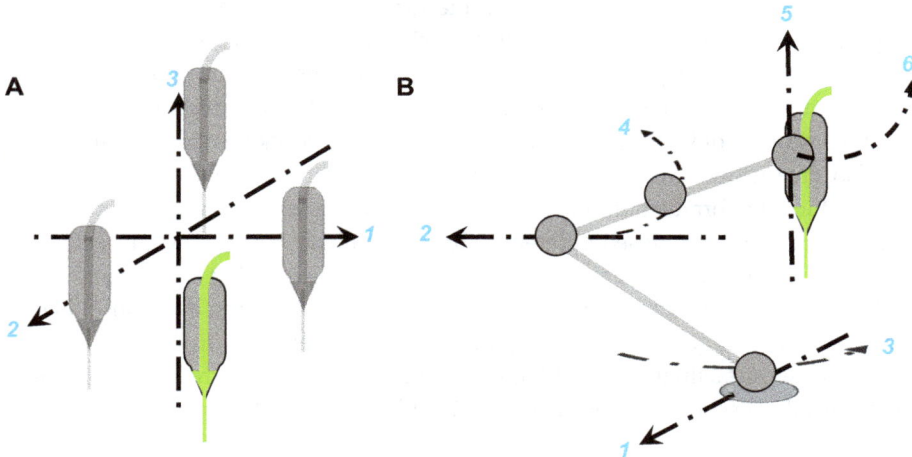

Figure 9.11: Comparison of the movement of freedom of (**A**) 3-axis printer with (**B**) 6-axis printer.

to have regenerative capabilities to improve the healing time and quality (Figure 9.12). This scenario is far from fiction, and technological proof-of-concept is being conducted. Beyond the engineering challenge, the outcome of such procedures is measured by the biological and medical benefits. To be adopted, automated surgical procedure must provide a cost, safety, and performance benefit for the patients [14, 24].

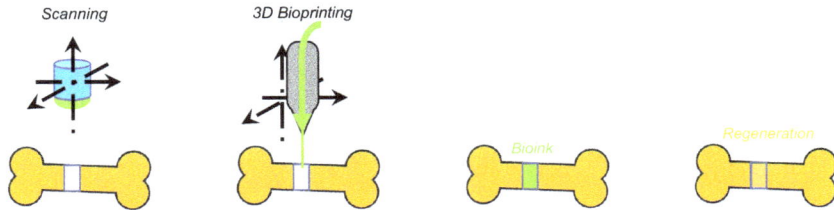

Figure 9.12: Proposed workflow for the introduction of bioprinting for medical procedures: first scanning of the pathology, deposition of a bioink, the bioink fits the pathology and provides a medical benefit such as better repair or regeneration of the tissue.

9.6 Pharmaceutical 3D printing

One of the applications for 3D printing and 3D bioprinting is the potential application for a patient-specific pill that could contain several active pharmaceutical ingredients (APIs). For a patient who must get several drugs per day, having a pill that contains different ingredients could be beneficial. When incorporating several drugs in one pill, not all the ingredients will be released at the same speed and amount. For instance, API positioned at the surface of the multi API pill will be released quicker than the one positioned inside of the pill. Here, the surface area and the material used to make the pill are critical to obtaining the required drug release profile. For instance, a high surface area means a quicker release, but biomaterials that quickly degrade in the stomach also give faster release. So, the geometry and materials need to be precisely adjusted for each formulation. Further application of this technology might be developed for drugs that are dosed differently for each patient. For instance, in pediatric, APIs doses are given as an amount per body mass. Because pediatric patients' mass greatly varies, their treatment is prepared by the in-house pharmacy in hospital settings. Using an AM approach has the potential to create an attractive shape with the required dose to facilitate the willingness of patients in taking the medicine while being versatile to be adapted for each dose [25, 26] (Figure 9.13).

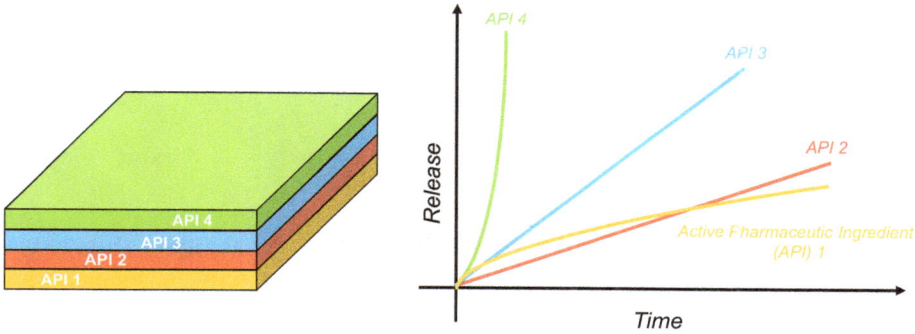

Figure 9.13: 3D printing of pharmaceutical pill containing multiple active pharmaceutical ingredients (API) to obtain multiple drug release profiles in one formulation.

9.7 4D Printing

Introducing the fourth dimension to the 3D printing process means that the part once printed will evolve with time upon exposure to an external stimulus. Using materials that are reactive to humidity, magnetic field, light, or pH, the shape of the printed object can evolve predictably. For instance, using hydrogel that responds to temperature objects that fold when exposed to a change of temperature can be designed. Such printed parts have the potential to be used for drug delivery when the drug carrier must change shape after implantation (Figure 9.14).

Because they are living materials, most of the 3D bioprinted objects are changing over time. Cells are replicating, secreting extracellular materials, or degrading the initial hydrogel matrices. Leveraging the predictable fate of cells, 4D printing concepts can be used to guide living cells into mature tissue replicate to stimulate

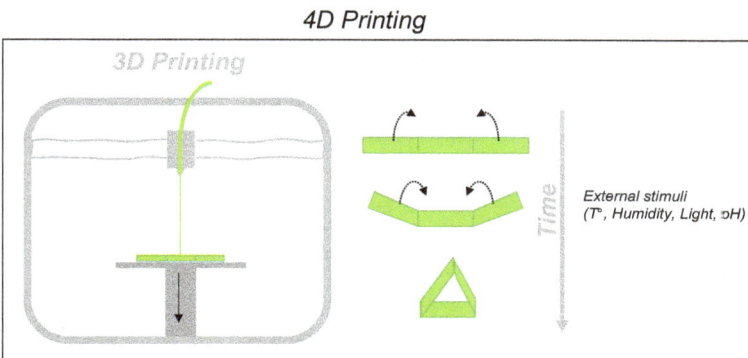

Figure 9.14: 4D printing concept. An object is 3D printed, and by getting exposed to external stimuli, the shape of the objects is changing.

their replication, differentiation, or matrices degradation capabilities. Yet, these concepts are in their infancy and more fundamental research needs to be carried out to transform these concepts into products [27, 28].

9.8 Imaging 3D object

Manufacturing 3D living objects by bioprinting requires the ability to observe the evolution of these living objects once printed. Yet, imaging tridimensional objects is challenging, but several techniques are now available (Figure 9.15).

9.8.1 Widefield microscopy

The widefield microscope uses a series of glass lenses to magnify the observed object. The light beam goes into a series of convex lenses that enlarge the image. Usual magnification ranges from 10x to 100x with a resolving power of about 200 nm. Unfortunately, this type of microscopy is not well suited for acquiring 3D images. So, 3D printed samples are usually sectioned into thin layers of materials deposited on a glass slide. If biological information needs to be obtained, then staining with antigen (immunohistology) or dye (histology) is needed. The usual dye that is used can target specifically the nuclei (Hoechst) or the actin skeleton (fluorescently - tagged phalloidin). While there is a constant improvement of fluorescent dye, most of them require crosslinking the proteins (fix) and permeabilizing the cell membrane to allow the dye to reach its target.

9.8.2 Laser microscopy

Using a laser instead of a light beam permits to selectively shine light on the sample and reduces the signals from unfocused parts. By using a point illumination method and a pinhole in front of the detector, out-of-focus signal can be reduced. It is then possible to gain a high resolution in the depth (height) of the sample so as to only acquire a signal from one focal plane. A 3D image of the sample can be gathered by scanning different focal planes. Here, only fluorescence information can be obtained and thus the sample needs to be stained with antigen (immunofluorescence) or fluorescently-tagged toxin such as phalloidin. While fluorescent dye that can be given to a living specimen without altering its physiology are available now, they are limited. If specific proteins are needed, the sample needs to be fixed and permeabilized to be marked with fluorescently-tagged antibodies. However, laser microscopy does not require sectioning of the specimen. Alternatively, genetically modified cells that expressed protein directly as a fluorescent protein like the green fluorescent protein can be used to follow the cell fate and function. Improvement of the laser microscopy technique has been made with a spinning disk that allows fastening the technique by using a rotating disk punched with holes that let the light go through at regular intervals.

9.8.3 Light-sheet microscopy

The light-sheet microscope uses a laser beam that is focused in one direction. Doing so illuminates only the observed section. Like the confocal, it reduces the photobleaching that can occur by the over exposition of the sample like in the widefield microscope. This is particularly appreciated when imaging living animals as it reduces the acquisition time and thus the stress on the animal. Instead of using points like in the confocal but a light sheet, the image acquisition is faster than the confocal while giving a good contrast. In biology, this technique is used with genetically modified animals that produces the protein of interest with an attached fluorescent protein.

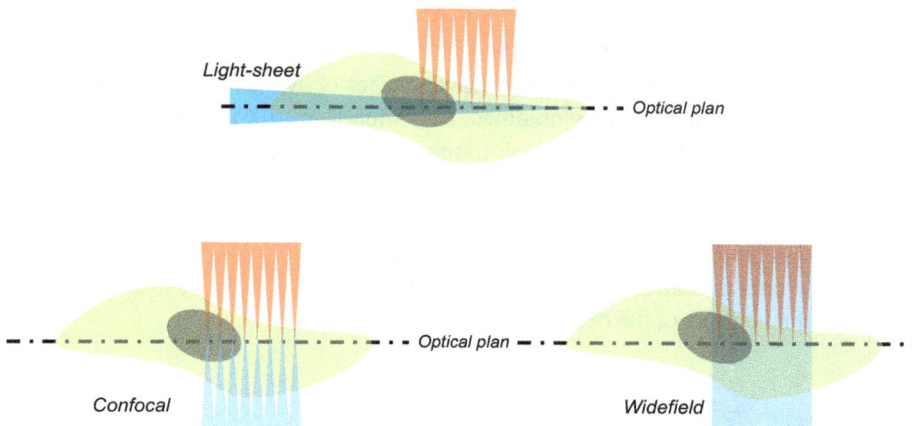

Figure 9.15: The three principle microscopic techniques used in biofabrication The **light-sheet** gives a rapid sampling time, the **confocal** gives high resolution but can take long, and the **widefield** microscopy has a contrast limitation as all the sample is illuminated at once.

9.8.4 Tomography

In light microscopy, to obtain a 3D image of an object, it is needed to physically section the object into thin slices. The slices are mounted on a glass slide, and images of the slices of the objects are acquired. The object can then be reconstructed into a digital 3D image by combining all the individual images together. With a technique that permits to go through the matter such as x-ray, MRI, and fluorescence, images of the inside of the object can be obtained. Using a tomography approach, imaging an object by section enables the reconstruction of a 3D digital image of the sampled object.

X-ray

X-rays are electromagnetic penetrating radiation with a wavelength between 10 pm and 10 nm. These radiations can go through many materials and are well suited to obtain anatomical information of living animals. For 3D printing and bioprinting, they permit to get precise 3D reconstruction of the object and obtain a 3D digital image that can be compared to the intended design. Initially restrained to imaging anatomy, the development of an instrument combining X-ray spectroscopy (to get chemical information) with imaging offers a promising potential to use X-ray for both anatomical and physiological characterization of the tested specimens. For example, in some medical applications of X-ray, contrast agents such as barium sulfate are ingested by the patient to increase the contrast of the digestive tract. Barium sulfates concentrate in the digestive tract, thus increasing the density of the tissue and permitting to obtain a better image.

Magnetic resonance imaging

Using magnetic field gradients and radio wave, images of living specimens anatomical and physiological process can be obtained. In comparison to X-ray, which is best for hard tissue, MRI is better suited for soft tissue imaging. Based on the magnetic resonance of hydrogen nuclei, a radio frequency pulse is used to excite the electrons of an atom, and the resulting signal of this excitation is collected with magnetic receiving coils. For some applications, contrast agents are required to improve the image. This consists of atoms that alter the electromagnetism of the surrounding water and thus obtain a better contrast of the observed tissue.

9.8.5 Scanning electron microscopy

Obtaining high-resolution imaging of an object often requires using electron microscopy. Instead of using photons, electrons are shot onto the specimen, and the scattered electrons are collected, giving an image of the sample. Additionally, if the instrument is equipped with an x-ray detector, information on the atomic composition of the sample can be obtained. In the scanning electron microscopy process, non-conductive samples must be coated with a thin conductive metal layer such as gold. The electron beam scans the surface of the sample, and a full image of the surface is obtained with depth contrast. The stage can be tilted, but the result is a planar 2D picture. Because the instrument operates in a high vacuum (to get the electron beam), the samples need to be crosslinked and dehydrated to avoid shrinking and change of dimension under vacuum.

Table 9.2: Commonly used imaging techniques for 3D samples.

Technique		Sample preparation	Resolution	Damage	Cost		Image
					Sample	Machine	
Microscopy	Light	Sectioning	µm	Yes	$	$	2D
		GMO/live staining		No			
	Confocal	Immunostaining	µm	Yes	$$	$$	3D
		GMO/live staining		No			
	Light-sheet	GMO	µm	No	$$	$$	3D
	Electronic	Fixed/coated	nm	Yes	$	$$$	2D
Tomography	X-ray	Contrast agent	µm	No	$	$$$	3D
	MRI	Contrast agent	µm	No	$	$$$	3D

9.9 Quiz

1. What was the first 3D printing technique developed, and when?
2. What was the first application of 3D printing in medicine?
3. Which 3D printing technique gives thermoset-made materials, which one a thermoplastic?
4. What polymeric materials are used for the 3D printing of implants?
5. What materials are used for bioprinting?
6. What is a bioink?
7. What are the main bioprinting technologies?
8. What are the main 3D printing technologies?
9. What is 4D printing?
10. What are the two main advantages that 3D printing of pharmaceutical p lls could bring?

References

[1] Müller, M.; Becher, J.; Schnabelrauch, M.; Zenobi-Wong, M. Nanostructured Pluronic Hydrogels as Bioinks for 3D Bioprinting. *Biofabrication*. **2015**, 7 (3), 035006. doi:https://doi.org/10.1088/1758-5090/7/3/035006.

[2] Moncal, K.K.; Ozbolat, V.; Datta, P.; Heo, D.N.; Ozbolat, I.T. Thermally-Controlled Extrusion-Based Bioprinting of Collagen. *J. Mater. Sci. Mater. Med.* **2019**, 30 (5). doi:https://doi.org/10.1007/s10856-019-6258-2.

[3] Osidak, E.O.; Karalkin, P.A.; Osidak, M.S.; Parfenov, V.A.; Sivogrivov, D.E.; Pereira, F.D.A.S.; Gryadunova, A.A.; Koudan, E.V.; Khesuani, Y.D.; Kasyanov, V.A., et al. V scoll Collagen Solution as a Novel Bioink for Direct 3D Bioprinting. *J. Mater. Sci. Mater. Med.* **2019**, 30 (3). doi:https://doi.org/10.1007/s10856-019-6233-y.

[4] Bertassoni, L.E.; Cecconi, M.; Manoharan, V.; Nikkhah, M.; Hjortnaes, J.; Cristino, A.L.; Barabaschi, G.; Demarchi, D.; Dokmeci, M.R.; Yang, Y., et al. Hydrogel Bioprinted

Microchannel Networks for Vascularization of Tissue Engineering Constructs. *Lab Chip.* **2014**, 14 (13), 2202–2211. doi:https://doi.org/10.1039/c4lc00030g.

[5] Forget, A.; Blaeser, A.; Messmer, F.; Köpf, M.; Campos, D.F.D.; Voelcker, N.H.; Blencowe, A.; Fischer, H.; Shastri, V.P. Mechanically Tunable Bioink for 3D Bioprinting of Human Cells. *Adv. Healthc. Mater.* **2017**, 1700255. doi:https://doi.org/10.1002/adhm.201700255.

[6] Annabi, N.; Tsang, K.; Mithieux, S.M.; Nikkhah, M.; Ameri, A.; Khademhosseini, A.; Weiss, A.S. Highly Elastic Micropatterned Hydrogel for Engineering Functional Cardiac Tissue. *Adv. Funct. Mater.* **2013**, 23 (39), 4950–4959. doi:https://doi.org/10.1002/adfm.201300570.

[7] Freeman, F.E.; Kelly, D.J. Tuning Alginate Bioink Stiffness and Composition for Controlled Growth Factor Delivery and to Spatially Direct MSC Fate within Bioprinted Tissues. *Sci. Rep.* **2017**, 7 (1), 1–12. doi:https://doi.org/10.1038/s41598-017-17286-1.

[8] Siqueira, G.; Kokkinis, D.; Libanori, R.; Hausmann, M.K.; Gladman, A.S.; Neels, A.; Tingaut, P.; Zimmermann, T.; Lewis, J.A.; Studart, A.R. Cellulose Nanocrystal Inks for 3D Printing of Textured Cellular Architectures. *Adv. Funct. Mater.* **2017**, 27 (12). doi:https://doi.org/10.1002/adfm.201604619.

[9] Groll, J.; Burdick, J.A.; Cho, D.-W.; Derby, B.; Gelinsky, M.; Heilshorn, S.C.; Jüngst, T.; Malda, J.; Mironov, V.A.; Nakayama, K., et al. A Definition of Bioinks and Their Distinction from Biomaterial Inks. *Biofabrication.* **2018**, 11 (1), 013001. doi:https://doi.org/10.1088/1758-5090/aaec52.

[10] Cleverley, C.; Robinson, M.; Willerth, S.M. 3D Printing Breast Tissue Models: A Review of past Work and Directions for Future Work. *Micromachines.* **2019**, 10 (8), 501. doi:https://doi.org/10.3390/mi10080501.

[11] Mironov, V.; Boland, T.; Truck, T.; Forgacs, G.; Markwald, R.R. Organ Printing: Computer-Aided Jet-Based 3D Tissue Engineering. **2003**, 21 (4), 157–161. doi:https://doi.org/10.1016/S0167-7799(03)00033-7.

[12] Xu, T.; Jin, J.; Gregory, C.; Hickman, J.J.; Boland, T. Inkjet Printing of Viable Mammalian Cells. *Biomaterials.* **2005**, 26 (1), 93–99. doi:https://doi.org/10.1016/j.biomaterials.2004.04.011.

[13] Blaeser, A.; Filipa, D.; Campos, D.; Puster, U.; Richter, W.; Stevens, M.M.; Fischer, H. Controlling Shear Stress in 3D Bioprinting Is a Key Factor to Balance Printing Resolution and Stem Cell Integrity. **2015**. doi:https://doi.org/10.1002/adhm.201500677.

[14] Keriquel, V.; Oliveira, H.; Rémy, M.; Ziane, S.; Delmond, S.; Rousseau, B.; Rey, S.; Catros, S.; Amédée, J.; Guillemot, F., et al. In Situ Printing of Mesenchymal Stromal Cells, by Laser-Assisted Bioprinting, for in Vivo Bone Regeneration Applications. *Sci. Rep.* **2017**, 7 (1), 1–10. doi:https://doi.org/10.1038/s41598-017-01914-x.

[15] Keriquel, V.; Guillemot, F.; Arnault, I.; Guillotin, B.; Miraux, S.; Amédée, J.; Fricain, J.-C.; Catros, S. In Vivo Bioprinting for Computer- and Robotic-Assisted Medical Intervention: Preliminary Study in Mice. *Biofabrication.* **2010**, 2 (1), 014101. doi:https://doi.org/10.1088/1758-5082/2/1/014101.

[16] Bertassoni, L.E.; Cardoso, J.C.; Manoharan, V.; Cristino, A.L.; Bhise, N.S.; Araujo, W.A.; Zorlutuna, P.; Vrana, N.E.; Ghaemmaghami, A.M.; Dokmeci, M.R., et al. Direct-Write Bioprinting of Cell-Laden Methacrylated Gelatin Hydrogels. *Biofabrication.* **2014**, 6 (2), 024105. doi:https://doi.org/10.1088/1758-5082/6/2/024105.

[17] Berlin, S.; Brown, G.; Lim, K.S.; Jung, T.; Boeck, T.; Blunk, T.; Tessmer, J.; Hooper, G.J.; Woodfield, T.B.F.; Groll, J. Thiol–Ene Clickable Gelatin: A Platform Bioink for Multiple 3D Biofabrication Technologies. *Adv. Mater.* **2017**, 29 (44). doi:https://doi.org/10.1002/adma.201703404.

[18] Gasperini, L.; Maniglio, D.; Motta, A.; Migliaresi, C. An Electrohydrodynamic Bioprinter for Alginate Hydrogels Containing Living Cells. *Tissue Eng. Part C Methods.* **2015**, 21 (2), 123–132. doi:https://doi.org/10.1089/ten.tec.2014.0149.

[19] He, J.; Zhang, B.; Li, Z.; Mao, M.; Li, J.; Han, K.; Li, D. High-Resolution Electrohydrodynamic Bioprinting: A New Biofabrication Strategy for Biomimetic Micro/Nanoscale Architectures and Living Tissue Constructs. *Biofabrication*. **2020**, 12 (4), 042002. doi:https://doi.org/10.1088/1758-5090/aba1fa.

[20] Mironov, V.; Visconti, R.P.; Kasyanov, V.; Forgacs, G.; Drake, C.J.; Markwald, R.R. Organ Printing: Tissue Spheroids as Building Blocks. *Biomaterials*. **2009**, 30 (12), 2164–2174. doi: https://doi.org/10.1016/j.biomaterials.2008.12.084.

[21] Arai, K.; Murata, D.; Verissimo, A.R.; Mukae, Y.; Itoh, M.; Nakamura, A.; Morita, S.; Nakayama, K. Fabrication of Scaffold-Free Tubular Cardiac Constructs Using a Bio-3D Printer. *PLoS One*. **2018**, 13 (12), 1–17. doi:https://doi.org/10.1371/journal.pone.0209162.

[22] Hinton, T.J.; Jallerat, Q.; Palchesko, R.N.; Park, J.H.; Grodzicki, M.S.; Shue H.J.; Ramadan, M.H.; Hudson, A.R.; Feinberg, A.W. Three-Dimensional Printing of Complex Biological Structures by Freeform Reversible Embedding of Suspended Hydrogels. *Sci. Adv.* **2015**, 1 (9). doi:https://doi.org/10.1126/sciadv.1500758.

[23] Forget, A.; Derme, T.; Mitterberger, D.; Heiny, M.; Sweeney, C.; Mudili, L.; Dargaville, T.R.; Shastri, V.P. Architecture-inspired Paradigm for 3D Bioprinting of Vessel-ike Structures Using Extrudable Carboxylated Agarose Hydrogels. *Emergent Mater*. **2019**, 2 (2), 233–243. doi:https://doi.org/10.1007/s42247-019-00045-5.

[24] Singh, S.; Choudhury, D.; Yu, F.; Mironov, V.; Naing, M.W. In Situ Bioprinting – Bioprinting from Beachside to Bedside?. *Acta Biomater*. **2019**. doi:https://doi.org/10.1016/j.actbio.2019.08.045.

[25] Khaled, S.A.; Burley, J.C.; Alexander, M.R.; Yang, J.; Roberts, C.J. 3D Printing of Five-in-One Dose Combination Polypill with Defined Immediate and Sustained-Release Profiles. *J. Control. Release*. **2015**, 217, 308–314. doi:https://doi.org/10.1016/j.jconrel.2015.09.028.

[26] Robles-Martinez, P.; Xu, X.; Trenfield, S.J.; Awad, A.; Goyanes, A.; Telford, R.; Basit, A.W.; Gaisford, S. 3D Printing of a Multi-Layered Polypill Containing Six Drugs Using a Novel Stereolithographic Method. *Pharmaceutics*. **2019**, 11 (6). doi:https://doi.org/10.3390/pharmaceutics11060274.

[27] Kirillova, A.; Maxson, R.; Stoychev, G.; Gomillion, C.T.; Ionov, L. 4D Biofabrication Using Shape-Morphing Hydrogels. *Adv. Mater*. **2017**, 29 (46), 1703443. doi:https://doi.org/10.1002/adma.201703443.

[28] Stoychev, G.; Puretskiy, N.; Ionov, L. Self-Folding All-Polymer Thermoresponsive Microcapsules. *Soft Matter*. **2011**, 7 (7), 3277. doi:https://doi.org/10.1039/c1sm05109a.

10 Applications of biofabrication

10.1 The start of an industry

With the growing range and complexity of available biofabrication techniques, we are getting closer to manufacturing synthetic organs. The initial aim of biofabrication was to address the transplantation organ shortage. But to achieve this, we need to be able to manufacture fully functional organs constituted of several tissues, which is a highly complex task. While the current advances in the field have permitted a significant breakthrough, new opportunities have emerged in unexpected areas. Now biofabrication methods are applied to generate tissue for drug and cosmetic testing, meat production, and living materials for architecture and design.

10.1.1 Tissue model

Three main factors are putting pressure on pharmaceutical drug development. The first one is an aging society. The United Nations predicts that in 2030 over 10% of the worldwide population will be over 65 years old. An aging population means more severe pathologies will occur and an increasing need for new medicines to sustain a comfortable life. In addition, there is legal pressure that will impact how pharmaceutical research and development is carried out. In 2009, the European Union banned animal testing for cosmetical research and development. In 2010, a directive followed up to draw a roadmap to replace laboratory animals with alternative methods, including engineered tissue. Finally, economic pressure is pushing to change how pharmaceuticals are developed. Indeed, drugs are not discovered by serendipity anymore, and the cost of the industrial research of pharmaceutical molecules has nearly doubled in the last decade. So, methods to reduce the cost of development are needed, and biofabricated tissues are one promising technology. Biofabricated tissue is expected to play a critical role in drug development and replace animal testing. Tissue models engineered with human physiology are expected to provide a closer model to human biology than animal models, thus providing an alternative to animal testing.

10.1.2 Meat production

With the exponential increase of the world population and the need to access cheap protein sources, animal farming has become industrialized, sometimes unethical, and a significant producer of greenhouse gases. One technological solution to the protein food supply currently in development is to grow meat in laboratories, the so-called cultured meat. Using current biofabrication techniques, cost reduction

https://doi.org/10.1515/9781501515736-010

in cultured meat production is in the making. But only growing the cells is not enough. To reproduce the texture and flavors of the natural meat, tissue organization must be mimicked. Indeed, natural muscle for meat consumption is marbled by fat tissue, which creates a particular taste upon cooking. Here biofabrication techniques are needed to organize the myocyte and adipocyte cells into the meat [1].

10.1.3 Living materials

To reduce animal farming, endeavors to create laboratory leather are explored. Two approaches are investigated: one to use animal cells to produce collagen and form animal-based leather [2], the second approach uses fungi and biomaterials to reproduce the materials properties of natural leather [3]. Using a combination of biomaterials with fungi or bacteria to engineer living materials [4] have been used to make packaging materials. The fungi are grown on a support. As the fungi develop, it creates a mycelium that forms strong materials [5]. A similar approach can be used to make cellulose using bacteria that naturally produce the polysaccharide. Upon inactivation of the bacteria, a strong and stable material is obtained. By shaping the growing support of the bacteria, complex shape and design of the bacterial cellulose can be achieved [6].

10.2 Industrial processes

For biofabrication to be utilized to manufacture products, it has to follow the regulation and guidelines to ensure the safety of biomedical materials. For the north-American market, medical devices and pharmaceuticals need to be produced following good manufacturing practices (GMP), which are a set of guidelines that provide for systems that assure proper design, monitoring, and control of manufacturing processes and facilities. It encounters the quality management system, personal training, facilities management, and product control. Next, automated systems utilized for manufacturing or quality control need to have an electronic signature that ensures that the electronic records are considered trustworthy and reliable as much as paper records. This limits the risk of falsification of electronic records. These two standards are brought to use to ensure the manufacturing process's safety and traceability. For medical devices (that could be biofabricated) a system to monitor the quality must be brought in place. These requirements are set into the ISO 13,485 norm. This norm is necessary for the obtention of approval for the commercialization of medical devices in the European Union and associated countries. Because there are rules and regulation that needs to be followed to ensure the safety and quality of medical devices, if biofabrication methods are required to manufacture such devices, these methods must follow the specific guidelines. Meaning that the computer and instruments utilized in the biofabrication process (electrospinning, 3D bioprinter as an example) must complain to the electronic documentation regulations, and the manufacturing process must follow a quality management system [7].

10.3 Sterilization

Finally, biofabricated medical devices must be produced without pathogens and microorganisms. Two manufacturing approaches can achieve this: manufacture in non-sterile conditions and then sterilize the device at the end of the process, or manufacture under aseptic conditions. Meaning the whole manufacturing process is done under sterile conditions. Usually, the first strategy is preferable for cost-associated reasons, but it is not possible to have final sterilization in some cases. For example, if living cells are incorporated into the medical device, no sterilization methods can be used. Three main techniques are available to sterilize medical devices: heat sterilization, chemical sterilization, and radiation sterilization (Figure 10.1).

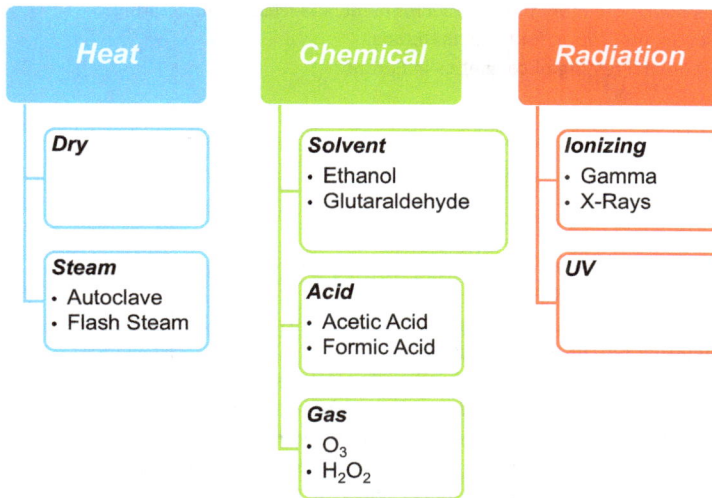

Heat
- **Dry**
- **Steam**
  - Autoclave
  - Flash Steam

Chemical
- **Solvent**
  - Ethanol
  - Glutaraldehyde
- **Acid**
  - Acetic Acid
  - Formic Acid
- **Gas**
  - O_3
  - H_2O_2

Radiation
- **Ionizing**
  - Gamma
  - X-Rays
- **UV**

Figure 10.1: Sterilization methods commonly used in biofabrication.

In contrast, heat and radiation are preferred because they can be easily scaled up and do not use toxic chemicals that must be removed. Heat sterilization usually consists of steam sterilization (autoclave) that heat the samples above 120 °C for several minutes. This technique might not be compatible with all polymers, especially those with melting temperature below the steam temperature. Radiation sterilization is the preferred method for poly(styrene) sterilization. The radiation does not induce heat and can go through packaging. Therefore, it allows for the sterilization of many manufactured and packaged products, especially plastic ware used for the *in vitro* cell culture. For some polymer systems, the radiation generated during the process can induce unwanted crosslinking of the polymers and modify the product's properties. In another case, the radiation can induce crosslinking of the polymer and further improve the mechanical properties. Finally, the efficiency of the sterilization process must be assessed and proven to ensure the safety and quality of the produced device.

10.4 Quiz

1. Why could tissue models reduce the use of animal models in pharmaceutical development?
2. What are the three main sterilization methods?
3. What are the three factors pushing toward a change in pharmaceutical development?
4. What is aseptically manufacturing?
5. What is the significant risk of radiation sterilization?
6. What is the major risk of heat sterilization?
7. What are the factors pushing toward laboratory meat production?
8. What are the three main regulatory frameworks for medical device manufacturing?
9. Give an example of an application for living materials.
10. When was banned animal testing of cosmetics in Europe?

References

[1] Chriki, S.; Hocquette, J.-F. The Myth of Cultured Meat: A Review. *Front. Nutr.* **2020**, 7. doi: https://doi.org/10.3389/fnut.2020.00007.
[2] Dance, A.;. Engineering the Animal Out of Animal Products. *Nat. Biotechnol.* **2017**, 35 (8), 704–707. doi:https://doi.org/10.1038/nbt.3933.
[3] Jones, M.; Gandia, A.; John, S.; Bismarck, A. Leather-like Material Biofabrication Using Fungi. *Nat. Sustain.* **2021**, 4 (1), 9–16. doi:https://doi.org/10.1038/s41893-020-00606-1.
[4] Nguyen, P.Q.; Courchesne, N.M.D.; Duraj-Thatte, A.; Praveschotinunt, P.; Joshi, N.S. Engineered Living Materials: Prospects and Challenges for Using Biological Systems to Direct the Assembly of Smart Materials. *Adv. Mater.* **2018**, 30 (19), 1–34. doi:https://doi.org/10.1002/adma.201704847.
[5] Abhijith, R.; Ashok, A.; Rajesh, C.R. Sustainable Packaging Applications from Mycelium to Substitute Polystyrene: A Review. *Mater. Today Proc.* **2018**, 5 (1), 2139–2145. doi:https://doi.org/10.1016/j.matpr.2017.09.211.
[6] Aswini, K.; Gopal, N.O.; Uthandi, S. Optimized Culture Conditions for Bacterial Cellulose Production by Acetobacter Senegalensis MA1. *BMC Biotechnol.* **2020**, 20 (1), 46. doi:https://doi.org/10.1186/s12896-020-00639-6.
[7] da Silva, L.R.R.; Sales, W.F.; Campos, F.; Dos, A.R.; de Sousa, J.A.G.; Davis, R.; Singh, A.; Coelho, R.T.; Borgohain, B. A Comprehensive Review on Additive Manufacturing of Medical Devices. *Prog. Addit. Manuf.* **2021**, 6 (3), 517–553. doi:https://doi.org/10.1007/s40964-021-00188-0.

Index

https://doi.org/10.1515/9781501515736-011

www.ingramcontent.com/pod-product-compliance
Lightning Source LLC
Chambersburg PA
CBHW081529220326
41598CB00036B/6375

Patrizio Raffa, Pablo Druetta

Chemical Enhanced Oil Recovery

Advances in Polymer Flooding and Nanotechnology

DE GRUYTER

Authors
Dr. Patrizio Raffa
University of Groningen
Faculty of Science and Engineering
Product Technology
Nijenborgh 4
9747 AG Groningen
The Netherlands
p.raffa@rug.nl

Dr. Pablo Druetta
University of Groningen
Faculty of Science and Engineering
Product Technology
Nijenborgh 4
9747 AG Groningen
The Netherlands
p.d.druetta@rug.nl

ISBN 978-3-11-064024-3
e-ISBN (PDF) 978-3-11-064025-0
e-ISBN (EPUB) 978-3-11-064043-4

Library of Congress Control Number: 2019941241

Bibliographic information published by the Deutsche Nationalbibliothek
The Deutsche Nationalbibliothek lists this publication in the Deutsche Nationalbibliografie;
detailed bibliographic data are available on the Internet at http://dnb.dnb.de.

© 2019 Walter de Gruyter GmbH, Berlin/Boston
Cover image: grandriver / E+ / Getty Images
Typesetting: le-tex publishing services GmbH, Leipzig
Printing and binding: CPI books GmbH, Leck

www.degruyter.com

Contents

List of Abbreviations

Abbreviation	Description
AA	Acrylic Acid
AIM	Adaptive Implicit
AM	Acrylamide
AMPDAC	(2-acrylamido-2-methylpropyl)dimethylammonium Chloride
AMPDAPS	3-(2-acrylamido-2-methylpropane-dimethylammonio)-1-propanesulfonate
AMPS	2-acrylamide-2-methylpropane sulfonic acid
AMR	Adaptive Mesh Refinement
API	American Petroleum Institute
ASP	Alkaline-Surfactant-Polymer Flooding
ATRP	Atom Transfer Radical Polymerization
BP	British Petroleum
CDC	Capillary Desaturation Curve
CDG	Colloidal Dispersion Gel
cEOR	Chemical Enhanced Oil Recovery
CMC	Carboxy methyl cellulose
cP	centiPoise
DLVO	Derjaguin, Landau, Verwey and Overbeek Theory
DPR	Disproportionate Permeability Reduction
EIA	Energy Information Administration
EIP	Emulsion Inversion Point
EOR	Enhanced Oil Recovery
EOS	Equation of State
E&P	Exploration and Production
FENE	Finitely Extensible Non-linear Elastic
FTUS	Forward in Time, Upwind Scheme
GPC	Gel Permeation Chromatography
HEC	Hydroxy ethyl cellulose
HPAM	Hydrolyzed Polyacrylamide
HLP	Hydrophobic and Lipophilic Polysilicon
IAPV	Inaccessible Pore Volume
IEA	International Energy Agency
IFT	Interfacial Tension
IMPEC	Implicit in Pressure, Explicit in Concentration
IMPES	Implicit in Pressure, Explicit in Saturation
IMPSAT	Implicit in Pressure and Saturation
IOR	Improved Oil Recovery

https://doi.org/10.1515/9783110640250-201

Abbreviation	Description
LCST	Lower Critical Solubility Temperature
LHP	Lipophobic and Hydrophilic Polysilicon
LTE	Local Truncation Error
MTOE	Million Tons of Oil Equivalent
NaAMB	Sodium 3-acrylamido-3-methylbutanoate
NWT	Neutral Wet Polysilicon
OOIP	Original Oil in Place
O/W	Oil/Water
PAM	Polyacrylamide
PDE	Partial Differential Equation
PIT	Phase Inversion Temperature
PNP	Polymer Coated Nanoparticle
PSNP	Polysilicon Nanoparticle
PVT	Pressure-Volume-Temperature
RAFT	Reversible Addition-Fragmentation Chain-Transfer Polymerization
REV	Representative Elementary Volume
RF	Oil Recovery Factor
RPM	Relative Permeability Modification
SG	Specific Gravity
SSCP	Smart-Covered Polymeric Particles
SP	Surfactant-Polymer Flooding
SSNa	Sodium Styrene Sulfonate
SWCNT	Single-Walled Carbon Nanotube
TDS	Total Dissolved Solids
TVD	Total Variation Dimishing
UCM	Upper Convected Maxwell
UNFCCC	United Nations Framework Convention on Climate Change
UVM	Unified Viscosity Model
WEO	World Energy Outlook
β-CD	β-cyclodextrin

List of Symbols

Symbol	Description
Ad_i	Adsorption of Component i
Bbl	Oil Barrel (approximately $0.159\,\text{m}^3$)
c	Volumetric Concentration
$c(r)$	Solubility around a particle of radius r
c_r	Rock Compressibility
Cr	Courant Number
$\underline{\underline{D}}$	Dispersion Tensor
d	Nanoparticle Diameter [nm]
dl	Longitudinal Dispersion Coefficient [m^2/s]
dm	Molecular Diffusion Coefficient [m^2/s]
dt	Transversal Dispersion Coefficient [m^2/s]
$\hat{f}$	Flow Efficiency Factor (nanoparticles)
$\bar{K}$	Absolute Permeability [m^2] or [Darcy]
$\bar{K}_{\text{MH}}$	Mark–Houwink Parameter
$\hat{k}_r$	Relative Permeability
$\hat{k}_f$	Constant for Fluid Seepage (nanoparticles)
M	Mobility Ratio
M_w	Molecular Weight
n	Power Law Exponent
n_2	UVM Parameter
N_c	Capillary Number
p	Pressure [Pa]
p_{wf}	Bottomhole Pressure [Pa]
q	Flowrate
R_f	Resistance Factor
r_w	Well Radius [m]
S	Phase Saturation
s	Skin Factor
T	Temperature
u	Darcy Velocity
V	Volumetric Concentration
z	Overall Concentration

Greek Letters

α_{MH}	Mark–Houwink Parameter
Γ	Boundary of the Domain
$\dot{\gamma}$	Shear Rate

https://doi.org/10.1515/9783110640250-202

Symbol	Description
γ	Interfacial Tension [mN/m]
γ_{ow}	Interfacial Tension of the Water-Oil System
δ_{ij}	Kronecker Delta
η	Intrinsic Viscosity
λ	Polymer Degradation Parameter
λ_2	UVM Parameter
λ^j	Phase Mobility [m^2/(Pa $\cdot$ s)]
μ	Dynamic Viscosity [Pa $\cdot$ s]
μ_{MAX}	UVM Parameter
Π	Disjoining Pressure
τ_2	UVM Parameter
τ_r	Critical Shear Rate (Carreau Model)
ϕ	Rock Formation Porosity
Ω	Reservoir Physical Model

Superscripts

a	Aqueous Phase
H	Water-Oil System (no Chemical)
j	Phase
[k]	Number of Iterations
$\langle n \rangle$	Time Step Number
o	Liquid Hydrocarbon Phase
r	Residual

Subscripts

0	Initial Condition
0sr	Zero Shear-Rate
c	Capillary, Chemical Component
cf	Carrier Fluid
i	Component
in	Injection
m, n	Grid Blocks
nf	Nanofluid
p	Petroleum Component
r	Residual
s	Salt Component
sp	Specific
t	Total
w	Wetting Phase, Water Component

1 An introduction to chemical enhanced oil recovery

1.1 Current trends in oil recovery

One of the main challenges of the current society is to replace fossil fuels with more sustainable and green resources [1–5]. The reasons for this need are simple: fossil fuels are 1) polluting and 2) finite. These issues are becoming more and more critical, and many experts agree that we have to focus all our efforts on finding ways to become independent from fossil fuels as soon as possible. The agreement recently negotiated in Paris by the United Nations Framework Convention on Climate Change (UNFCCC) [6], aims at contrasting climate change by significantly reducing emissions of greenhouse gases (mainly CO_2) generated by human activity worldwide in a relatively short time. In Europe, the energy-focused scenarios recently developed by Shell [7], denominated respectively the *Sky*, *Oceans*, and *Mountains* scenario, aimed at responding to the challenges posed by the Paris agreement, and describe a technologically, industrially, and economically possible way forward, consistent with limiting the global average temperature rise to well below 2 °C from pre-industrial levels.

Fossil fuels are nowadays our main source of both energy and platform chemicals, thus the challenge is actually twofold: the use of fossil fuels for the production of energy should be increasingly replaced by renewable resources (solar, wind, hydroelectric, geothermal, biofuels, etc.); and, platform chemicals should be obtained by bio-based resources, according to the concept of bio-refineries, which has become a familiar term in recent times [5]. Some worrying questions comes naturally to mind when thinking about this resource problem: how far are we from becoming independent from fossil fuels? Do we have enough of them to keep going until we manage to do so? Are we already late?

Of course, the answers to these questions are not straightforward, but we can have a look at some recent data. Figure 1.1 shows the global primary energy consumption (in million tonnes oil equivalent) from 1991 to 2016, according to the BP (British Petroleum) statistical review of world energy 2017 [8].

The BP report shows that the total world primary energy consumption grew by 1.0% in 2016, well below the last 10-year average of 1.8% and the third consecutive year at or below 1%. All fuels except oil and nuclear power grew at below-average rates. Oil provided the largest increment to energy consumption at 77 million tonnes of oil equivalent (mtoe), followed by natural gas (57 mtoe) and renewable power (53 mtoe). On a very simple level, it appears very clearly that even though the trend is positive (Fig. 1.2), renewable sources are not gaining much ground on fossil fuels yet and a "fossil fuel free" world is still far off.

https://doi.org/10.1515/9783110640250-001

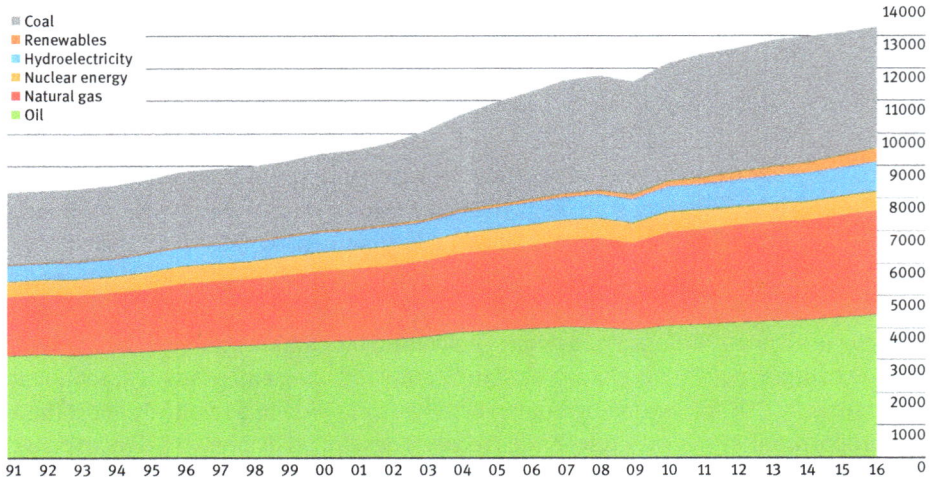

Fig. 1.1: Global primary energy consumption between 1991 and 2016 (in million ton oil equivalent) [8].

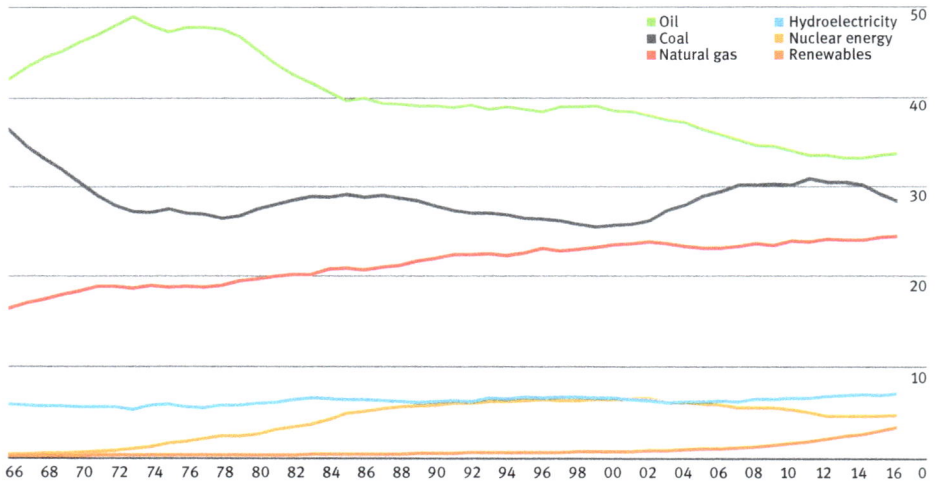

Fig. 1.2: Shares of global primary energy consumption (in %) between 1966 and 2016 [8].

Looking more specifically at the oil situation, Fig. 1.3 illustrates the world oil demand and supply over the last few years, as reported by the International Energy Agency (IEA) in 2017, including predictions for 2018 [9]. Of course, the overall scenario of demand and supply is extremely complicated, due to all the political and economic fac-

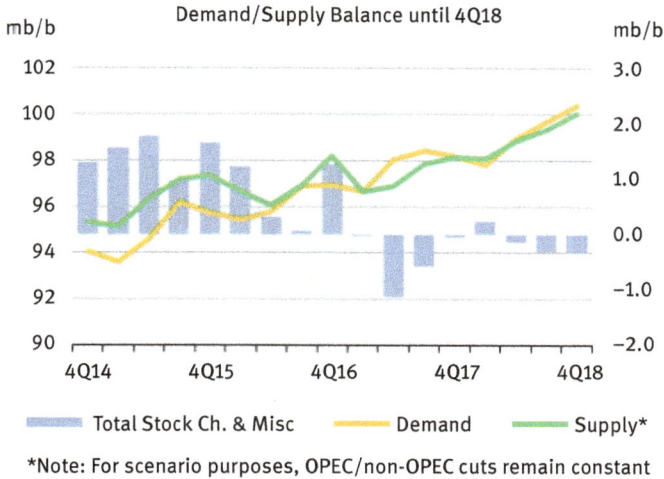

Fig. 1.3: Balance between world supply and demand of oil (in million barrel per day) until 2017, with predictions for 2018, according to [9].

tors in play, but it is again clear to the non-expert eye, that both are increasing and are predicted to keep on doing so.

Oil is obviously a limited resource and it will eventually become unavailable, when there will be no new reservoirs to be discovered and the existing one will be no longer exploitable. It is practically impossible to foresee such a moment, but several theories, the first rigorous one being proposed by geophysicist Hubbert (as early as in 1956), have predicted the existence of a "peak oil" [10–13]. According to the theory, the current increase in oil production will reach a peak, followed by a steady decline until full depletion of economically exploitable oil resources. This follows the evolution of every oil reservoir, which generally experiences a rise, peak, decline and depletion. Several studies have tried to place the position of the peak in a specific year. This will depend on many factors, such as the global economy, the discovery of new reservoirs, the development of new technologies or improvement of existing ones, which are so complex they make such an estimate very impractical. The most recent estimates places the "oil peak" around 2020, and in any case it seems very unlikely that it will be reached after 2030 [13].

The current and prospective worldwide energy demand has led either to exploiting more difficult and costly unconventional oil reserves (oil shale, tar sands, etc.), or to maximizing the exploitation of conventional oil sources. The latter triggered and still drives the development of new techniques aimed at improving the efficiency and

Non-recoverable
Oil 4120 Gbbl

Recoverable Oil
3000 Gbbl

Produced so far
1311,13 Gbbl

Remaining as
Proved Reserved
1687,9 Gbbl

World Total Conventional
Oil Resources = 7120 Gbbl

Non-recoverable
Oil 6390 Gbbl

Recoverable Oil
2210 Gbbl

Remaining as
Proved Reserved
2120 Gbbl

Produced so far
90 Gbbl

World Total Unconventional
Oil Resources = 8600 Gbbl

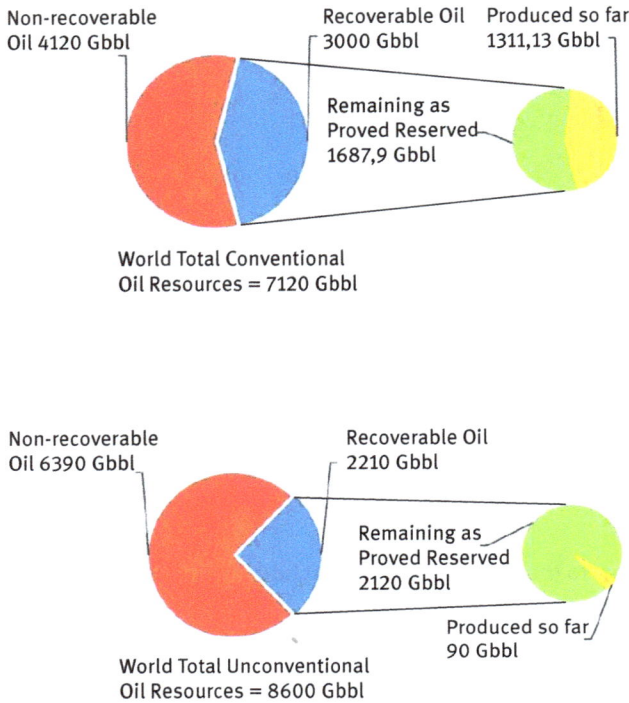

Fig. 1.4: Conventional (top) and unconventional (bottom) oil resources expressed in Giga barrels [14].

lifetime of mature oil fields. These techniques usually go under the collective term of enhanced oil recovery (EOR). Figure 1.4 gives an estimate of total conventional and unconventional oil made by BP and IEA [14].

1.2 Enhanced oil recovery

Oil recovery processes can be divided into three main stages: primary, secondary and tertiary recovery. The latter is also known as enhanced oil recovery, (EOR) [15].

Primary recovery uses natural forces to produce the oil, through three different mechanisms: the aquifer drive, the gas cap drive and the gravity flow. The aquifer drive, according to which the pressure that is exerted on the oil by the aquifer represents the driving force for extraction, is the most efficient mechanism. The production of oil leads to a decrease in pressure of the reservoir, and the aquifer moves towards the production well. The oil cut decreases as more and more water is produced along with the oil. The gas cap drives the oil in a similar fashion as the aquifer drive. Gas production (along with the oil) is not seen as a disadvantage, since it also can be used as an energy source. Finally, gravity is the important factor in the gravity flow, for which the well placement is obviously relevant. The use of this method is limited and is heavily dependent on the geology of the reservoir. The primary techniques recover, depending on the oil reservoir, on average between 5 and 25% of the original oil in place (OOIP) [15].

The secondary method involves the injection of either water or gas to increase the pressure in the reservoir, which in turn drives the oil out. After a given time, the injected water breaks through in the production wells. As the production well ages, after the water breakthrough, the water cut increases. The use of the secondary methods enables the extraction of additional 5–30% of the OOIP depending on the reservoir after primary extraction.

At most 55% of the OOIP can be recovered (in most cases this value is much lower) using the primary and secondary techniques. Therefore, a large portion of the OOIP remains embedded in the reservoir.

Since the 1970s, many different methods have been developed to increase the oil recovery as a response to the oil crisis. These all belong to the category improved oil recovery (IOR). Improved oil recovery implies improving the oil recovery by any means, such as operational strategies. Enhanced oil recovery, a subgroup of IOR, is different in that the objective is to reduce the oil saturation below the residual oil saturation (the latter being defined as the oil saturation after a prolonged waterflood). Figure 1.5 shows schematically an oil well with implemented chemical EOR techniques [14]. Injection wells have to be built around the main production well, which of course means a significant economic investment.

Looking at the evolution of EOR in the USA, the number of projects kept growing until reaching a peak in the mid-80s (Fig. 1.6) [16], when the drop in oil prices discouraged further investments. Recent times, however, experienced again a growth of EOR techniques, as the resources become scarcer and fewer new reservoirs are discovered.

The term EOR includes a series of techniques, summarized in Fig. 1.7 [17]. They can roughly be divided into two groups, thermal and non-thermal. Thermal methods have been tested since the 1950s, and they are the most advanced EOR methods, as far as field experience and technology are concerned. They are best suited for heavy oils (10–20°API) and tar sands ($\leq 10°$API).

Thermal methods supply heat to the reservoir and vaporize some of the oil. The major mechanisms include a large reduction in viscosity, and hence mobility ratio. Thermal methods have been highly successful in Canada, USA, Venezuela, Indonesia and other countries [15].

While thermal methods rely on the introduction of energy into the well in the form of heat, the non-thermal methods involve the injection of substances to ensure better oil extraction. These can be miscible with the oil (organic solvents, gases) or non-

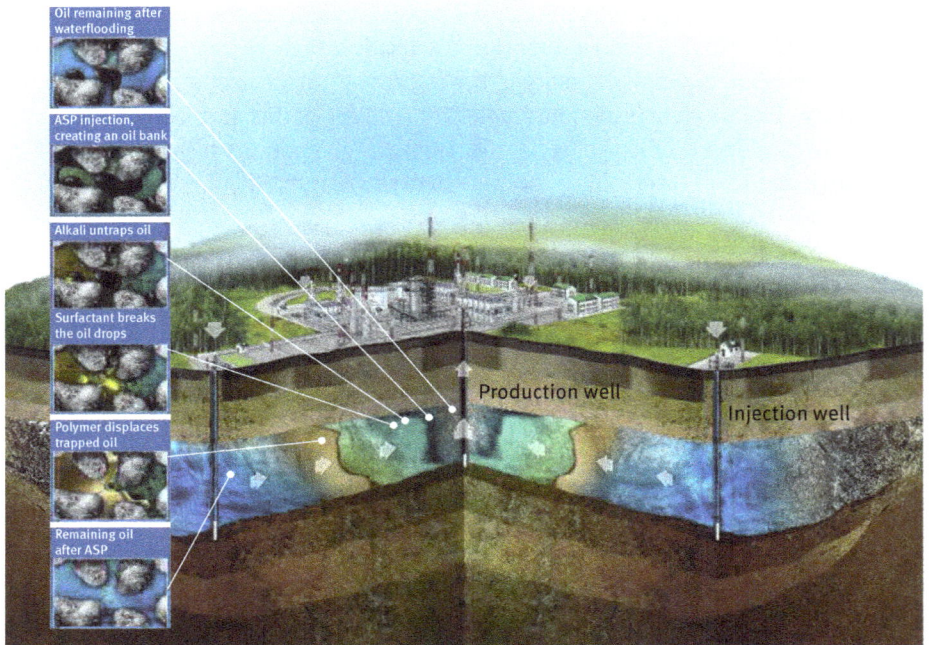

Fig. 1.5: Schematic representation of a chemical EOR process [14].

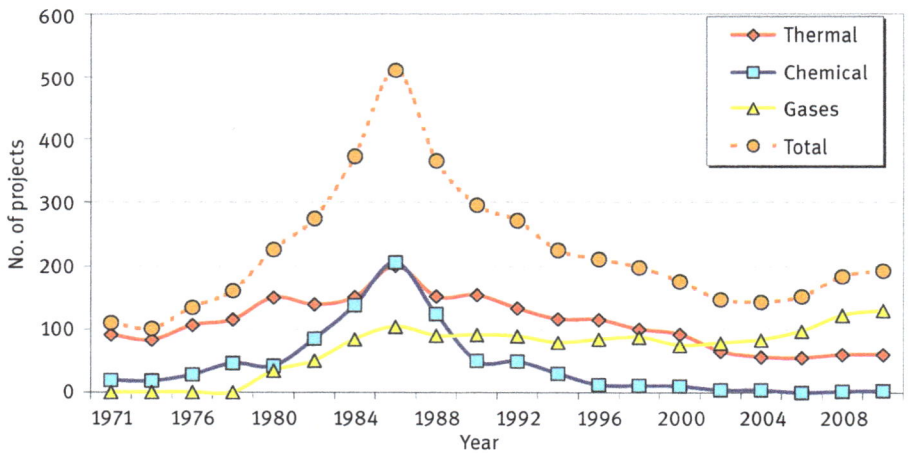

Fig. 1.6: Evolution of EOR projects in the USA [16].

miscible. The latter include all the chemical EOR methods, for brevity cEOR, which are the focus of this book. One non-thermal and non-chemical EOR method worthy of

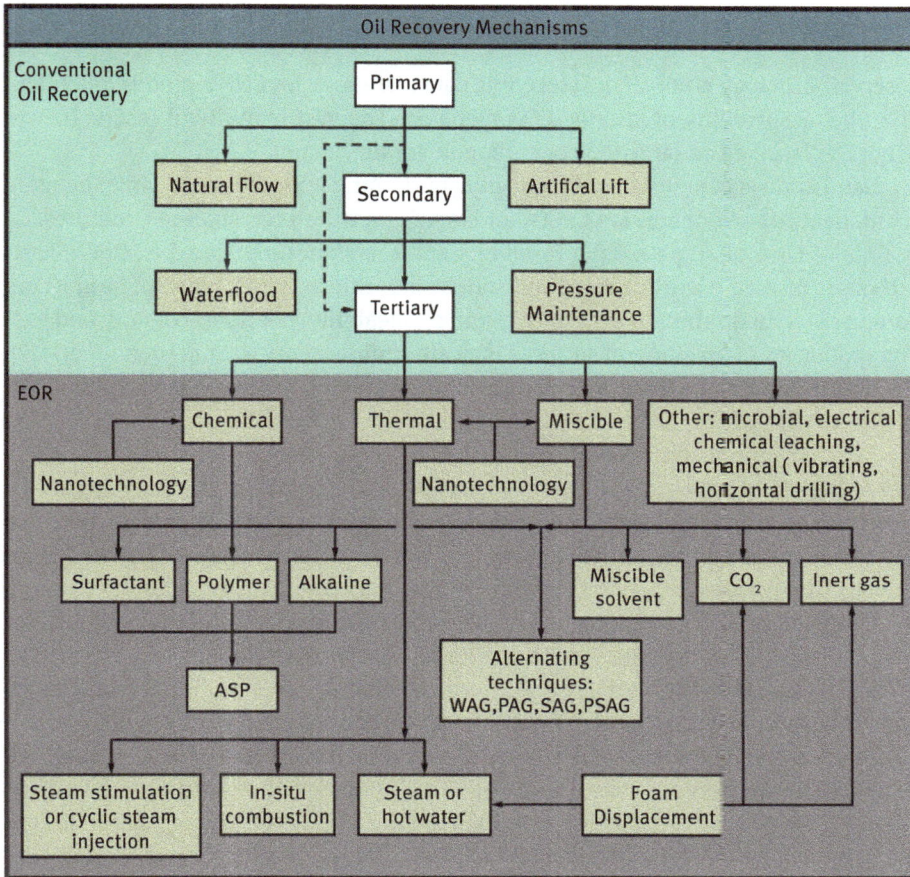

Fig. 1.7: Summary of chemical EOR techniques [17].

mention is CO_2 flooding, which is gaining a lot of popularity [15, 16]. The use of CO_2 is attractive for both economic and environmental reasons: it is cheap and readily available and when used for oil extraction it can be trapped under the soil, reducing its emission into the atmosphere.

The vast majority of cEOR methods are constituted by mainly three of them, and their combinations: polymer flooding, surfactant flooding and alkaline flooding [18]. Combined surfactant-polymer flooding (usually abbreviated to SP flooding) and al-kali-surfactant-polymer flooding (ASP flooding) have been also intensively investigated [19–21].

Which EOR method is more suitable for a particular case depends on many factors, including reservoir lithology and the type of oil, which will be discussed in the following sections.

1.3 Reservoir lithology

Reservoir lithology is one of the screening considerations for EOR methods, often limiting the applicability of specific EOR methods. The vast majority of reservoirs can either be classified as sandstone or carbonate (or limestone).

Sandstone is composed mainly of silicates (quartz and feldspar), while the main components of carbonates are $CaCO_3$ and $MgCO_3$ in various crystalline structures.

About 60% of all petroleum reservoirs are sandstones; outside the Middle East, carbonate reservoirs are less common, and the percentage is even higher. The most important reservoir properties are porosity and permeability, but pore geometry and wetting properties of the mineral surfaces may also influence petroleum production [22]. It is generally accepted that oil recovery is more efficient in water-wet rocks rather than in oil-wet ones [23].

Most EOR applications have been in sandstone reservoirs, as derived from a collection of over 1500 international EOR projects [16]. EOR thermal and chemical projects are the most frequently used in sandstone reservoirs compared to other lithologies (e.g., carbonates and turbiditic formations) [16]. Sandstones usually possess higher porosity and permeability and favorable wetting properties (water-wet).

However, a considerable portion of the world's hydrocarbon endowment is in carbonate reservoirs, which poses more challenges. Carbonate reservoirs usually exhibit lower porosity and may be fractured. These two characteristics along with rock wetting properties, usually result in lowered hydrocarbon recovery rates [24, 25].

Another problem associated with carbonate reservoirs is related to their chemical composition: they are composed mainly by $CaCO_3$ and $MgCO_3$, thus they contain significant amounts of Ca^{2+} and Mg^{2+} ions in their structure. As we will see later, this limits the implementation of chemical EOR methods, due to precipitation of most chemicals employed in the processes (polymers and surfactants) in the form of calcium and magnesium salts.

The recoverable reserves of carbonate reservoirs are huge and thus it is considered that the exploitation of them has tremendous economic value, so people are shifting attention from traditional sandstone reservoirs to carbonate reservoirs which are widely distributed around the world, in North America, North Europe, China and Australia (Fig. 1.8) [26, 27].

Applying EOR technology to improve oil recovery is critical. Worldwide EOR projects for carbonate reservoir development have been undertaken since the 1970s. The main techniques to enhance oil recovery for carbonate reservoirs include air injection, thermal recovery and chemical flooding. The injected gas can be carbon dioxide, nitrogen or air, and hydrocarbon gas where carbon dioxide flooding is the dominant EOR process used in the US because of the high availability of low-cost CO_2. Steam injection in viscous oil carbonates has been piloted in the Middle East and Canada because it can change reservoir behaviors, leading to production enhancement. Be-

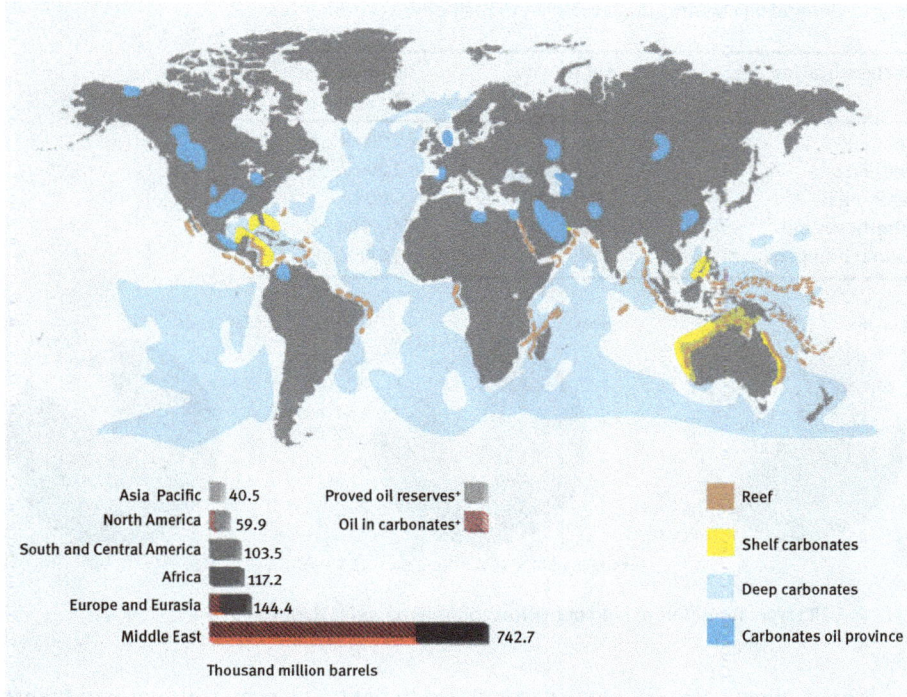

Fig. 1.8: Geographical distribution of large carbonate reservoirs worldwide [27].

sides, SiO_2 nanoparticles and microbial biomass have also been investigated for EOR of carbonate reservoirs [26].

1.4 Types of crude oil

The most common distinction made when referring to crude oil is between light, heavy and extra-heavy oil. The latter are also called bitumen, tar or oil sands. Although the distinction can be made somewhat arbitrarily, it is generally accepted to define as heavy any oil with a gravity of 20°API or lower [28]. It is worth mentioning here that the API gravity is an inverse measure of a petroleum liquid's density relative to that of water. Another distinction can be made based on viscosity. Table 1.1 gives an overview of a general classification of oils based on these physical properties. The viscosity of heavy oil is typically higher than 100 cP, while extra-heavy oils are those which do not flow significantly in ordinary conditions. Since these cannot be extracted using conventional methods (the ones discussed in this book), they are also referred to as "unconventional oils".

Tab. 1.1: General oil classification according to their physical properties.

oil classification	API Gravity °API	Dynamic Viscosity cP
Light oils	> 31.1	< 100 cP
Medium oils	22.3 < °API < 31.1	< 100 cP
Heavy oils	10 < °API < 22.3	< 100 cP
Extra-heavy oils	< 10	< 10.000 cP
Natural bitumen (or tar sands)	< 10	> 10.000 cP

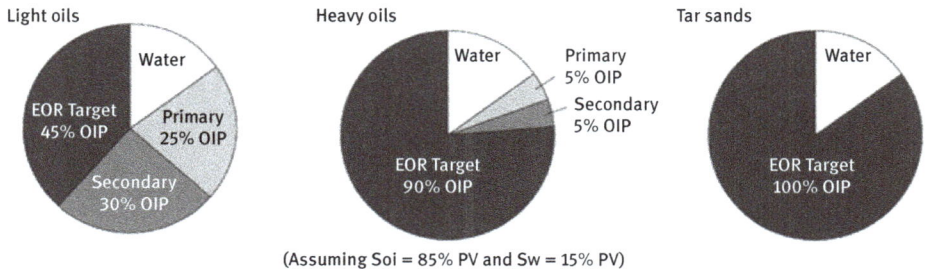

(Assuming Soi = 85% PV and Sw = 15% PV)

Fig. 1.9: EOR target for different kind of hydrocarbons. From ref [15].

From a chemical point of view, compared to light oils, heavy oils contain higher amounts of aromatic hydrocarbons, simple (benzene derivatives) or polycondensate (asphaltenes), as well as higher amounts of heteroatoms (sulfur, nitrogen, and heavy metals).

Due to their higher density and viscosity, heavy oils and tar sands respond poorly to primary and secondary recovery methods, and the bulk of the production from such reservoirs come from EOR methods (Fig. 1.9) [15].

1.5 Chemical EOR

Chemical EOR methods had their best times in the 1980s, most of them in sandstone reservoirs.

The total number of active projects peaked in 1986 with polymer flooding as the most important chemical EOR method. However, since 1990s, oil production from chemical EOR methods has dropped down significantly. Nowadays the biggest cEOR projects are running in China. Nevertheless, chemical flooding has been shown to be sensitive to the volatility of oil markets despite recent advances (e.g., low surfactant concentrations) and lower costs of chemical additives.

Chemical EOR includes various interesting techniques. The most important among them are polymer flooding, which is the most mature and used, surfactant flooding, alkaline flooding and combinations of the above, such as surfactant-polymer (SP) flooding and alkali-surfactant-polymer (ASP) flooding [18, 20, 21, 29]. Besides these, microbial flooding [30], and more recently, nanofluids [31] flooding are also relevant cEOR methods. Polymeric surfactants, as an alternative to polymer and SP flooding have also been investigated [32].

Polymer flooding can date its origins back to the 1950s, when the importance of fluid mobility for oil recovery was proposed and proved experimentally shortly after [33]. It was observed that increasing the viscosity of water used in waterflooding by dissolving a small amount of high molecular weight soluble polymer, would greatly improve the macroscopic sweep efficiency. Polymer flooding will be discussed in detail in the next chapter. Surfactants are known to decrease the interfacial tension between water and oil. This allows the formation of emulsions when a surfactant solution is flooded into an oil reservoir, also known as micellar flooding. This causes a significant decrease in residual oil saturation, with subsequently better microscopic oil displacement. The role of an alkali in alkaline flooding, is basically the same as the surfactant, but it is achieved by a different mechanism. The alkali (or base), is able to generate surfactant species *in situ*, simply by neutralizing fatty acids present in the crude oil, turning them in amphiphilic molecules. Alkaline flooding is generally used in combination with the previous ones, in what is called ASP flooding (from alkali-surfactant-polymer). Besides the role already mentioned, the chemicals used for cEOR can also alter the rocks wettability, which can also result in better recovery.

Important phenomena to take into account in cEOR are adsorption and retention of the chemicals onto the porous rock systems. Adsorption is generated by all kind of interactions (Van der Waals, electrostatic, and hydrogen bonding) between the chemicals and the rock surface. Polymers generally display high adsorption and retention, due to their high molecular weight, and this can cause a significant change in rocks permeability and porosity. Both polymers and surfactants are very often constituted by charged molecules that able to adsorb onto the rock surfaces via electrostatic interactions with the ions present in the rock crystalline structure.

Another relevant aspect is the environmental impact of the chemicals used for cEOR. Underground injection of huge amounts of chemicals can have a significant and often unpredictable effect on ecosystems. Water flooding should be performed after cEOR methods are implemented, in order to remove residual chemicals from the reservoir.

The use of chemical products such as polymers, surfactants, alkali, microorganisms, nanomaterials and nanofluids, has attracted a lot of interest and poses several scientific challenges, which have been tackled by many researchers. The chemicals used should be designed to withstand the harsh conditions present in the reservoir (e.g., dissolved salts, pH, temperature, presence of bacteria) and increase the efficiency of the process. One of the key factors in this development is the (macro)molecules architecture and its influence on the physical properties of the fluids being injected: from linear to branched polymers, and from monomeric to gemini surfactants.

The most recent advancements in cEOR will be treated in the rest of the book, with particular focus on academic studies aimed at designing new polymers for EOR, with the aid of both experimental work and mathematical simulations.

References

[1]   Dresselhaus, M. S. and Thomas, I. L. Alternative energy technologies. *Nature*, 414:332–337, 2001.
[2]   Hoffert, M. I., Caldeira, K., Benford, G., Criswell, D. R., Green, C., Herzog, H., Jain, A. K., Kheshgi, H. S., Lackner, K. S., Lewis, J. S., Lightfoot, H. D., Manheimer, W., Mankins, J. C., Mauel, M. E., Perkins, L. J., Schlesinger, M. E., Volk, T., and Wigley, T. M. L. Advanced technology paths to global climate stability: energy for a greenhouse planet. *Science*, 298:981–987, 2002.
[3]   York, R. Do alternative energy sources displace fossil fuels? *Nat. Clim. Chang.*, 2:441–443, 2012.
[4]   McGlade, C. and Ekins, P. The geographical distribution of fossil fuels unused when limiting global warming to 2 °C. *Nature*, 517:187–190, 2015.
[5]   Cherubini, F. The biorefinery concept: Using biomass instead of oil for producing energy and chemicals. *Energy Convers. Manag.*, 51:1412–1421, 2010.
[6]   UNFCCC. The Paris Agreement. Available online: https://unfccc.int/process-and-meetings/the-paris-agreement/the-paris-agreement (accessed on Jun 6, 2018).
[7]   Shell Global. Sky Scenario. Available online: https://www.shell.com/energy-and-innovation/the-energy-future/scenarios/shell-scenario-sky.html (accessed on Feb 15, 2019).
[8]   BP. Statistical Review of World Energy | Energy economics. Available online: https://www.bp.com/en/global/corporate/energy-economics/statistical-review-of-world-energy.html (accessed on May 23, 2018).
[9]   OMR. OMR Public. Available online: https://www.iea.org/oilmarketreport/omrpublic/ (accessed on May 23, 2018).
[10]   Bardi, U. Peak oil: The four stages of a new idea. *Energy*, 34:323–326, 2009.
[11]   De Almeida, P. and Silva, P. D. The peak of oil production—Timings and market recognition. *Energy Policy*, 37:1267–1276, 2009.
[12]   Kerr, R. A. Peak Oil Production May Already Be Here. *Science (80-.).*, 331:1510–1511, 2011.
[13]   Madureira, N. L. Oil Reserves and Peak Oil. In *Key Concepts in Energy*, pp. 101–130. Springer International Publishing, Cham, 2014.
[14]   Druetta, P. D., Picchioni, F., and Rijksuniversiteit Groningen. *Numerical simulation of chemical EOR processes*. University of Groningen, 2018.
[15]   Thomas, S. Enhanced oil recovery – An overview. *Oil Gas Sci. Technol. L Inst. Fr. Du Pet.*, 63:9–19, 2008.
[16]   Alvarado, V. and Manrique, E. Enhanced Oil Recovery: An Update Review. *Energies*, 3:1529–1575, 2010.

[17] Druetta, P., Raffa, P., and Picchioni, F. Plenty of Room at the Bottom: Nanotechnology as Solution to an Old Issue in Enhanced Oil Recovery. *Appl. Sci.*, 8:2596, 2018.

[18] Lake, L. W. *Enhanced Oil Recovery.* New Jersey, Prentice Hall, 1989.

[19] Li, G. Z., Mu, J. H., Li, Y., and Yuan, S. L. An experimental study on alkaline/surfactant/polymer flooding systems using nature mixed carboxylate. *Colloids Surfaces A Physicochem. Eng. Asp.*, 173:219–229, 2000.

[20] Sheng, J. J. A comprehensive review of alkaline-surfactant-polymer (ASP) flooding. *Asia-Pacific J. Chem. Eng.*, 2014.

[21] Olajire, A. A. Review of ASP EOR (alkaline surfactant polymer enhanced oil recovery) technology in the petroleum industry: Prospects and challenges. *Energy*, 77:963–982, 2014.

[22] Bjørlykke, K. and Jahren, J. *Petroleum Geoscience*, chapter Sandstones and Sandstone Reservoirs, pp. 113–140. Springer Berlin Heidelberg, Berlin, Heidelberg, 2010.

[23] El-hoshoudy, A. N. N., Desouky, S. E. M. E. M., Elkady, M. Y. Y., Al-Sabagh, A. M. M., Betiha, M. A. A., and Mahmoud, S. Hydrophobically associated polymers for wettability alteration and enhanced oil recovery – Article review. *Elsevier*, 26:757–762, 2017.

[24] Manrique, E. J., Muci, V. E., and Gurfinkel, M. E. EOR Field Experiences in Carbonate Reservoirs in the United States. *SPE Reserv. Eval. Eng.*, 10:667–686, 2007.

[25] Chilingar, G. V. and Yen, T. F. Some Notes on Wettability and Relative Permeabilities of Carbonate Reservoir Rocks, II. *Energy Sources*, 7:67–75, 1983.

[26] Bai, M., Zhang, Z., Cui, X., and Song, K. Studies of injection parameters for chemical flooding in carbonate reservoirs. *Renew. Sustain. Energy Rev.*, 75:1464–1471, 2017.

[27] Schlumberger. Carbonate Reservoirs. Available online: https://www.slb.com/services/technical_challenges/carbonates.aspx (accessed on Feb 19, 2019).

[28] Saboorian-Jooybari, H., Dejam, M., and Chen, Z. Heavy oil polymer flooding from laboratory core floods to pilot tests and field applications: Half-century studies. *J. Pet. Sci. Eng.*, 142:85–100, 2016.

[29] Chang, H. L. ASP Process and Field Results. In *Enhanced Oil Recovery Field Case Studies.* Gulf Professional Publishing (Elsevier), Waltham, MA, 2013.

[30] Sen, R. Biotechnology in petroleum recovery: The microbial EOR. *Prog. Energy Combust. Sci.*, 34:714–724, 2008.

[31] Peng, B., Zhang, L., Luo, J., Wang, P., Ding, B., Zeng, M., and Cheng, Z. A review of nanomaterials for nanofluid enhanced oil recovery. *RSC Adv.*, 7, 2017.

[32] Raffa, P., Broekhuis, A. A., and Picchioni, F. Polymeric surfactants for enhanced oil recovery: A review. *J. Pet. Sci. Eng.*, 145, 2016.

[33] Thomas, A. Polymer Flooding. In Romero-Zeron, L., ed., *Chemical Enhanced Oil Recovery (cEOR) – a Practical Overview*, p. Ch. 02. InTech, Rijeka, 2016.

2 Polymer flooding

2.1 Introduction

Polymer flooding is the most established chemical EOR method. It has a long history and demonstrated effectiveness. This technology outnumbers other chemical technologies, because it is easier to implement, it's applicable in a large range of reservoir conditions and it has low risks when compared to others.

The environmental aspects are seldom considered and the long-term effects on the ecosystems of injection of tons of polymers underground is still unclear. However, polymer flooding can be expected to have a lower environmental impact when compared, for example, to surfactants or microbial flooding. From this point of view, it could be considered a positive sign that after polyacrylamides, the most used polymers for EOR are biopolymers such as Xanthan gum and cellulose. However, this is not necessarily better for the environment, because the introduction of huge amount of these materials, which are nutrients for microorganisms, could affect the underground flora and fauna, with unpredictable consequences.

Nonetheless, the application of polymer flooding has increased over the past years, especially in high temperature and high salinity reservoirs [1–8]. Polymer flooding consists of injecting a polymer water solution into a subterranean oil formation in order to improve the sweep efficiency in the reservoir. The increased viscosity of the water due to the dissolved polymer causes a better mobility control between the injected fluid and the hydrocarbons within the reservoir, allowing to recover an additional 12 to 15% of the original oil in place (OOIP) [3, 9]. Usually polymer flooding follows water flooding, but it can be implemented also directly after primary recovery.

The early pioneering work on polymer flooding dates back to the 60s, and the first large commercial uses of polymers to increase oil recovery were performed in the United States, during a crude oil price control period between the 60s and the 70s [1]. After peaking during the 80s, the number of projects abruptly decreased for several economic and technological reasons. However, research has continued over the years and polymer flooding regained interest, especially in China, in the 90s. The largest polymer flooding project is currently running in the Daqing oilfield, after being started in 1996. Another example of successful polymer injection in the 1990s was in Courtenay, France, where extra oil recoveries from 5 to 30% have been reported after the technology was conducted in a secondary recovery mode as augmented water flooding. Germany, Austria, Oman and North Kuwait have also running EOR projects [4].

As of 2004, more than 30 commercial projects were implemented, involving approximately 2427 injection wells and 2916 production wells according to [1]. Polymer injection in the Shengli and Daqing oilfields yielded incremental oil recoveries ranging from 6 to 12%, contributing to 250.000 barrels per day in 2004 [1].

https://doi.org/10.1515/9783110640250-002

An extensive recent update from researchers in Norway [7] reports results from 72 documented polymer projects (including pilots). Among these, 66 projects were implemented onshore and only 6 offshore. Of the reported projects, 40 were classified as successes and 6 were assessed discouraging. 46 projects have been performed in the U.S.A., followed by 6 in Canada, 6 in China and 4 in Germany. Partially hydrolysed polyacrylamide (HPAM) was used as polymer in 92% of the cases and the rest were using biopolymers. Only one project used a hydrophobic associative polymer [7].

Several polymers are commercially available for chemical EOR applications, HPAM being the most widely used. Water soluble polymers for chemical EOR applications have been reviewed several times. The main contributions from the last years include general reviews on polymer flooding [1, 3–7, 10–12] as well as reviews focusing on particular classes of polymers for EOR such as hydrophobically associating polymers [13, 14], or particular aspects such as oil displacement mechanisms [9] and injection parameters [15, 16]. Our own group also recently contributed with reviews focusing on solution properties [2] and on polymeric surfactants [17].

Currently used polymers suffer various limitations and cannot generally be used in severe conditions of salinity and temperature or in low porosity rock formations. Therefore, considering the still high and actually increasing demand of oil and its decreasing availability, which is expected to cause the oil price to rise again in the near future, research of new systems for polymer flooding is still active. The number of patents filed by many companies, even in recent times, about new polymers for EOR certainly supports this statement [17].

In our opinion, along with elucidating the oil recovery mechanisms, establishing proper structure-properties relationships for water-soluble polymer is the key to design new successful systems.

This chapter will shortly illustrate the underlying theory of polymer flooding and will then describe from a chemical point of view the polymers currently used and proposed as flooding agents, with particular emphasis on structure-properties relationship and how those could determine possible advantages and disadvantages in chemical EOR processes.

2.2 Theory

The overall oil recovery efficiency in oil production processes is generally divided in two distinct contributions: macroscopic and microscopic recovery efficiency. The macroscopic recovery efficiency refers to the volume that the flooding agents are able to sweep and it is also referred to as volumetric sweep efficiency; the microscopic recovery efficiency is a measure of the effectiveness of the displacing fluid(s) in mobilizing the oil trapped at pore scale by capillary forces. In other words, any mechanism that can improve either macroscale or microscale oil recovery efficiency is beneficial for increasing oil production [3].

The mechanisms of polymer flooding have been elucidated and discussed thoroughly. It is normally suggested that polymer flooding can only improve the volumetric sweep efficiency without any effect on the microscopic displacement efficiency.

The well-established relationship between capillary number and oil recovery indicates that a substantial increase in oil recovery at the pore level (microscale) can be obtained only when the capillary number is increased by at least three orders of magnitude. For polymer flooding, the capillary number is normally increased less than 100 times. This can be visualized considering that the capillary number can be expressed as the viscous-to-capillary forces, according to Eq. (2.1):

$$N_c = \frac{u\mu_c}{\gamma} \tag{2.1}$$

Where u is the velocity gradient, μ_c is the viscosity of the displacing fluid (or continuous phase) and γ the interfacial tension between displacing fluid and oil.

Therefore, as the polymer can increase water viscosity by two to three orders of magnitude, the effect on capillary number will be of the same order of magnitude. Microscopic efficiency is much more affected by reduction of interfacial tension operated by the presence of surfactants.

However, many reported values of increased oil recovery seem to be higher than those achievable only by sweep efficiency improvement [3]. This fact made researchers to revisit the oil recovery mechanisms occurring in polymer flooding. It is now proposed that incremental oil recovery by polymer flooding can also be explained by the simultaneously increased microscopic displacement efficiency due to the distinctive flow characteristic of polymer solutions [3].

In any case, the main mechanism of oil recovery achieved by polymer flooding is the increase of macroscopic efficiency by increasing water viscosity, in order to achieve better mobility control. The mobility can be also altered by changing the permeability of the rock formation. The second most important mechanism of oil recovery is permeability reduction. Viscoelastic behaviour of polymer solutions might also play a role, as discussed later.

Mobility control is discussed in the next section.

2.2.1 Mobility control

Mobility control remains one of the most important concepts in any enhanced oil recovery process. According to the most general consensus, mobility control can be achieved through injection of chemicals to change displacing fluid viscosity or to preferentially reduce specific fluid relative permeability through injection of foams, or even through injection of chemicals, to modify wettability [18].

The existing concept of mobility control is that the displacing fluid mobility should be equal to or less than the (minimum) total mobility of displaced multiphase fluids. This statement has been recently challenged [18].

The mobility of a fluid in a porous formation (λ) is usually defined as the ratio between the effective permeability of the fluid (which is the product of the absolute permeability K and the relative permeability of the phase k_r) and its viscosity (μ), as shown in Eq. (2.2):

$$\lambda = \frac{Kk_r}{\mu} \tag{2.2}$$

What matters in an EOR process, is the ratio between the mobility of the displacing fluid (water, or as in the case of polymer flooding, polymer solution) and the displaced fluid (oil), called the mobility ratio (M), as expressed as in Eq. (2.3) (the superscripts w and o, stand for water phase and oil phase respectively):

$$M = \frac{\lambda^w}{\lambda^o} = \frac{\frac{k^w}{\mu^w}}{\frac{k^o}{\mu^o}} = \frac{k^w}{k^o}\frac{\mu^o}{\mu^w} \tag{2.3}$$

Mobility ratio influences the microscopic (pore level) and macroscopic (areal and vertical sweep) displacement efficiencies. A value of M much higher than one, which results from a high difference in viscosity between the water and the oil phases (this is can be easily seen by Eq. (2.2)) is considered unfavourable, because it indicates that the displacing fluid flows more readily than the displaced fluid (oil), and it can cause channeling of the displacing fluid, and as a result, bypassing some of the residual oil. Under such conditions, and in the absence of viscous instabilities, more displacing fluid is needed to obtain a given residual oil saturation. The effect of mobility ratio on displaceable oil is shown in Fig. 2.1 [8]. The three curves represent one, two and three pore volumes of total fluid injected, respectively. Displacement efficiency is increased when $M = 1$ and is denoted a "favourable" mobility ratio.

It has been established that during a typical flooding, for a mobility ratio close to one, the displacing fluid will push the oil in a "piston-like" fashion while, as the value becomes higher, the phenomena of viscous fingering and channeling will become relevant (Fig. 2.2) [19, 20], with subsequent lowered sweep efficiency. Simply put, as the viscosity of the oil phase is always bigger than that of water, increasing the water viscosity should have beneficial effects in sweep efficiency and thus in oil recovery. This consideration is at the basis of the use of water thickeners, particularly water-soluble polymers, in chemical enhanced oil recovery. In the next section, the viscosity of polymer water solutions and rheological behaviour, will be discussed in the context of enhanced oil recovery.

It has been recently proposed by Sheng [9, 18] that the mobility ratio M value used to determine the most favourable viscosity conditions for the displacing fluid, should be corrected by a multiplicative factor, which he called "normalized movable oil saturation".

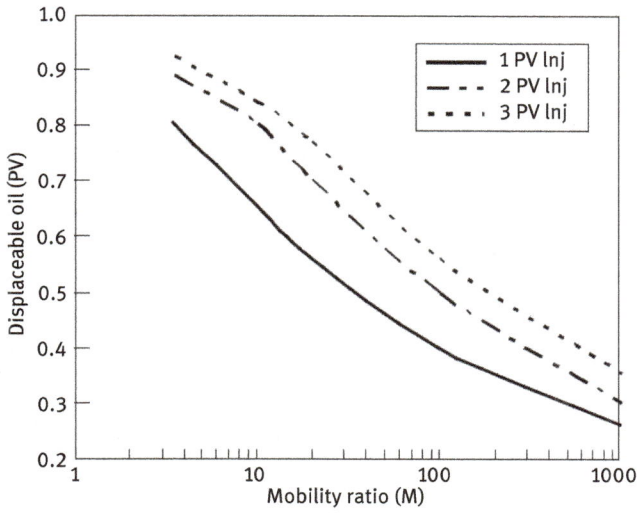

Fig. 2.1: Effect of mobility ratio on displaceable oil (simulated) [8].

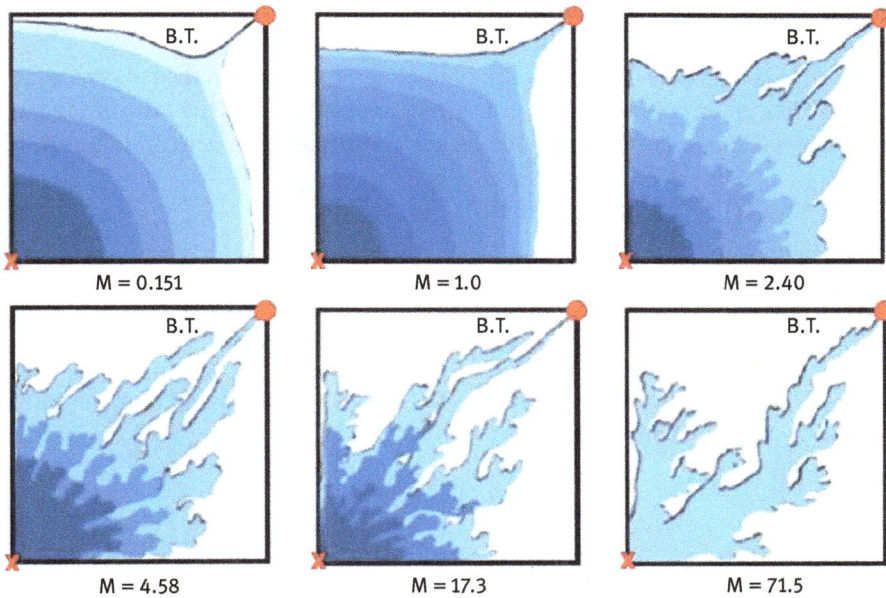

Fig. 2.2: Water flooding at different mobility ratios, showing viscous fingering [20].

2.2.2 Permeability reduction

Viscosity is not the only relevant phenomena affecting mobility in polymer flooding. Another mechanism, known as disproportionate permeability reduction (DPR), or simply permeability reduction, can facilitate polymer flooding to improve the macroscopic sweep efficiency. This refers to a reduction of water permeability (k^w, Eq. (2.3)) by the polymer, while the oil permeability remains basically unaffected [3]. The reasons for DPR are varied and include mainly segregation of flow pathways of oil and water and adsorption of the polymer, but also wettability alteration and polymer swelling phenomena, to a minor extent [3]. By looking at Eq. (2.3), it is apparent that the effect achieved by DPR is the same as the increase of viscosity of the water phase with respect to the oil phase: a decrease of the mobility ratio.

The permeability reduction is also known as the residual resistance factor [1] and mathematically it can be expressed as the mobility ratio of water (brine) before and after polymer flooding:

$$R_{Rf} = \frac{\lambda_{bp}}{\lambda_{ap}} = \frac{\frac{k_{ap}}{\eta_{ap}}}{\frac{k_{bp}}{\eta_{bp}}} \cong \frac{k^w}{k^p} \tag{2.4}$$

where the symbols have the meanings as defined above, and the indexes bp and ap refer respectively to before and after polymer flooding, while k^p and k^w are the permeabilities of the polymer solution and water respectively. The k^w/k^p ratio is called the permeability reduction factor and is often used for the quality estimation of polymer solutions [19].

The residual resistance factor R_{Rf} should not be confused with the resistance factor R_f, another important parameter in EOR, which is defined as the ratio of the mobility of water (brine) to that of the polymer solution under the same conditions [19]:

$$R_f = \frac{\lambda^w}{\lambda^p} \tag{2.5}$$

The residual resistance factor can be used to estimate the thickness layer of adsorbed polymer χ_p, according to the formula [21]:

$$\chi_p = d_p \left(1 - R_{Rf}^{-\frac{1}{4}}\right) \tag{2.6}$$

where d_p is the average pore diameter that, in turn, can be estimated from the water permeability and rock porosity Φ [21]:

$$d_p = 1.15 \left(\frac{8k^w}{\Phi}\right)^{1/2} \tag{2.7}$$

2.2.3 Flow resistance

Flow resistance is another mechanism believed to improve the macroscopic sweep efficiency during polymer flooding. Some authors reported a positive contribution of polymer elasticity in EOR and they connected it to higher resistance to flow through porous media, compared to less elastic polymers, at identical shear viscosities [3].

Since elastic behaviour is related to the polymer structure, this will play a relevant role in the design of polymers for polymer flooding. Indeed, it has been proposed that branched PAM would perform better in polymer flooding compared to linear structures, at comparable shear viscosity. Control over the polymer structure, however, might prove to be challenging and expensive from a synthetic point of view. Synthetic polymers traditionally used for EOR are prepared by methods based on free-radical polymerization, which allows hardly any control. The structures obtained are usually high-molecular weight polymers, with large distributions of molecular weights and mostly linear structures, containing random branching. Later, synthetic strategies aimed at controlling polymer structures and branching will be discussed.

In the next sections, the effect of polymer structure's viscoelastic properties in solution, will be discussed.

2.2.4 Polymer solution rheology

Based on the previous considerations, solution rheology, particularly viscosity, has a relevant role in polymer flooding. Many well-known water-soluble polymers are effective thickening agents. These includes natural ones such as polysaccharides or proteins and synthetic ones such as polyethylene glycol and polyacrylamide. Generally speaking, the viscosity of a polymer solution increases with the hydrodynamic volume of the polymer chains, and therefore with the molecular weight, according to the well-known Mark–Houwink–Sakurada Eq. (2.8):

$$[\eta] = K_{MH} M_w^{\alpha_{MH}} \tag{2.8}$$

Where $[\eta]$ is the intrinsic viscosity, M_w is an average molecular weight and K_{MH} and a_{MH} are semi-empirical parameter characteristics for a polymer-solvent pair, related to the Flory–Huggins interaction parameter and the conformation of polymer in solution. In particular, the value of α can vary from 0.5 for a polymer in a theta solvent, to 0.7–0.8 for a polymer in a good solvent, to values higher than one for polymers in extended conformation [22]. This latter case can present itself in water for polymers bearing charges along the backbones, also known as polyelectrolytes, due to Coulombic and osmotic effects [23].

These considerations suggest that high molecular weight polyelectrolytes would be the most effective water viscosifiers and this is indeed often the case. In fact, the most used polymers in EOR are HPAM, which are high-molecular weight polyelectrolytes. Their chemical and physical characteristics will be described later.

However, when talking about polymer flooding, we must consider that the solution is forced to flow through a porous media, which causes shear forces to act on the system. Therefore, the most representative rheological parameter is the shear viscosity; polymer solutions generally present a non-Newtonian behaviour, therefore effects such as shear thinning and shear thickening, should be taken into account [18]. The typical viscosity profile of a polymer solution in a porous media is shown in Fig. 2.3 [20]. At low shear rates, the solution behaves as a typical polymer solution, with a Newtonian plateau followed by a shear thinning region as the shear increases. As the shear increases even further, a shear thickening effect is observed, attributed to a transition from shear flow to elongational flow occurring in the porous media. In the high shear region, further shear thinning can occur, due to polymer degradation causing a decrease in molecular weight.

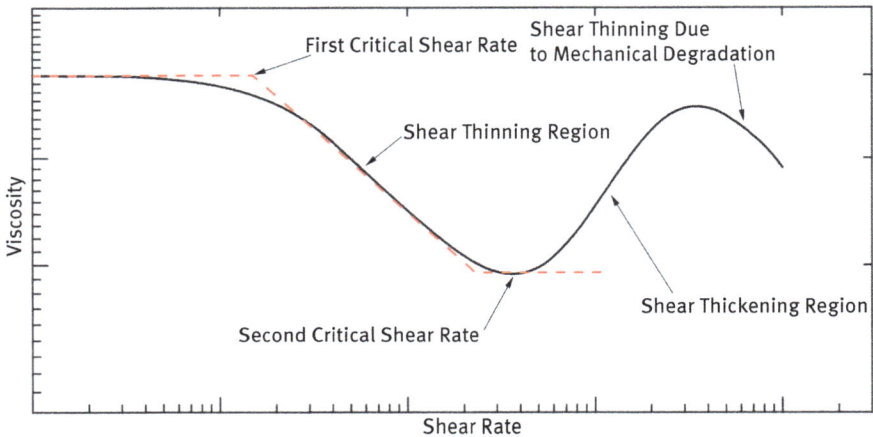

Fig. 2.3: Schematic illustration of viscosity profile in porous media [20].

The role of viscoelasticity, rather than just viscosity, has been often emphasized [2, 9]. This has been used to justify oil recovery values higher than the ones expected purely by an increase in macroscopic sweeping efficiency [3]. It has been suggested by Wang et al. [24–26] that if a fluid with elastic properties flows over dead ends, normal stresses between the oil and polymer solution are generated in addition to the shear stresses resulting from the long molecular chains. Thus, polymer imposes a larger force on oil droplets and pulls them out of dead ends. The amount of residual oil pulled out from dead ends is proportional to the elasticity of the driving fluid; these observations are presented in Fig. 2.4, where the effect is shown comparing a HPAM solution to a viscous liquid without an elastic component, such as glycerine [18, 26].

It was also shown in core flood experiments that after glycerine flooding, polymer flooding further increases the recovery by an extra 6% OOIP [18]. Since viscosity and interfacial tension were comparable, the increased recovery was explained in terms of elasticity of the polymer solution.

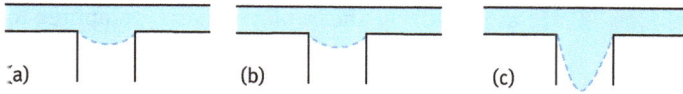

Fig. 2.4: Schematic representation of residual oil in "dead ends" after (a) water, (b) glycerine, and (c) HPAM floods.

Various displacing mechanisms have been proposed and studied via numerical simulations, based on "pulling" or "stripping" from dead ends, formation of oil threads, and shear thickening [18].

2.3 Criteria for polymer selection

Knowing the theoretical background can help in screening polymers and conditions for optimal recovery. However, there are several empirical factors to take into account.

In chemical EOR, given the harsh conditions present in most oil reservoirs, some problems and limitations arise with the use of water-soluble polymers. For example, high ionic strength drastically reduces polyelectrolytes viscosity due to the charge screening effect. Moreover, high salt concentration can cause "salting out" effects or flocculation, especially in the presence of polyvalent cations such as calcium and magnesium. High temperatures greatly accelerate polymer chemical degradation and/or transformations, the porosity of the rocks favour obstruction and polymer adsorption, shear forces cause mechanical degradation. all these aspects will be discussed later.

The aforementioned factors can seriously limit the choice of polymers for oil recovery. Moreover, and this is often the most stringent requirement, the polymer should be cheap and easily available in large quantities, in order to make the recovery process cost-effective.

The selection criteria, based on those recently summarized by Rellegadla et al. [6], are presented and further elaborated as follows.

2.3.1 Costs

Costs are obviously one of the main criteria for the selection of polymers for chemical EOR. These include not only the cost of the chemicals employed, but also process costs, operation costs, transportation, etc. For synthetic polymers, the price is strictly related to oil price, which in turn can be affected by EOR processes in use, creating a paradoxical situation, like a dog chasing its tail.

Polyacrylamides, belonging to this category, are the cheapest polymers used for EOR, but their price can indeed fluctuate depending on oil prices, while biopolymers such xanthan gum, have higher but more stable prices. Hydroxy ethyl cellulose (HEC) and carboxy methyl cellulose (CMC) are relatively cheap biopolymers, compared to xanthan gum [6].

When new polymers are proposed for chemical EOR, one should always consider if the improved recovery surpasses or at least balances out the increased costs for production, which is normally very difficult to evaluate.

2.3.2 Filtration properties

Suitable polymers for chemical EOR should not cause wellbore plugging. This is why polymers proposed for EOR are usually subject to filtration tests of various types [6].

Issues can be generated by a not proper dissolution of the polymer: if the hydration is not efficient, structures such as "fish eyes" or "gel balls" can form, potentially causing near wellbore plugging. This can be avoided by slow addition and vigorous agitation. Polymer cross-linking can be a further cause of plugging. Problems can be minimized by using good quality water and filtering the solution prior to injection, even though this increases operation costs. Enzyme clarification and diatomaceous earth filtration are now being implemented for treatment process of polymer solutions prior to flooding.

2.3.3 Viscosity

The general theoretical background about viscosity is given in Section 2.2.3. Polymer flooding has proven particularly effective for oils with medium viscosity, typically lower than 150 cP [4, 7]. It is generally believed that polymer flooding is only suitable for oils with a viscosity not higher than 100 cP, while for heavy or extra-heavy oils, thermal methods or miscible solvents methods seem to be better [11]. However, applications of the latter to many reservoirs is restricted by technical, economical, and environmental issues. For this reason, and because polymer flooding still remains the most practical EOR method, it is of particular research interest in the investigation of possible implementation of polymer flooding processes for heavy oils. Core flood experiments have been performed to evaluate the effectiveness of polymer flood for oils with high viscosities, in the range 200–8400 cP, which has been recently reviewed [11]. A very important finding from a set of experiments performed by Wang and Dong on oils with viscosities from 430 to 5500 cP [27], is the existence of a S-shape region in the viscosity/recovery profile (Fig. 2.5). This defines a range within which tertiary oil recovery by polymer flooding increases significantly with an increase in the viscosity of the polymer solution. On the other hand, for the polymer viscosity outside of this region, incremental oil recovery changes slightly with polymer viscosity.

A master curve has been built with this data by Guo et al. [28], where the dependence on the oil viscosity is normalized via the mobility ratio M (Fig. 2.6).

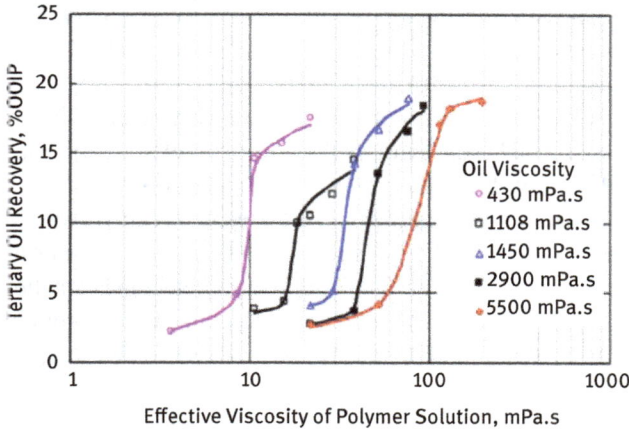

Fig. 2.5: Tertiary oil recovery as a function of polymer solution viscosity at different oil viscosities. Adapted from [27].

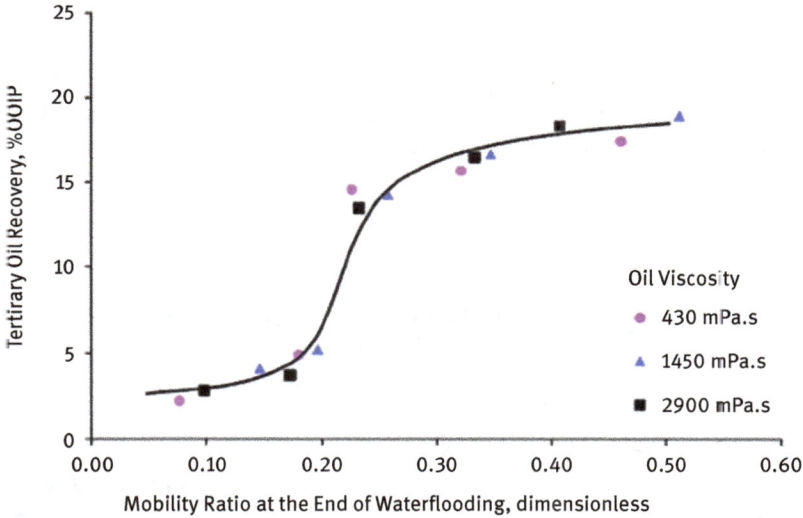

Fig. 2.6: Normalized relationship between tertiary oil recovery and oil/water mobility ratio. Reproduced from [28].

Analogous observations were made by other researchers for oils with viscosity as high as 8400 cP. The main overall conclusion is that in order to have an effective polymer flooding, a threshold value of viscosity (and thus of polymer concentration) must be reached [11].

As it is known, the polymer solution viscosity can be increased by increasing polymer concentration, polymer molecular weight, or reducing salinity. However, increasing the polymer solution viscosity can have severe negative effects in injectivity. As a matter of fact, real field polymer flooding operations in heavy oil reservoirs have been conducted with solution viscosities in the range of 12–100 cP [11].

Keeping all other parameters constant, a higher molecular weight polymer achieves a higher viscosity polymer solution, more permeability reduction, and consequently better mobility control. But if the molecules are too large, they may block the pores. In other words, there exists an optimum molecular weight of polymer for a particular formation that must be determined from laboratory test results. As a rule of thumb, the experiments suggests that if the radius of a polymer molecule is smaller than one-fifth of the mean pore throat size of a reservoir, pore plugging will not occur [11, 29]. This emphasizes the importance of the polymers structures in determining their efficiency in enhanced oil recovery.

2.3.4 Interactions with surfactants

This is relevant for SP and ASP flooding, and will not be considered in this book. Some comprehensive reviews of these EOR methods can be found in recent literature [19, 30]. Interactions between polymers and surfactants during a chemical flooding process can be really complicated. In general, the surfactant can influence polymer adsorption on the rocks, polymer solubility, and the solution viscosity.

2.3.5 Salinity resistance

Salinity is known to have a big influence in polymer flooding. This is particularly true for polymers containing charged moieties, as in the case of HPAM, which is *defacto* a copolymer of acrylamide and acrylic acid [2]. Since acrylic acid is a weak acid, it exists in water solution in both associated and dissociated forms, which is negatively charged. HPAM is therefore a polyelectrolyte.

Due to Columbic repulsion and osmotic effects, the charged acrylate groups cause the polymer chains to extend in solution, which has a positive effect on its hydrodynamic volume and therefore it increases the viscosity [31]. The presence of salts dissolved in water has a shielding effect, which reduces the viscosity. The higher the degree of hydrolysis of the HPAM, the more pronounced this effect [6]. The chemical nature of the ions also has an influence on polymer flooding. At the same concentration, ions with a higher multiplicity of charges have a more pronounced effect, due to higher ionic strength. Hard cations such as Ca^{2+} and Mg^{2+} can also cause the precipitation of the polymer, for example in case of HPAM, with subsequent loss of viscosity and negative effects on EOR. On the other hand, some polyvalent cations can act as binders for polysaccharides, causing partial plugging, that can have positive effects for EOR [6]. Generally speaking, in an EOR process, for low salinity/hardness brine it is recommended the use of HPAM, while for high salinity/hardness the use of biopolymers is preferred [1].

Salinity can also affect adsorption of anionic polymers on reservoir rock surface, that occur by several types of interactions, such as hydrogen bonding, hydrophobic interaction, ion binding, electrostatic interaction, and Van der Waals forces. Rocks are themselves constituted by ionic species, with the possibility of increased interaction and adsorption.

2.3.6 Thermal and mechanical stability

High temperatures can be reached in a reservoir. As a result, all chemical reactions are accelerated, including hydrolysis. Hydrolysis of HPAM causes amide group conversion to carboxylic acid groups, with increased salt sensitivity, which causes a loss of viscosity (see the previous section). The glyosidic bond of polysaccharides is also susceptible to hydrolysis, which in this case causes a drastic reduction in molecular weight and then viscosity.

The rate of hydrolysis for HPAM seems to be reasonably low at 50 °C, with polymer solutions being stable for months, even in the presence of a high concentration of divalent cations, but it increases rapidly above the 60–70 °C. At 90 °C the hydrolysis is very fast and polymer solutions lose viscosity and precipitate very quickly in presence of divalent cations [6].

Another important factor is the mechanical stability of polymeric chains. When polymer solutions are forced through the rock pores they experience strong shear forces, which can cause mechanical degradation of the polymer. Intuitively higher molecular weight polymers are more sensitive to this kind of degradation and this is confirmed by experiments [6].

2.3.7 Injectivity, retention and polymer adsorption

Injectivity can be an issue in EOR process, especially for low-permeability reservoirs, where high molecular weight polymers are likely to cause plugging of the pore throats and further reduce permeability, causing lower injectivity. This, in the end, makes it harder to maintain pressure in low-permeability reservoirs.

The effective permeability can be reduced also by chemical adsorption of the polymers in the pore throats.

It can be noted here, however, that polymer adsorption can also vary the wettability of the rocks. This can increase oil mobility, having a positive effect on oil recovery [6].

2.3.8 Microbial and chemical degradation

Besides the already mentioned hydrolysis reactions – favoured by high temperatures – and mechanical degradation, a series of biological and chemical transformations can occur in the polymers employed for EOR processes.

Microbial degradation is an issue particularly for biopolymers such as xanthan gum. These are obviously susceptible to degradation by bacteria, fungi, yeasts and other microorganisms possibly living in the reservoir. When the polymer is stocked as a solution, biological degradation can start occurring even at this stage. To prevent this, biocides such as formaldehyde or sodium azide are added to the solutions. However, this causes environmental concerns and the use of biocides should be avoided in field applications.

Synthetic polymers such as HPAM are much less affected by microbial degradation, but some microorganisms can still use them as a source of nutrients.

Chemical degradation can occur mostly by the action of oxygen or by non-neutral pH levels.

Oxygen, especially in presence of catalysts such as Fe^{2+} ions, can trigger formation of radicals which eventually may lead to de-polymerization or other side reactions.

Acidic or basic pH can accelerate hydrolysis and alter the ionization degree of polyelectrolytes causing more rock adsorption, with related problems as discussed before.

2.3.9 Conclusions on selection criteria

Many of the factor discussed in this paragraph are inter-connected. Often one structural parameter influences others, adding extra challenges in the design of polymers for EOR applications. For example, salinity can simultaneously affect time viscosity, polymer adsorption and chemical degradation. all this can have an effect on polymer injectivity, and so on.

Also the reservoir characteristics (homogeneity, sandstone versus carbonate, reservoir temperature, etc.) are important in the selection of the right polymer to use. For example, for sandstone and clay reservoirs, which are negatively charged, the injection of anionic macromolecules is obviously preferred to limit ionic interactions [1]. Here we have decided to focus on the polymers, thus we will not examine the reservoir characteristics, but they will be mentioned in relation to polymer structure where necessary.

When polymers are evaluated by researchers for EOR applications, these aspects should be taken into account. In the following sections we present the most used polymers for EOR and the most recently proposed ones, emphasizing their chemical characteristics, in relation to their applicability to EOR, based on the discussed selection criteria.

2.4 Currently used polymers for chemical enhanced oil recovery

Despite the many possible water-soluble polymers that are potentially viable for EOR applications, once the previously mentioned selection criteria are applied, only few candidates remain. As mentioned several times, the most used polymer for EOR is by far HPAM. A recent survey shows that HPAM is the polymer used in the vast majority of running chemical EOR projects [7].

As it will be shown later, many of the new polymers proposed for EOR are derivatives of HPAM. The next Section (2.5) describes in detail PAM and HPAM, while Section 2.6 includes their main derivatives proposed as improved systems for EOR.

After HPAM, the most used polymer for EOR application is a biopolymer, in particular Xanthan gum, a polysaccharide. Section 2.7 will illustrate Xanthan gum and other biopolymers used or proposed for EOR. The final section of this chapter will present miscellaneous alternative polymers proposed for EOR, not belonging to the previously mentioned classes.

2.5 PAM and HPAM

Partially hydrolysed polyacrylamide (HPAM) is a synthetic copolymer constituted by acrylamide (AM) and acrylic acid (AA), which is its hydrolysis product. The AA units are neutralized (often as sodium salt), thus the resulting polymer is anionic in nature and its pH in solution is slightly basic, due to the weak acid nature of the carboxylic groups. The chemical structure of HPAM is reported in Fig. 2.7. In the polymers used for EOR applications, the typical amount of AA units, which is referred to as the degree of hydrolysis, is between 25 and 35% and the molecular weight ranges between 4 and 30 million g/mol, as determined by intrinsic viscosity measurements [1]. The largest producer worldwide of HPAM is SNF in France.

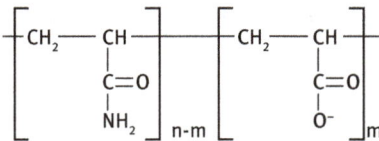

Fig. 2.7: Chemical structure of HPAM.

2.5.1 Synthesis

At the industrial level, HPAM can be produced either by free radical polymerization of polyacrylamide, followed by partial hydrolysis, or directly by co-polymerization of acrylamide and sodium acrylate [2].

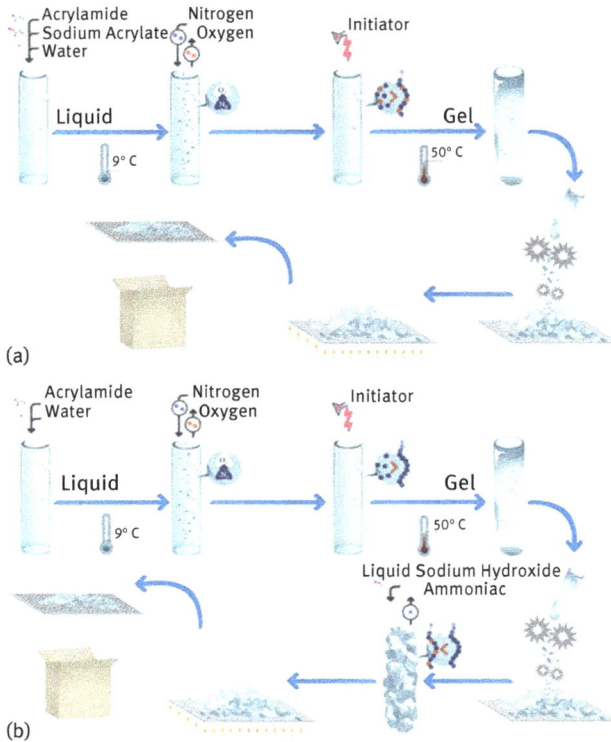

(a)

(b)

Fig. 2.8: Synthesis of HPAM by a) post-hydrolysis and b) co-polymerization. Adapted from [1].

In the first case (Fig. 2.8a), acrylamide is polymerized in water, upon removal of oxygen and the introduction of a radical initiator. The obtained gel is ground up, and the hydrolysis step is performed with sodium hydroxide or ammonia. The product is then dried and stored. When the polymer is produced by co-polymerization, the hydrolysis step is not necessary, thus the polymerization is followed by grinding of the obtained gel and drying (Fig. 2.8b).

2.5.2 Molecular structure

Depending upon the manufacturing process, the structure of the polymer will differ, resulting in polymers with different properties. For example, a different charge distribution can be expected. Post-hydrolyzed polyacrylamides are composed of a wide range of chains, some being highly charged and others less charged. The copolymerization method produces a polymer with a more even charge distribution along the backbone [1]. These properties are of paramount importance for the behaviour of the polymers in an aqueous solution, especially in the presence of divalent cations such as calcium and magnesium.

Not much can be said about the molecular weight distribution of commercial PAM and HPAM. Values of molecular weight can be obtained by intrinsic viscosity measurements, but more direct analysis such as gel permeation chromatography (GPC) are not possible or reliable in these cases [1]. Note that free-radical polymerization processes do not allow any control over the polymer structure and architecture. Most likely, the obtained polymers possess very large molecular weight distributions, and extensive branching and cross-linking can be expected.

Laboratory scale studies are customarily performed with the aim of evaluating the effects of the HPAM structure and composition on the recovery efficiency [11]. The general approach used for these studies is the preparation of copolymers with different structures, the measurement of the relevant properties of the obtained polymers, and the evaluation of the performances in core flood experiments, possibly simulating typical reservoir conditions. A recent good example of this kind of investigation is represented by the quite extensive work of Riahinezhad et al. [32–37]. They systematically studied the influence of several parameters in the kinetics of aqueous free radical copolymerization of AA and AM, such as monomer concentration, pH and ionic strength, with the aim of providing the tools for the synthesis of polymers with tailored characteristics for EOR applications. By carefully evaluating the reactivity ratio for the monomers, the authors were able to produce HPAM with tailored molecular weight, degree of hydrolysis and monomers distribution [37]. In follow-up works, they evaluate the relevant rheological properties [35] and performances in core flood experiments with heavy oil [36]. The studied polymers had molecular weights between 4 and 9 million g/mol and degree of hydrolysis between 8 and 33%. The main finding seems to be that the polymers with higher molecular weights experience higher retention in the sand-pack cores, with subsequent worse performance in oil recovery. This study is quite empirical in nature and, even though the differences in molecular characteristics and resulting properties are evaluated, it does not attempt to correlate them with oil recovery efficiency.

Molecular weight and the degree of hydrolysis affect the most relevant property of HPAM for its use in EOR, which is its thickening ability (see Section 2.2). The correlation between molecular weight and viscosity in solution is quite straightforward: with other parameters being constant, viscosity increases with molecular weight [12]. As typical for high molecular weight polymers, HPAM solutions show non-Newtonian viscosity profiles. although the HPAM solutions display pseudoplastic behavior (shear thinning) in simple viscometers, it has been demonstrated that these solutions show pseudodilatant characteristics (shear thickening) in porous media as well as in viscometers at relatively high shear rates [2]. The pseudoplastic behaviour is related to uncoiling of polymer chains and the disentanglement between separate polymer coils due to shear, thus it is expected to be more pronounced for higher molecular weight HPAM [2] and possibly the ones with more branched structures. However, accurate structural information about the polymer is usually not available, thus no direct ob-servations can be made.

As we have also observed in Section 2.2.4, it is often speculated that the viscoelas-ticity of polymer solutions might play a relevant role in the recovery mechanisms, by affecting *in-situ* viscosity or even by reducing residual oil saturation [38]. It is well known that viscoelastic behaviour is influenced by the molecular characteristics of the polymer, including molecular architecture. Since viscoelastic behaviour arises from molecular entanglement in solution, linear polymeric chains will exhibit different vis-coelasticity than more branched ones. This makes it interesting to study the effect of the molecular architecture of polymers in EOR processes. Unfortunately, these kind of studies are hampered by the fact that the vast majority of PAM and HPAM eval-uated for EOR applications are prepared via free-radical polymerization, which pro-duces polymers with undefined and non-homogeneous architectures. Wever et al. per-formed the synthesis of PAM with well-defined non-linear architectures, achieved via controlled radical polymerization methods, and tried to correlate their molecular ar-chitecture with viscoelasticity of water solutions [39–41]. They compared a series of star-like PAM with comparable arm lengths and observed that polymers with a low number of branches have lower viscosities than the analogous linear ones, but the more highly branched polymers possess a more pronounced viscosifying effect and more pronounced shear thinning. More recently, various long chain branched PAM were synthesized by RAFT (reversible addition-fragmentation chain-transfer polymer-ization) and the influence of viscoelastic effects on oil recovery was systematically in-vestigated [42, 43]. It was found that branched polymers possess higher robustness under shear and higher salt tolerance than their linear analogues. They perform ap-proximately 3–5 times better than their linear analogues of similar molecular weight.

The use of controlled radical polymerization methods for large-scale production of polymers is still quite limited [44], however they represent a great tool for the study of the structure-properties relationship of polymeric products, enabling the design of new products for desired applications, such as EOR [45–47].

2.5.3 Degree of hydrolysis

Besides the molecular architecture, one of the most important features in determin-ing the polymer's properties is the degree of hydrolysis. The presence of hydrolysed units in HPAM introduces charges into the structure, which influences several charac-teristics. In line with the general theory of polyelectrolytes solutions, the presence of electrostatic charges along a polymer backbone is responsible for stretching (due to electric repulsion) of the polymeric chains in water which results in a higher viscosity than the corresponding uncharged analogue, as a consequence of the increased hy-drodynamic radius [2]. It has indeed been observed that increasing the relative amount of AA units in the polymer with respect to the neutral AM units, causes an increase in shear viscosity. However, higher amounts of AA in the monomer feed ratio also causes a decrease in the molecular weight of the polymer. As a result, there seems to be an optimum viscosity at around a 30% content of AA [35].

It is known that the viscosifying effect of polyelectrolytes is greatly influenced by the type and amount of salts dissolved in the water used to prepare the solution. In particular, the presence of salt causes a decrease in viscosity of polyelectrolytes solutions. This is attributed to the shielding effect of the charges leading to a decreased electrostatic repulsion and therefore to a reduced expansion of the polymer coils in the aqueous solution. This results in a lower hydrodynamic volume and consequently a lower viscosity. Therefore, to reduce the extent of the salt sensitivity, the charge density on the polymer backbone cannot be too high.

A further complication comes from the presence of divalent cations such as Ca^{2+} and Mg^{2+} or other polyvalent cations. These are known to cause aggregation and eventually precipitation of HPAM, because of their ability to bridge carboxylic groups. This is of course a problem because it causes a loss of viscosity and pore plugging.

As briefly discussed before, the presence of charged moieties in the polymer also influences its adsorption on the rock surface during flooding. HPAM being anionic, it will be preferentially adsorbed on those surfaces bearing an excess of positive charges, present in both limestone and sandstone reservoirs [4, 17, 19]. However, adsorption of HPAM is very high in carbonate reservoirs when compared to sandstone reservoirs, because of the strong interaction between carboxylate anion and Ca^{2+} cations [4].

The degree of hydrolysis also has an influence on the shear-thinning region, with this interval being reduced by the increase in the degree of hydrolysis. This can be attributed to the increased rigidity of the polymer structure with increased charge density [2].

Another effect of the high shear experienced by HPAM during EOR processes is possible mechanical degradation. HPAM is sensitive to shear degradation because of its long and flexible chains [4]. The incorporation of more AA groups increases chain stiffness due to charge repulsion. Thus, HPAM become less sensitive to shear degradation as the degree of hydrolysis increases. also, longer chains are more flexible and more susceptible of degradation, thus high molecular weight HPAM are more sensitive to shear degradation when compared to low molecular weight ones.

2.5.4 Degradation

The viscosity of HPAM decreases with temperature, as it is the case for most polymer solutions. However, the main effect of temperature on HPAM solution viscosity during polymer flooding, derives from the significant influence of temperature on polymer stability, in particular towards further hydrolysis, as discussed before (see Section 2.2).

The hydrolysis rate of HPAM is dependent on pH, salinity and temperature. Up until 50 °C HPAM is rather stable towards hydrolysis, but the rate increases significantly at 70 °C and becomes very rapid at 90 °C [4]. Especially when high salinity brines are used, this almost always results in a decrease in performance, due to a lowering of viscosity and partial precipitation of the polymer.

As observed, a certain degree of hydrolysis is desirable, but if the hydrolysed fraction increases too much during polymer flooding, salting out effects and adsorption on the rock surface become the predominant phenomena, with negative effects on the recovery.

Hydrolysis is the most relevant chemical degradation for HPAM, but it is not the only one. Another cause of chemical degradation is the presence of free radicals, which can cause chain scission, depolymerisation, and chain transfer reactions, ultimately leading to a decrease in molecular weight [1, 48].

Free radicals can be generated from various sources, including molecular oxygen. Again, high temperatures favour the formation of free-radical, accelerating degradation.

Overall, considering all the mentioned effects, application of HPAM is limited to 75 °C in the presence of divalent cations and can be extended to 100 °C in the presence of negligible amounts of divalent cations (< 200 ppm). Unfortunately, most of the reservoirs with residual oil have more hostile levels of temperature and salinity [4].

Mechanical degradation is another relevant phenomena for PAM and HPAM [1]. Shear forces and flow through narrow pores can cause chain scission due to mechanical stress, proportional to the molecular weight of the polymer. Shear forces are strong near injection, thus mechanical degradation occurs already in the first stages of polymer flooding.

Biological degradation is generally not considered relevant for PAM and HPAM, but it is for biopolymers, and will be treated later in this chapter.

2.6 Modified PAM and HPAM

From the overview given in the previous section, it is apparent that, even though HPAM is the most used polymer in EOR processes, it suffers from several limitations, especially for high-temperature, high-salinity reservoirs. This is why the research into new polymers for EOR is still very active. Due to the prominent role of HPAM in this respect, the proposed alternatives are mostly based on modifications of PAM or HPAM structures, designed to cope with some of the presented limitations. Many approaches have been used. The most common method to extend the application of PAM and HPAM is copolymerization of AM and AA with suitable monomers that can introduce some advantages. For example, incorporation of N-vinylpyrrolidone up to 50% in HPAM provides better shear and thermal stability [1, 49, 50].

The studied systems can be roughly divided into three categories, which will be treated separately in the following sections. The first category consists of those copolymers or ter-polymers synthesized by incorporating another charged unit. These monomers, compared to AA, are more resistant to chemical degradation, more resistant to charge shielding, and sometimes can sterically hinder the polymer chain to keep the hydrodynamic radius at a reasonable value at high salinity [4]. The second category includes those polymers synthesized by incorporating a hydrophobic monomer, which provides intermolecular association with subsequent enhancement of the viscosity. The third category consists of PAM-based thermo-viscosifying polymers having a thermo-responsive monomer on the main chain. Many modified polymers based on HPAM (acrylamide, acrylic acid plus one or more acrylate or vinyl monomers) have been tested or proposed for EOR. Extensive lists can be found in previously published reviews [2, 4]. Since the method used to prepare these polymers is based on free-radical polymerization, the monomer distribution is most likely random and there is little or no control over the structure and molecular weight of the produced polymers. The synthesis is easy to implement but it usually increases costs because the additional monomer is more expensive than AA or AM so the benefits should obviously compensate for the increased production expenses.

2.6.1 HPAM modified with charged monomers

Even though the presence of AA in HPAM provides a higher viscosity because of electrostatic repulsions, it also introduces problems, because the carboxylic group has the ability to strongly bind cationic species such as Ca^{2+} and Mg^{2+}, causing precipitation and increased adsorption on the rock surfaces, especially in carbonate reservoirs.

In order to minimize this effect without losing the advantages of using polyelectrolytes, AA can be replaced (partially or completely) with another charged monomer. This strategy has been used in the design of polymers for EOR. A list of charged monomers typically incorporated in PAM for EOR applications is shown in Fig. 2.9. For a more complete list and related references the reader can refer to the review by Wever et al. [2].

The incorporation of sulfonate monomers such AMPS, (also abbreviated as ATBS) or styrene sulfonate to the polymer backbone improves its tolerance to salinity, especially calcium cations, because of the lower tendency of sulfonate groups to bind such cations. In most cases, the temperature stability is also improved [1, 3, 4, 6].

The problem of adsorption in carbonate reservoirs can be solved by using cationic PAMs. Synthesis of HPAM with incorporated cationic monomers or cross-linker, and their performances in core flood in chalk pack have been reported [51]. Often, the monomers used are amphiphilic in nature (so called "surfmers") and the resulting polymers are also surface active, which has beneficial effects in terms of wettability alteration and oil emulsification [51, 52].

Various zwitterionic monomers have been incorporated in PAM or HPAM. The presence of charged moieties is beneficial because it increases water solubility and viscosity, the latter due to the electrostatic effects discussed previously. The presence

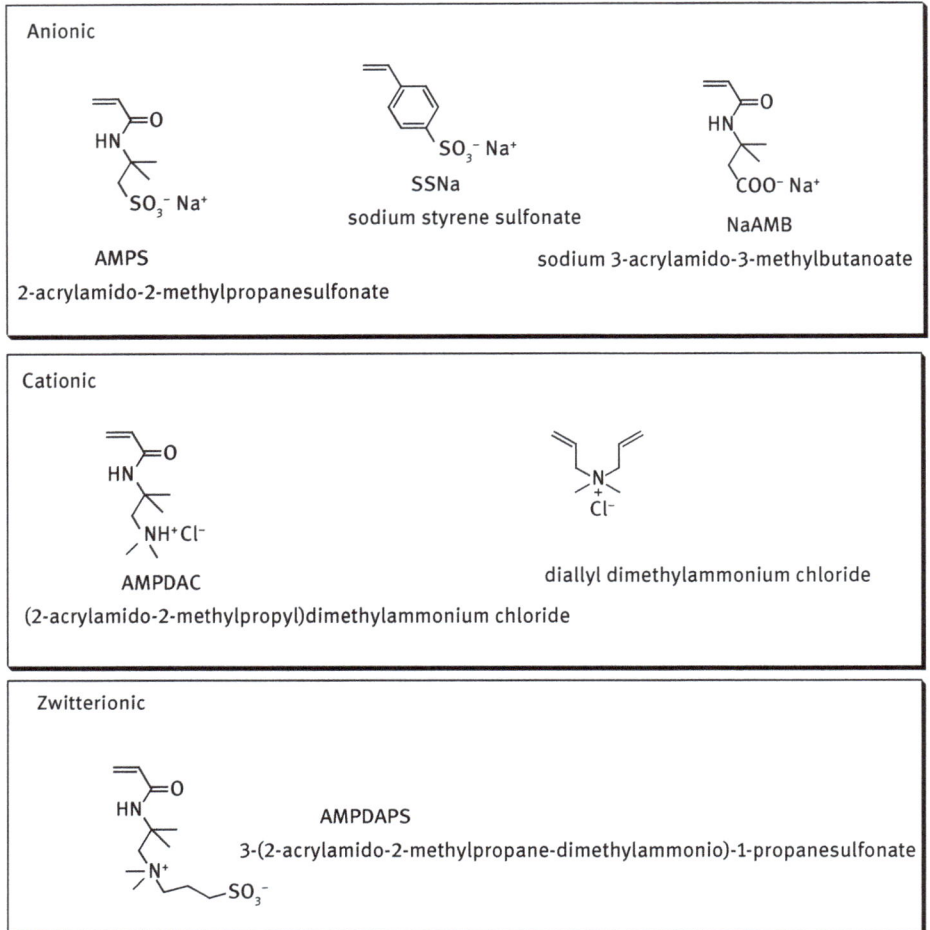

Fig. 2.9: Charged co-monomers typically incorporated in HPAM for EOR applications.

of zwitterionic monomers in the polyelectrolyte chain should provide a better resistance to salinity and pH, with potential beneficial effects for EOR applications. The addition of a zwitterionic monomer into PAM can also provide thermo-thickening behaviour [2].

2.6.2 Hydrophobically associating HPAM

Various hydrophobically associating (or hydrophobically modified) HPAM also received attention as polymers with improved characteristics for EOR [2, 13, 14, 17, 53]. This class of polymers is probably the most investigated for EOR purposes.

~ HPAM chain
● hydrophobic group

Fig. 2.10: Intermolecular hydrophobic association in HPAM.

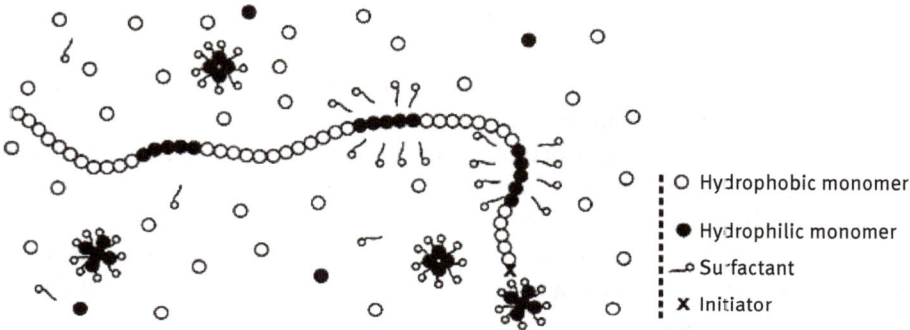

O Hydrophobic monomer
● Hydrophilic monomer
—o Surfactant
✗ Initiator

Fig. 2.11: Schematic representation of synthesis of hydrophobically modified HPAM in emulsion.
Adapted from [2].

The main idea behind the use of such polymers is that the presence of hydrophobic groups allows intermolecular aggregation [2], as schematically shown in Fig. 2.10, with a subsequent thickening effect. Moreover, due to the amphiphilic character of these polymers, they have the ability to lower the surface tension of water and stabilize emulsions of the oil. This can contribute to better performances in oil recovery [17].

The synthesis of hydrophobically modified PAM or HPAM can be performed via free radical polymerization, analogously to the non-modified versions of the polymer [2]. However, while PAM and HPAM can be prepared in homogeneous aqueous conditions, this is obviously not possible when hydrophobic monomers (which are, by definition, water-insoluble) are incorporated in the HPAM structure. In these cases the polymers can still be prepared in homogeneous conditions in the presence of a co-solvent, such an alcohol, or more frequently, via emulsion polymerization. The latter makes use of a surfactant to generate micelles containing the hydrophobic monomer, while the hydrophilic monomers and the initiator are in water. The final structure of the polymer is obviously affected by the polymerization method used. Polymerization in homogeneous solution affords random copolymers, while emulsion polymerization gives a "multi-block" structure. This is inherent to the method itself, as shown schematically in Fig. 2.11: chains of the hydrophilic monomers grow in water, generating hydrophilic blocks; when the growing radical meets a droplet of hydrophobic monomer, the polymerization proceeds in the new environment, where mostly hydrophobic monomers are incorporated, until the monomer is consumed and the radical keeps growing in water with the hydrophilic monomers.

It has been reported that hydrophobically modified HPAM prepared with different methods (solutions versus emulsion), present sensibly different rheological properties [2, 54].

Various hydrophobic groups have been used to modify PAM and HPAM. Many of them are based on AM and/or AA monomers, substituted with alkyl chains or aromatic groups. Reference [2] provides a quite complete list of hydrophobically modified HPAM prepared and proposed for EOR. Associating polymers have been prepared by incorporating hydrophobic groups into the polymer after the polymerization process. The advantage of this approach is that commercially available polymers can be used as a starting material. A disadvantage is that reactions involving viscous polymer solutions are not easily carried out because of problems associated with the mixing and reaction homogeneity [13]. This approach is used mostly for the preparation of hydrophobically modified polysaccharides, that will be treated later.

A quite general approach found in the literature is to use a combination of AM with charged monomers (AA, AMPS) and hydrophobic ones (alkyl, aryl, perfluoroalkyl). The polymers are usually synthesized via free-radical emulsion polymerization, with the composition optimized for EOR purposes (mostly aimed at maximize the viscosity), then the rheological properties and performances in core flood experiments are evaluated [2, 17].

The great popularity of hydrophobically modified HPAM for EOR applications, especially for high-salinity, high-temperature reservoirs, is due to their unique properties, which can be summarized as follows [14]:

- in aqueous solutions, above a critical association concentration (C*), their hydrophobic groups develop inter-molecular hydrophobic associations in nano-domains, leading to a build up of a 3D-transient network structure in high ionic strength medium, providing excellent viscosity building capacity, remarkable rheological properties and better stability with respect to salts than the unmodified HPAM precursors;
- reduced interfacial tension at the solid/liquid interface, since hydrophobic moieties associate forming aggregates or micelles;
- wettability alteration due to an unusual adsorption isotherm;
- improved resistance to mechanical degradation under high shear stress such as those encountered in pumps and near the well bore area, since the physical links between chains are disrupted before any irreversible degradation occurs, also they reform and retain their viscosity upon shear decreasing;
- high resistance to physicochemical conditions (temperature, pH, and ion content) prevailing around the wells. Hydrophobically modified HPAM can show anti-polyelectrolyte behaviour (increase of viscosity with salinity, instead of decrease).

Several studies confirmed experimentally that hydrophobically modified HPAM possess higher viscosity, shear and temperature stability than HPAM [53, 55–57]. It has been proposed that the resistance to mechanical degradation comes from the fact that the hydrophobic Van der Waals interactions between chains are reversibly disrupted before any irreversible degradation can occur, and they reform when shear forces cease, thus high viscosity can be regained upon decreasing the shear [14, 58].

It has been observed in flooding experiments that hydrophobically modified HPAM present a high resistance factor (R_f) and residual resistance factor (R_{Rf}). This can be explained by the increased polymer adsorption, due to the presence of hydrophobic groups, which can cause permeability reduction with beneficial effects on mobility [58].

Particularly relevant is the amphiphilic character of hydrophobically modified PAM and HPAM, which makes them suitable as polymeric surfactants. The idea of using polymeric surfactants for EOR is interesting because it can combine the advantages of solution thickening and interfacial tension (IFT) reduction, without suffering the problems of SP and ASP flooding, such as differential adsorption, polymer-surfactant interactions and chromatographic separations in the porous rocks [17, 18].

It is worth noticing that it is normally reported that in order to have an effect on oil recovery, a surfactant should give ultralow interfacial tension (IFT) values (on the order of magnitude of 10^{-3} mN/m), while polymeric surfactants are usually only able to achieve moderate IFT reduction (in the best cases around 0.1 mN/m) [17]. Nonetheless, it has been recently shown that an surface active modified HPAM solution with a measured IFT of 0.1 mN/m can give ~5% more oil recovery than a conventional HPAM in a core-flood experiment, at even lower viscosity [59]. The authors of this work attribute the better performance to the emulsification properties of the polymer and claim that, even though the recovery is not as good as for an SP flooding, the addition of a small percentage of the hydrophobic monomer (which is not specified) could still be competitive because of the low costs of implementation. In some cases they proved to be even more effective than a SP formulation in sandpack core flood experiments, showing excellent temperature and shear resistance [60].

Interfacial properties of hydrophobically modified PAM and HPAM proposed for chemical EOR have been studied. As already mentioned, the IFT values obtained are not as low as for traditional low-molecular weight surfactants, but the polymer solutions are able to form stable emulsions with crude oil, which should result in better microscopic displacement efficiency [17].

In conclusion the improved performance in oil recovery by hydrophobically modified HPAM, compared to the non-modified version, can most likely be attributed to both rheological and interfacial effects [17].

2.6.3 Thermo-thickening HPAM

One of the main drawbacks of PAM and HPAM for polymer flooding is their sensitivity to temperature, which limits their use in high temperature reservoirs. As we have already discussed, hydrolysis, oxidations and various degradation reactions occur much faster at high temperature, causing a sharp decrease in viscosity. Moreover, the viscosity of water-soluble polymers it is known to generally decrease with temperature and HPAM it is not an exception to that (thermo-thinning behaviour) [61].

However, the high temperature can be turned into an advantage by making use of thermo-thickening (or thermo-viscosifying polymers), that are polymers which, contrarily to the usual behaviour, experience an increase in viscosity with temperature. Thermo-viscosifying polymers for chemical EOR based on PAM and HPAM have indeed been developed and tested [61–64]. In order to give thermo-thickening behaviour to PAM, chains of a polymer exhibiting a lower critical solubility temperature (LCST) have been grafted to it.

Below LCST the polymer is completely hydrophilic and exhibits the usual thickening ability of PAM solutions, but as the temperature increases above the LCST value, the grafted chains become insoluble and aggregate, causing an increase in viscosity in the same fashion as hydrophobically modified PAM (Fig. 2.12) [2]. Conveniently, various polymers from N-alkyl substituted acrylamide possess a LCST above room temperature [65].

In comparative core flood experiments, thermo-thickening PAM showed better recovery than PAM both at 45 and 85 °C [61]. However, the thermal stability can be an issue for these kind of polymers. Aging experiments showed that the viscosity of the

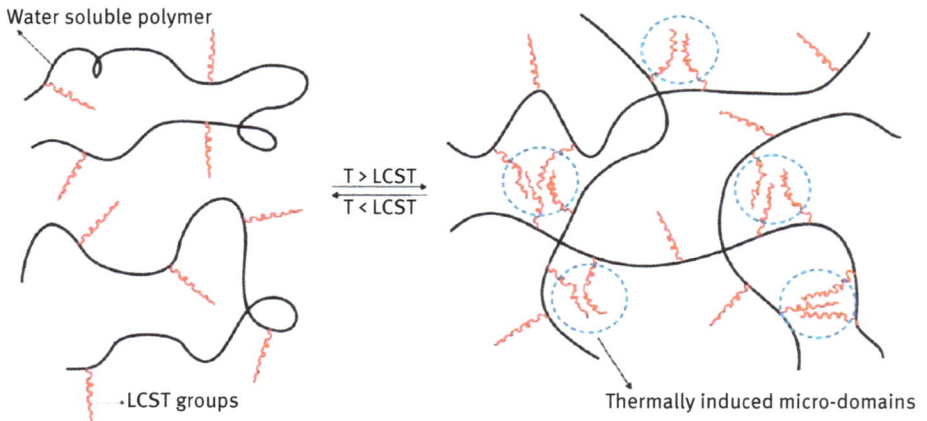

Water soluble polymer

T > LCST
T < LCST

·LCST groups

Thermally induced micro-domains

Fig. 2.12: Thermally induced aggregation at LCST. Adapted from [2].

thermo-thickening polymer decreases very quickly, even though the retained viscosity of the modified polymer is still high than the unmodified PAM after one month at both 45 and 85 °C.

2.6.4 Other PAM-derived polymers proposed for EOR

Even though the vast majority of the PAM-derived polymers used or proposed for EOR are included in one of the previously discussed categories, the search for new and more efficient polymers is still quite active. The recent scientific literature presents few other systems based on PAM as potential candidates for chemical EOR, some of which are worth mentioning because they introduce new concepts to improve oil recovery. The fact that the polymers are still based on PAM presents the clear advantage that they could be easily implemented in the already existing production processes.

One interesting concept, used recently by Zou et al., is the introduction of β-cyclodextrin (β-CD) as side group in the PAM-based polymer [66–69]. β-CD are cyclic oligosaccharides, characterized by a toroidal ring structure. The interior cavity of β-CD is hydrophobic, while the exterior shell is hydrophilic. Due to its specific architectural conformation, β-CD can selectively incorporate hydrophobic molecules of appropriate size as guests into its cavity to generate host/guest inclusion complexes [70].

The polymers proposed for EOR by Zou et al. are characterized by the incorporation of a vinyl monomer containing β-CD (which needs to be synthesized) on HPAM [66] or hydrophobically modified PAM [68]. These systems are able to form a network-like system in solution, alone or in combination with surfactants (Fig. 2.13). The authors suggest that when this polymer is used in combination with surfactants, the surfactant is trapped into the β-CD cavities during flooding, which prevents loss of surfactant by adsorption on the rocks; when oil is struck, because of the competition between oil molecule and surfactant in the cyclodextrin cavity, surfactant would be released, resulting in the increase of local surfactant concentration. This would cause a reduction of viscosity, due to disruption of intermolecular association and a decrease in interfacial tension and emulsification of the oil, with a subsequent improved recovery (Fig. 2.13) [68]. This mechanism is of course difficult to prove, nonetheless the author found an improved oil recovery in sand pack flood tests.

Slightly different approaches have been used by other researchers [69, 70]. It these cases β-CD are used to induce viscosity enhancement by association in two ways: 1) incorporating β-CD and adamantane (hydrophobic) moieties [69], obtaining a system similar to the one illustrated in Fig. 2.13; 2) mixing β-CD with HPAM and surfactants or hydrophobically modified HPAM, obtaining intermolecular association as described in Fig. 2.14 [70]. also these systems are able to give better performances in core flood experiments, when compared to HPAM.

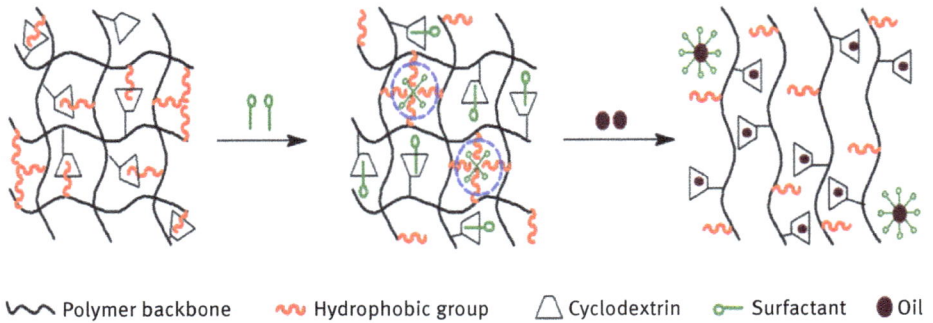

∿∿ Polymer backbone   ∿∿ Hydrophobic group   △ Cyclodextrin   ⊶ Surfactant   ● Oil

Fig. 2.13: Interaction of cyclodextrin-containing polymers with surfactant and oil. Reproduced with permission from [68].

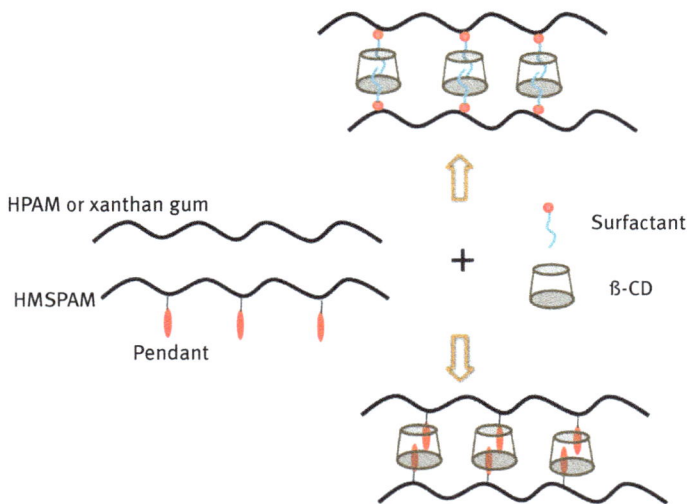

HPAM or xanthan gum

HMSPAM

Pendant

Surfactant

ß-CD

+

Fig. 2.14: Intermolecular association via cyclodextrin. Reproduced with permission from [70].

A very interesting concept introduced recently is the use of smart-covered polymeric particles (SCPP) [71]. The systems proposed here is consists of core-shell nanoparticles, with a core of high-molecular weight PAM covered by a thin layer of low-molecular weight polystyrene (Fig. 2.15). The particles are prepared by inverse emulsion polymerization of AM, followed by styrene, affording basically a nanoparticles colloid. The idea proposed by the author is that the smart polymer (SCPP) dissolves at the water-oil interface in three successive steps: 1) due to the hydrophobic behaviour of the shell layer during polymer flooding, the SCPP remains insoluble in the injected water until transported to the water-oil interface, thus the viscosity

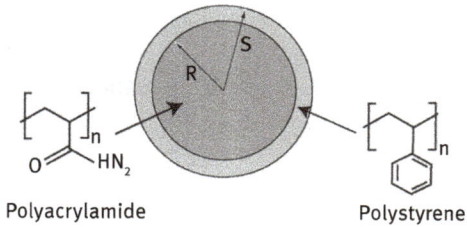

Fig. 2.15: Nanocapsules of PAM coated with PS obtained by emulsion polymerization. Adapted from [71].

Polyacrylamide

Polystyrene

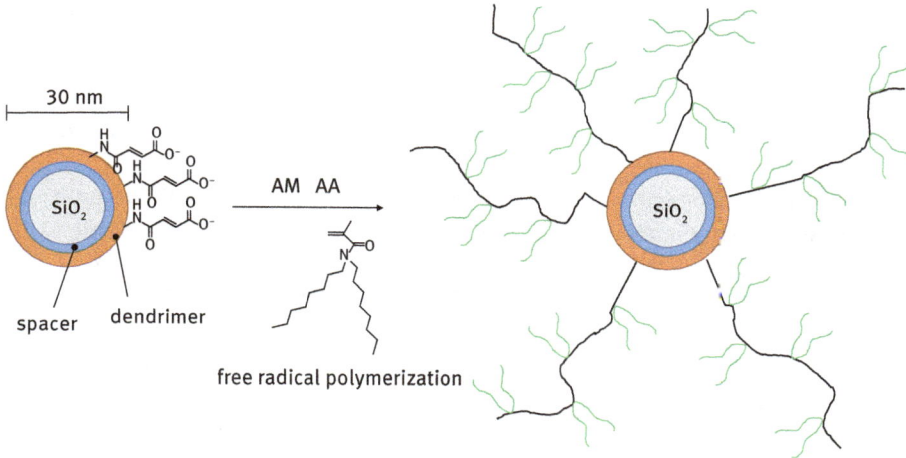

Fig. 2.16: Silica nanoparticles-grafted hydrophobically modified PAM [76].

remains low; 2) at the interface, the oil-soluble shell layer starts dissolving into the oil phase; 3) after full dissolution of the coating, the main hydrophilic core polymer (i.e., polyacrylamide) diffuses and dissolves in the aqueous phase, increasing its viscosity, particularly in the region of the interface. The concept has been tested in a glass-etched micromodel, with good results.

Another popular recent approach to new systems for chemical EOR, is the use of nano-hybrids constituted by opportunely modified HPAM grafted to various nanoparticles [72–76]. One recent interesting example is illustrated in Fig. 2.16 [76]. In this case, the silica nanoparticle core is surrounded by hydrophobically associative PAM grafts. This system shows a much higher viscosity than a corresponding HPAM at the same concentration, better stability, shear thickening behaviour and better recovery in core flood experiments.

However, in general these cannot be considered strictly as systems for polymer flooding, as the polymers here are used mainly as a means to stabilize the nanoparticles, which are the ones responsible for the oil recovery. Somewhat border-line systems are constituted by nano- or micro-spheres of cross-linked polymers made by inverse emulsion polymerization [77].

A later chapter will illustrate in detail the use of nanoparticles and nanotechnology, in EOR.

2.7 Biopolymers

The use of biopolymers for chemical EOR is certainly attractive for economic and environmental reasons. In fact, after polyacrylamides, the most used polymer for EOR is xanthan gum, a polysaccharide. High molecular weight polysaccharides have good viscosifying properties in water solutions, with the advantage over PAM of being bio-derived and biocompatible.

However, the biocompatibility of these polymers is not only their strength but also their weakness: microorganisms are always present in water, and are able to digest them, quickly reducing their effectiveness as water thickeners. To overcome this problem, biocides (e.g., formaldehyde, azides) are often added to the polymer solutions [1, 6]. Interestingly, utilization of biopolymer by resident bacteria inside the reservoir can lead to bio-plugging and increased oil recovery. The idea of using bacteria to improve oil recovery it is a separate EOR technique, which goes under the name of microbial EOR [78]. Besides xanthan gum, other polysaccharides have been proposed as biopolymer for EOR. They will be discussed in a separate section.

2.7.1 Xanthan gum

Xanthan gum is a polysaccharide produced through fermentation of glucose or fructose by different bacteria, such as *Xanthomonas campestris* [1–4, 13]. The addition of "gum" to the name refers to the fact that these polysaccharides are able to form highly viscous solutions at relatively low concentration, hence their use in EOR [4]. To date it's the only biopolymer significantly used in field applications [7]. The structure of xanthan is shown in Fig. 2.17 [2]. The backbone consists of a glucose unit, not much different from cellulose; however, the side chain consists of β-D-mannose-1,4-β-D-glucuronic acid-1,2-α-D-mannose and the terminal mannose is normally in the form of a 4,6- linked pyruvic acid ketal [4]. X-ray diffraction studies showed that the xanthan backbone has a helical structure where the side chains fold down along the helix. The average molecular weight of xanthan gum used in EOR processes is from 1 to 15 million [3].

Xanthan gum has a more rigid structure than PAM and it is more resistant to shear degradation.

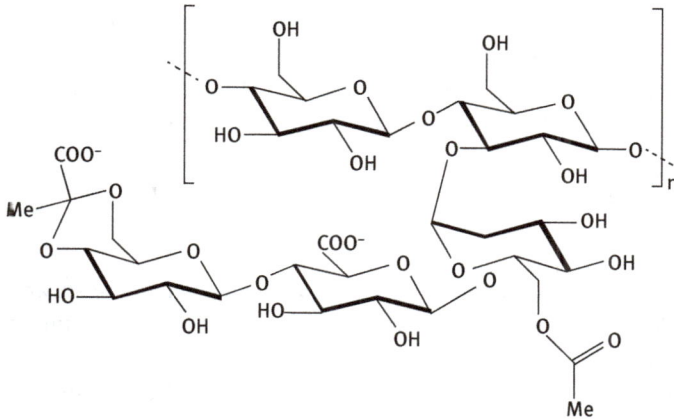

Fig. 2.17: Chemical structure of Xanthan gum.

In water solutions, xanthan gum has non-Newtonian pseudo-plastic behaviour, showing shear thinning. The higher shear resistance, when compared to HPAM, can be attributed to the rigid or rod-like structure of xanthan. The viscoelastic behaviour of HPAM is due to chain entanglement, and on applying high shear HPAM chains are unable to relax fast and breakage may occur. On the contrary, at high shear rigid rod-like chains of xanthan tend to align with the flow field with minimum breakage. A highly viscous solution is obtained due to the helical conformation and it is stabilized by the formation of hydrogen bonds between the backbone and the side chains. At high concentrations xanthan solutions can even form a liquid crystalline phase [4].

Moreover, Xanthan is much less sensitive to salinity and particularly to the presence of divalent cations with respect to HPAM. Screening effects typical for polyelectrolytes are present also in this case, due the presence of charged groups on the side chains, but to a much limited extent.

These constitute clear advantages over HPAM, especially in high salinity and low permeability reservoirs. Disadvantages of xanthan are higher cost, high susceptibility to biodegradation and potential for injectivity problems due to cellular debris remaining from the manufacturing process. Xanthan is used extensively in drilling fluids because it is not subject to shear degradation. In general, polyacrylamide is efficient at increasing viscosity up to about one mass percent sodium chloride, while xanthan gum is more efficient at higher salt concentrations [13]. It is relatively compatible with most surfactants and other injection fluid additives used in tertiary oil recovery formulations [5].

In general, xanthan gum seems to be more temperature stable than PAM and HPAM: its stability seem to be excellent up to 70 °C and still acceptable at 80 °C, while degradation starts to be significant above 90 °C [2–4]. Xanthan gum adsorption on rocks is salinity and pH dependent, similarly to HPAM.

Overall, xanthan gum injection is more effective than HPAM under higher salinity reservoir conditions, as it has proven in comparative flooding experiments [79].

2.7.2 Other biopolymers

Xanthan gum is not the only polysaccharide proposed for EOR. Scleroglucan and schyzophyllan, non-ionic analogous polysaccharides obtained by the fermentation of fungi, have been tested for EOR [1, 4]. The non-ionic nature of the polymers makes them salt resistant, and their helix structures are semi-rigid in aqueous solution. As a consequence, the viscosity, shear resistance and thermal resistance of these polymers are high in aqueous solution, which makes them suitable candidates for polymer flooding. Welan gum and guar gum are similar systems also proposed for EOR [1].

Hydroxyethlcellulose (HEC) has actually been used in at least one EOR project [3]. It is produced by the reaction of cellulose with ethylene glycol, to obtain a more water-soluble form [3, 4]. It actually gives less injectivity problems than xanthan gum. HEC has good salinity resistance, due to its non-ionic nature, but it does not have a helix structure in solution, therefore its viscosity decreases significantly with temperature and its thermal stability is in general lower than xanthan gum. For these reasons, HEC is suitable for polymer flooding in high salinity but low temperature formations. Chemically cross-linked nanoparticles of HEC, obtained by reaction of the biopolymer with divinyl sulfone, have also been proposed for EOR [80].

Hydrophobic modification of HEC and other polysaccharides has also been proposed for EOR [2, 13]. The hydroxyl side groups present on the polysaccharides chains can be modified rather easily by reaction with organic halides. The purpose is logically the same as for hydrophobically modified HPAM, to increase viscosity via association in solution. However, these systems present many drawbacks, such as low viscosity at low concentrations and poor thermal stability, and have not been exploited significantly [13].

Polysaccharides bearing charged groups, such as carrageenan (an anionic polysaccharide containing sulfonate groups) [81] and cationic starch [82, 83] have also been proposed as thickeners for EOR.

Another approach, based on grafting of PAM chains on polysaccharide backbones has also been attempted for EOR purposes [84–86]. For example starch-graft-poly(AM-co-AMPS) was synthesized and tested for EOR, showing better results than HPAM in oil recovery, attributed to much better resistance to temperature and shear [84].

2.8 Other miscellaneous polymers

Besides PAM-derived polymers and biopolymers, other synthetic polymers have been proposed for chemical EOR. Not much remains after excluding the previous mentioned categories, however the scientific literature still contains few systems that we will mention in this paragraph. Most of these systems can be classified as polymeric surfactants and have been reviewed recently [17]. The general idea is to obtain good viscosifying properties and interfacial activity reduction, in the same direction of what it is accomplished with hydrophobically modified PAM. Many of these systems are found in patent literature [17] and include statistic copolymers of styrene and maleimide derivatives, as well as block copolymers of styrene and acrylic acid. Some examples from open literature are copolymers of carboxymethylcellulose and an alkyl acrylate with a blocky structure [87], low molecular weight nonionic co-polyester surfactants [88]. The research group of the authors of this book recently proposed block copolymers prepared by ATRP (atom transfer radical polymerization) as a potential surface active viscosifiers for EOR [47].

2.9 Conclusions on polymer flooding

Polymer flooding is a very well established technique for chemical EOR. Even though it experienced a decrease of interest after booming in the 80s, connected with a drop in crude oil prices, it might become more popular again in the future, as the global energy demand become more and more difficult to handle. For this reason, parallel to the search for more sustainable sources of energy and chemicals, the scientific investigation of polymer flooding is still active, as described in this chapter. Here, a survey of old and new polymers proposed for chemical EOR, with their problems and advantages have been illustrated. The main target of the current research is to find suitable polymers for high-temperature high-salinity reservoirs, carbonate and low permeability rock formations. Studies aimed at elucidating more details of the mechanisms of polymer flooding are frequently performed, also with the aid of mathematical simulation, which will be illustrated in a following chapter of this book.

References

[1]   Thomas, A. Polymer Flooding. In Romero-Zeron, L., ed., *Chemical Enhanced Oil Recovery (cEOR) – a Practical Overview.* InTech, Rijeka, 2016. Ch. 02.
[2]   Wever, D. A. Z., Picchioni, F., and Broekhuis, A. A. Polymers for enhanced oil recovery: A paradigm for structure-property relationship in aqueous solution. *Prog. Polym. Sci.*, 36:1558, 2011.
[3]   Wei, B. Advances in Polymer Flooding. In El-Amin, M. F., ed., *V. and V. M.* InTech, Rijeka, 2016. Ch. 01.

[4] Kamal, M. S., Sultan, A. S., al Mubaiyedh, U. A., and Hussein, I. A. Review on Polymer Flooding: Rheology, Adsorption, Stability, and Field Applications of Various Polymer Systems. *Polym. Rev.*, 55:491–530, 2015.

[5] Abidin, A. Z., Puspasari, T., and Nugroho, W. A. Polymers for Enhanced Oil Recovery Technology. *Procedia Chem.*, 4:11–16, 2012.

[6] Rellegadla, S., Prajapat, G., and Agrawal, A. Polymers for enhanced oil recovery: fundamentals and selection criteria. *Appl. Microbiol. Biotechnol.*, 101:4387–4402, 2017.

[7] Standnes, D. C. and Skjevrak, I. Literature review of implemented polymer field projects. *J. Pet. Sci. Eng.*, 122:761–775, 2014.

[8] Thomas, S. Enhanced oil recovery – An overview. *Oil Gas Sci. Technol. L Inst. Fr. Du Pet.*, 63:9–19, 2008.

[9] Wei, B., Romero-Zerón, L., and Rodrigue, D. Oil displacement mechanisms of viscoelastic polymers in enhanced oil recovery (EOR): a review. *J. Pet. Explor. Prod. Technol.*, 4, 2014.

[10] Sheng, J. J., Leonhardt, B., and Azri, N. Status of polymer-flooding technology. *J. Can. Pet. Technol.*, 54:116–126, 2015.

[11] Saboorian-Jooybari, H., Dejam, M., and Chen, Z. Heavy oil polymer flooding from laboratory core floods to pilot tests and field applications: Half-century studies. *J. Pet. Sci. Eng.*, 142:85–100, 2016.

[12] Morgan, S. E. and McCormick, C. L. Water-soluble polymers in enhanced oil recovery. *Prog. Polym. Sci.*, 15:103–145, 1990.

[13] Taylor, K. C. and Nasr-El-Din, H. A. Water-soluble hydrophobically associating polymers for improved oil recovery: A literature review. *J. Pet. Sci. Eng.*, 19:265–280, 1998.

[14] El-hoshoudy, A. N. N., Desouky, S. E. M. E. M., Elkady, M. Y. Y., al Sabagh, A. M. M., Betiha, M. A. A., and Mahmoud, S. Hydrophobically associated polymers for wettability alteration and enhanced oil recovery – Article review. *Elsevier*, 26:757–762, 2017.

[15] Bai, M., Zhang, Z., Cui, X., and Song, K. Studies of injection parameters for chemical flooding in carbonate reservoirs. *Renew. Sustain. Energy Rev.*, 75:1464–1471, 2017.

[16] Seright, R. S. How Much Polymer Should Be Injected During a Polymer Flood? Review of Previous and Current Practices. *SPE J.*, 22:001–018, 2017.

[17] Raffa, P., Broekhuis, A. A. A. A. A., and Picchioni, F. Polymeric surfactants for enhanced oil recovery: A review. *J. Pet. Sci. Eng.*, 145:723–733, 2016.

[18] Sheng, J. J. Modern Chemical Enhanced Oil recovery: Theory and Practice. *Elsevier*, 2010.

[19] Olajire, A. A. Review of ASP EOR (alkaline surfactant polymer enhanced oil recovery) technology in the petroleum industry: Prospects and challenges. *Energy*, 77:963–982, 2014.

[20] Druetta, P. D., Picchioni, F., and Rijksuniversiteit Groningen. *Numerical simulation of chemical EOR processes.* University of Groningen, 2018.

[21] Mishra, S., Bera, A., and Mandal, A. Effect of Polymer Adsorption on Permeability Reduction in Enhanced Oil Recovery. *J. Pet. Eng.*, 2014:1–9, 2014.

[22] Teraoka, I. Dynamics of Dilute Polymer Solutions. In *Polymer Solutions*, pp. 167–275. Wiley, New York, USA.

[23] Dobrynin, A. V., Colby, R. H., and Rubinstein, M. Scaling Theory of Polyelectrolyte Solutions. *Macromolecules*, 28:1859–1871, 1995.

[24] Wang, D., Wang, G., Wu, W., Xia, H., and Yin, H. The Influence of Viscoelasticity on Displacement Efficiency–From Micro to Macro Scale. In *SPE Annual Technical Conference and Exhibition*. Society of Petroleum Engineers, 2007.

[25] Wang, D., Wang, G., and Xia, H. Large Scale High Visco-Elastic Fluid Flooding in the Field Achieves High Recoveries. In *SPE Enhanced Oil Recovery Conference*. Society of Petroleum Engineers, 2011.

[26] Wang, D. M. Development of new tertiary recovery theories and technologies to sustain Daqing oil production. *P.G.O.D.D.*, 20:1–7, 2001.

[27] Wang, J. and Dong, M. Optimum effective viscosity of polymer solution for improving heavy oil recovery. *J. Pet. Sci. Eng.*, 67:155–158, 2009.

[28] Guo, Z., Dong, M., Chen, Z., and Yao, J. A fast and effective method to evaluate the polymer flooding potential for heavy oil reservoirs in Western Canada. *J. Pet. Sci. Eng.*, 112:335–340, 2013.

[29] Wang, D., Liu, H., Niu, J., and Chen, F. Application Results and Understanding of Several Problems of Industrial Scale Polymer Flooding in Daqing Oil Field. In *SPE International Oil and Gas Conference and Exhibition in China*. Society of Petroleum Engineers, 1998.

[30] Sheng, J. J. A comprehensive review of alkaline-surfactant-polymer (ASP) flooding. *Asia-Pacific J. Chem. Eng.*, 2014.

[31] Raffa, P., Brandeburg, P., Wever, D. A. Z., Picchioni, F., and Broekhuis, A. A. Polystyrene-Poly(sodium methacrylate) amphiphilic block copolymers with different molecular architectures synthesized by ATRP: the effect of structure, pH and ionic strength on their rheological properties in water. *Macromolecules*, 46:7106–7111, 2013.

[32] Riahinezhad, M., Kazemi, N., McManus, N., and Penlidis, A. Optimal estimation of reactivity ratios for acrylamide/acrylic acid copolymerization. *J. Polym. Sci. Part A Polym. Chem.*, 51:4819–4827, 2013.

[33] Riahinezhad, M., Kazemi, N., McManus, N., and Penlidis, A. Effect of ionic strength on the reactivity ratios of acrylamide/acrylic acid (sodium acrylate) copolymerization. *J. Appl. Polym. Sci.*, 131:40949, 2014.

[34] Riahinezhad, M., McManus, N., and Penlidis, A. Effect of Monomer Concentration and pH on Reaction Kinetics and Copolymer Microstructure of Acrylamide/Acrylic Acid Copolymer. *Macromol. React. Eng.*, 9:100–113, 2015.

[35] Riahinezhad, M., McManus, N., and Penlidis, A. Shear Viscosity of Poly (Acrylamide/Acrylic Acid) Solutions. *Macromol. Symp.*, 360:179–184, 2016.

[36] Riahinezhad, M., Romero-Zerón, L., McManus, N., and Penlidis, A. Evaluating the performance of tailor-made water-soluble copolymers for enhanced oil recovery polymer flooding applications. *Fuel*, 11:269–278, 2017.

[37] Riahinezhad, M., Romero-Zerón, L., McManus, N., and Penlidis, A. Design of Tailor-Made Water-Soluble Copolymers for Enhanced Oil Recovery Polymer Flooding Applications. *Macromol. React. Eng.*, 11, 2017.

[38] Sandengen, K., Melhuus, K., and Kristoffersen, A. Polymer "viscoelastic effect"; does it reduce residual oil saturation*. *J. Pet. Sci. Eng.*, 153:355–363, 2017.

[39] Wever, D. A. Z., Picchioni, F., and Broekhuis, A. A. Branched polyacrylamides: Synthesis and effect of molecular architecture on solution rheology. *Eur. Polym. J.*, 49:3289–3301, 2013.

[40] Wever, D. A. Z., Polgar, L. M., Stuart, M. C. A., Picchioni, F., and Broekhuis, A. A. Polymer Molecular Architecture As a Tool for Controlling the Rheological Properties of Aqueous Polyacrylamide Solutions for Enhanced Oil Recovery. *Ind. Eng. Chem. Res.*, 52:16993–17005, 2013.

[41] Wever, D. A. Z., Riemsma, E., Picchioni, F., and Broekhuis, A. A. Comb-like thermoresponsive polymeric materials: Synthesis and effect of macromolecular structure on solution properties. *Polymer (Guildf).*, 54:5456–5466, 2013.

[42] Klemm, B., Picchioni, F., van Mastrigt, F., and Raffa, P. Starlike Branched Polyacrylamides by RAFT Polymerization—Part I: Synthesis and Characterization. *ACS Omega*, 3:18762–18770, 2018.

[43] Klemm, B., Picchioni, F., Raffa, P., and van Mastrigt, F. Star-Like Branched Polyacrylamides by RAFT polymerization – Part II: Performance Evaluation in Enhanced Oil Recovery (EOR). *Ind. Eng. Chem. Res.*, p. acs.iecr.7b03368, 2018.

[44] Destarac, M. Controlled Radical Polymerization: Industrial Stakes, Obstacles and Achievements. *Macromol. React. Eng.*, 4:165–179, 2010.

[45] Raffa, P., Wever, D. A. Z., Picchioni, F., and Broekhuis, A. A. Polymeric surfactants: Synthesis, properties, and links to applications. *Chem. Rev.*, 115, 2015.

[46] Raffa, P., Stuart, M. C. A., Broekhuis, A. A., and Picchioni, F. The effect of hydrophilic and hydrophobic block length on the rheology of amphiphilic diblock Polystyrene-b-Poly(sodium methacrylate) copolymers prepared by ATRP. *J. Colloid Interface Sci.*, p. 428, 2014.

[47] Raffa, P., Broekhuis, A. A. A. A., and Picchioni, F. Amphiphilic copolymers based on PEG-acrylate as surface active water viscosifiers: Towards new potential systems for enhanced oil recovery. *J. Appl. Polym. Sci.*, 133, 2016.

[48] Lake, L. W. *Enhanced Oil Recovery*. Prentice Hall, New Jersey, 1989.

[49] Zaitoun, A., Makakou, P., Blin, N., al Maamari, R. S., al Hashmi, A. A. R., and Abdel-Goad, M. Shear Stability of EOR Polymers. *SPE J.*, 17:335–339, 2012.

[50] Gaillard, N., Giovannetti, B., Favero, C., Caritey, J. P., Dupuis, G., and Zaitoun, A. New Water Soluble Anionic NVP Acrylamide Terpolymers for Use in Harsh EOR Conditions. In *SPE Improv. Oil Recover. Symp.*, 12–16 April 2014, Tulsa, Oklahoma, USA.

[51] El-hoshoudy, A. N. Quaternary ammonium based surfmer-co-acrylamide polymers for altering carbonate rock wettability during water flooding. *J. Mol. Liq.*, 250:35–43, 2018.

[52] Hussein, I. A., Ali, S. K. A., Suleiman, M. A., and Umar, Y. Rheological behavior of associating ionic polymers based on diallylammonium salts containing single-, twin-, and triple-tailed hydrophobes. *Eur. Polym. J.*, 46:1063–1073, 2010.

[53] Dastan, S., Hassnajili, S., and Abdollahi, E. Hydrophobically associating terpolymers of acrylamide, alkyl acrylamide, and methacrylic acid as EOR thickeners. *J. Polym. Res.*, 23:175, 2016.

[54] Hill, A., Candau, F., and Selb, J. Properties of Hydrophobically Associating Polyacrylamides – Influence of the Method of Synthesis. *Macromolecules*, 26:4521–4532, 1993.

[55] El-hoshoudy, A. N. N., Desouky, S. E. M. E. M., al Sabagh, A. M. M., Betiha, M. A. A., M.Y., E. K., and Mahmoud, S. Evaluation of solution and rheological properties for hydrophobically associated polyacrylamide copolymer as a promised enhanced oil recovery candidate. *Egypt. J. Pet.*, 26:779–785, 2017.

[56] Sarsenbekuly, B., Kang, W., Fan, H., Yang, H., Dai, C., Zhao, B., and Aidarova, S. B. Study of salt tolerance and temperature resistance of a hydrophobically modified polyacrylamide based novel functional polymer for EOR. *Colloids Surfaces A Physicochem. Eng. Asp.*, 514:91–97, 2017.

[57] Viken, A. L., Spildo, K., Reichenbach-Klinke, R., Djurhuus, K., Skauge, T., Løbø Viken, A., Spildo, K., Reichenbach-Klinke, R., Djurhuus, K., and Skauge, T. Influence of Weak Hydrophobic Interactions on in Situ Viscosity of a Hydrophobically Modified Water-Soluble Polymer. *Energy & Fuels*, 32:89–98, 2018.

[58] Zhang, P., Wang, Y., Chen, W., Yu, H., Qi, Z., and Li, K. Preparation and Solution Characteristics of a Novel Hydrophobically Associating Terpolymer for Enhanced Oil Recovery. *J. Solution Chem.*, 40:447–457, 2011.

[59] Co, L., Zhang, Z., Ma, Q., Watts, G., Zhao, L., Shuler, P. J., and Tang, Y. Evaluation of functionalized polymeric surfactants for EOR applications in the Illinois Basin. *J. Pet. Sci. Eng.*, 134:167–175, 2015.

[60] Li, F., Luo, Y., Hu, P., and Yan, X. Intrinsic viscosity, rheological property, and oil displacement of hydrophobically associating fluorinated polyacrylamide. *J. Appl. Polym. Sci.*, 134, 2017.

[61] Li, X., Xu, Z., Yin, H., Feng, Y., and Quan, H. Comparative Studies on Enhanced Oil Recovery: Thermoviscosifying Polymer Versus Polyacrylamide. *Energy & Fuels*, 31:2479–2487, 2017.

[62] Chen, Q., Wang, Y., Lu, Z., and Feng, Y. Thermoviscosifying polymer used for enhanced oil recovery: Rheological behaviors and core flooding test. *Polym. Bull.*, 2013.

[63] Kamal, M. S., Sultan, A. S., al Mubaiyedh, U. A., Hussein, I. A., and Feng, Y. Rheological Properties of Thermoviscosifying Polymers in High-temperature and High-salinity Environments. *Can. J. Chem. Eng.*, 93:1194–1200, 2015.

[64] Akbari, S., Mahmood, S., Tan, I., Ghaedi, H., and Ling, O. Assessment of Polyacrylamide Based Co-Polymers Enhanced by Functional Group Modifications with Regards to Salinity and Hardness. *Polymers (Basel).*, 9:647, 2017.

[65] Hocine, S. and Li, M. H. Thermoresponsive self-assembled polymer colloids in water. *Soft Matter*, 9:5839–5861, 2013.

[66] Zou, C., Wu, H., Ma, L., and Lei, Y. Preparation and Application of a Series of Novel Anionic Acrylamide Polymers with Cyclodextrin Sides. *J. Appl. Polym. Sci.*, 119:953–961, 2011.

[67] Zou, C., Zhao, P., Ge, J., Lei, Y., and Luo, P. β-Cyclodextrin modified anionic and cationic acrylamide polymers for enhancing oil recovery. *Carbohydr. Polym.*, 87:607–613, 2012.

[68] Zou, C., Zhao, P., Hu, X., Yan, X., Zhang, Y., Wang, X., Song, R., and Luo, F. β-Cyclodextrin-functionalized hydrophobically associating acrylamide copolymer for enhanced oil recovery. *Energy and Fuels*, 27:2827–2834, 2013.

[69] Pu, W. F., Yang, Y., Wei, B., and Yuan, C. D. Potential of a β-Cyclodextrin/Adamantane Modified Copolymer in Enhancing Oil Recovery through Host–Guest Interactions. *Ind. Eng. Chem. Res.*, 55:8679–8689, 2016.

[70] Wei, B. beta-Cyclodextrin associated polymeric systems: Rheology, flow behavior in porous media and enhanced heavy oil recovery performance. *Carbohydr. Polym.* 134:398–405, 2015.

[71] Ashrafizadeh, M., S. A., A. R., and Sadeghnejad, S. Enhanced polymer flooding using a novel nano-scale smart polymer: Experimental investigation. *Can. J. Chem. Eng.*, 95:2168–2175, 2017.

[72] Zhu, D., Wei, L., Wang, B., and Feng, Y. Aqueous hybrids of silica nanoparticles and hydrophobically associating hydrolyzed polyacrylamide used for EOR in high-temperature and high-salinity reservoirs. *Energies*, 7, 2014.

[73] Rezaei, A., Abdi-Khangah, M., Mohebbi, A., Tatar, A., and Mohammadi, A. H. Using surface modified clay nanoparticles to improve rheological behavior of Hydrolized Polyacrylamid (HPAM) solution for enhanced oil recovery with polymer flooding. *J. Mol. Liq.*, 222:1148–1156, 2016.

[74] Cheraghian, G. and Hendraningrat, L. A review on applications of nanotechnology in the enhanced oil recovery part A: effects of nanoparticles on interfacial tension. *Int. Nano Lett.*, 6:129–138, 2016.

[75] Bera, A. and Belhaj, H. Application of nanotechnology by means of nanoparticles and nanodispersions in oil recovery – A comprehensive review. *J. Nat. Gas Sci. Eng.*, 34:1284–1309, 2016.

[76] Liu, R., Pu, W., Sheng, J. J., and Du, D. Star-like hydrophobically associative polyacrylamide for enhanced oil recovery: Comprehensive properties in harsh reservoir conditions. *J. Taiwan Inst. Chem. Eng.*, 80:639–649, 2017.

[77] Pu, W., Zhao, S., Wang, S., Wei, B., Yuan, C., and Li, Y. Investigation into the migration of polymer microspheres (PMs) in porous media: Implications for profile control and oil displacement. *Colloids Surfaces A*, 540:265–275, 2018.

[78] Sen, R. Biotechnology in petroleum recovery: The microbial EOR. *Prog. Energy Combust. Sci.*, 34:714–724, 2008.

[79] Jang, H. Y., Zhang, K., Chon, B. H., and Choi, H. J. Enhanced oil recovery performance and viscosity characteristics of polysaccharide xanthan gum solution. *J. Ind. Eng. Chem.*, 21:741–745, 2015.

[80] al Manasir, N., Kjoniksen, A. L., and Nystrom, B. Preparation and Characterization of Cross-Linked Polymeric Nanoparticles for Enhanced Oil Recovery Applications. *J. Appl. Polym. Sci.*, 113:1916–1924, 2009.

[81] Iglauer, S., Wu, Y., Shuler, P., Tang, Y., and Goddard, W. A. Dilute iota- and kappa-Carrageenan solutions with high viscosities in high salinity brines. *J. Pet. Sci. Eng.*, 75:304–311, 2011.

[82] Haicun, Y., Weiqun, Z., Ning, S., Yunfei, Z., and Xin, L. Preliminary study on mechanisms and oil displacement performance of cationic starch. *J. Pet. Sci. Eng.*, 65:188–192, 2009.

[83] Yang, S., Dai, C., Wu, X., Liu, Y., Li, Y., Wu, Y., and Sun, Y. Novel investigation based on cationic modified starch with residual anionic polymer for enhanced oil recovery. *J. Dispers. Sci. Technol.*, 38:199–205, 2017.

[84] Song, H., Zhang, S. F., Ma, X. C., Wang, D. Z., and Yang, J. Z. Synthesis and application of starch-graft-poly(AM-co-AMPS) by using a complex initiation system of CS-APS. *Carbohydr. Polym.*, 69:189–195, 2007.

[85] Singh, R. and Mahto, V. Synthesis, characterization and evaluation of polyacrylamide graft starch/clay nanocomposite hydrogel system for enhanced oil recovery. *Pet. Sci.*, 14:765–779, 2017.

[86] Gou, S., Li, S., Feng, M., Zhang, Q., Pan, Q., Wen, J., Wu, Y., and Guo, Q. Novel Biodegradable Graft-Modified Water-Soluble Copolymer Using Acrylamide and Konjac Glucomannan for Enhanced Oil Recovery. *Ind. Eng. Chem. Res.*, 56:942–951, 2017.

[87] Cao, Y. and Li, H. L. Interfacial activity of a novel family of polymeric surfactants. *Eur. Polym. J.*, 38:1457–1463, 2002.

[88] al Sabagh, A. M. Surface activity and thermodynamic properties of water-soluble polyester surfactants based on 1,3-dicarboxymethoxybenzene used for enhanced oil recovery. *Polym. Adv. Technol.*, 11:48–56, 2000.

3 Numerical simulation of chemical EOR

3.1 Oil extraction and fluid flow equations

3.1.1 Introduction

The era of discovery and subsequent exploitation of the denominated "easy oil" (part of the so-called conventional reserves) is over [1–5]. After primary and secondary recovery stages, the oil companies have shifted to more complex enhanced oil recovery (EOR) processes, involving the injection of different fluids or chemical agents causing a modification of the physical properties of fluids and/or rock formation. Thus, these techniques require a greater understanding of the phenomena taking place in the porous medium. In addition to these complications, there are non-technical related factors that might increase the risk of an investment [6]. An example of this may be the developing of off-shore platforms, plants located in remote areas of the world, or the exploitation of unconventional deposits (heavy oil, shale oil, tar sands). All these projects require a prior investment (on the order of several tens of millions of dollars in high risk projects), so companies make use of several techniques in the stage of feasibility analysis to limit this risk and increase the chances of success of the operation (Figs. 3.1 and 3.2) [7, 8].

Along with the steps necessary during the exploration, explcitation and subsequent abandonment of an oil field, any EOR process demands a number of previous steps in order to guarantee the success of the operations (Fig. 3.2). Primary and secondary stages may be used to achieve a better determination of the physical dimensions and properties of the rock formation before the application of the more expensive EOR processes. These determine a number of requisites for the chemicals to be injected, which are then tested at a minor scale in the reservoir, providing the necessary feedback to optimize them. Finally, after all these steps are met, the large-field application is implemented, which increases the operational life of the reservoir (Fig. 3.3).

These characterization techniques, used to predict and optimize the exploitation, include laboratory and field tests (e.g., seismic 2D/3D/4C before starting operations and 4D to follow up the changes during exploitation, geostatistics) to give an idea of the conditions of the porous medium and its production performance [12–17]. However, these tools have proven to be insufficient [14, 18, 19] since there is a gap in the reservoir resolution than current techniques cannot resolve, and thus are complemented with geostatistics tools in order determine the reservoir characteristics (Fig. 3.4). Therefore, oil companies have begun using computational tools to predict and optimize their projects and production facilities, which is known as *reservoir*

https://doi.org/10.1515/9783110640250-003

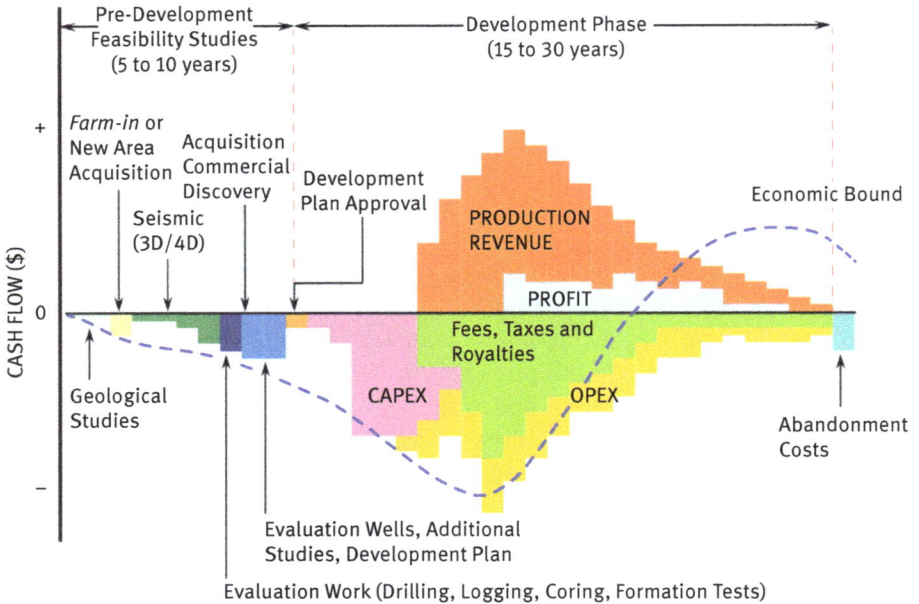

Fig. 3.1: A typical cash-flow of an exploration and production project, where the blue dashed line represents the instantaneous total cash-flow [9].

simulation. The latter consists of numerically solving the differential equations describing the fluid flow in porous media, which have no analytical solution, by taking into account all geological, physical, chemical and/or mechanical phenomena occurring during operation so as to analyze and predict behavior as a function of time. The reservoir simulation can be used as well in inverse engineering problems for optimizing existing numerical models and to couple the dynamic/historic data (production) in the simulation [19–22].

Generally speaking, reservoir simulation consists of three main parts: the physical characterization of a geological model describing the rock formation; a model characterizing the fluid flow; and finally "well models" which describe the conditions under which fluids are injected or extracted from the reservoir [23]. The latter, along with the wellbore and the primary surface facilities, constitutes what is known as *upstream.* During the last 30 years numerous theoretical and practical advances have been developed due to the appearance of new numerical techniques and increased computational power, respectively [19, 20, 24–26]. This led to a new generation of more complex and detailed models. The accurate representation of the reservoir and the fluid contained in it is an issue that still needs to be more carefully addressed in order to

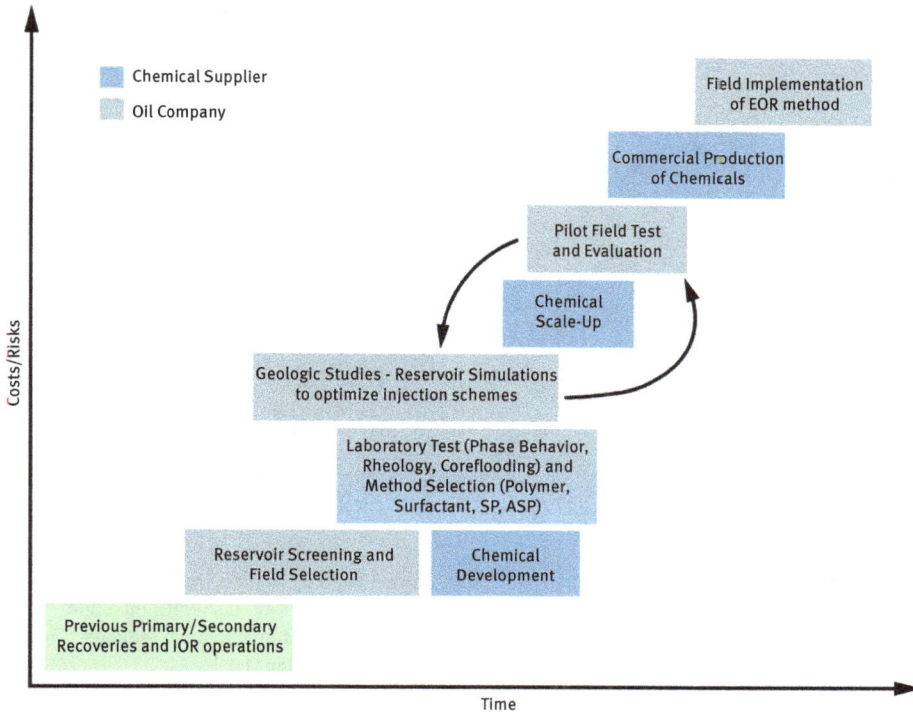

Fig. 3.2: Typical phases of a chemical EOR project, including previous primary and secondary recoveries (adapted from Jürgensson [10]).

reduce risks in exploration and production (E&P) projects. One of the most important points is related to the scales of the models: current grids are still large when compared to the geological characterization, description of chemical processes or fluid flow. Most of the current numerical models used in reservoir simulation may contain from 10^5 to 10^8 grid blocks, depending on the model type, complexity, computational power available and fluid behavior.

The development of increasingly complex and detailed models requires the use of numerical techniques to solve these within reasonable times. Moreover, the representation of the properties of the porous medium and the characteristics of crude oil and natural gas at high pressures and temperatures may differ from laboratory tests, causing differences between the results and simulation. Another important topic is how to assess and properly estimate the properties of the rock formation. Geologists use statistical techniques in order to recreate the model properties of a porous medium, which are determined by several tests (e.g., seismic studies, drill core sam-

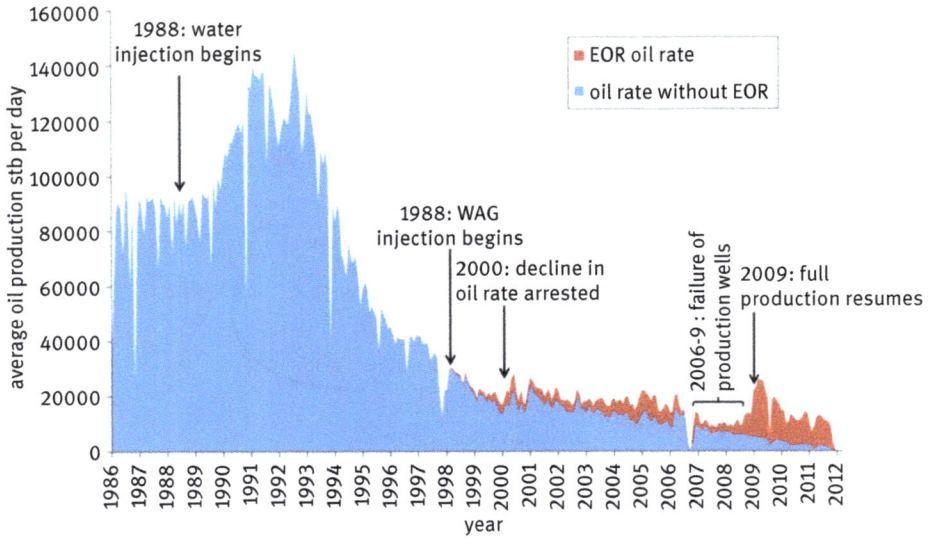

Fig. 3.3: Daily oil production rate (in STB – stock tank barrels per day) from the Magnus oilfield (North Sea – UK) since the start of the recovery in 1983. Water-alternating-gas (WAG) injection started in 2002 and by 2005 it was clear that the decline in oil production had been reduced. The oil rate expected without EOR was estimated using numerical simulation [11].

ples and even production data). However, other numerical tools are required (e.g., Monte-Carlo or stochastic processes) to take into account the effects of uncertainties in the model [26–28].

3.1.2 Reservoir characterization and formation

Origin of the oil

A reservoir is an underground trap where different fluids (water, oil and gas) have accumulated due to a migration from the source rock where they were originated. The porous medium is generally considered of sedimentary origin and consists of a series of microchannels (about 1–100 μm diameter) interconnected where these fluids can flow (Fig. 3.5). These formations may have some tens of meters of thickness, but may extend several kilometers in the lateral directions [29].

The origin of crude oil prior to the migration and deposit of hydrocarbons in the porous media is also a long and complex process [30–32]. The source of hydrocarbons consists of a series of phenomena, both organic and inorganic, taking place over long periods of time (in the order of million years) [33]. Most of these hydrocarbons originate in organic decomposition processes. The first stage (called diagenesis) involves the sedimentation of remains of dead plants and animals. Under these condi-

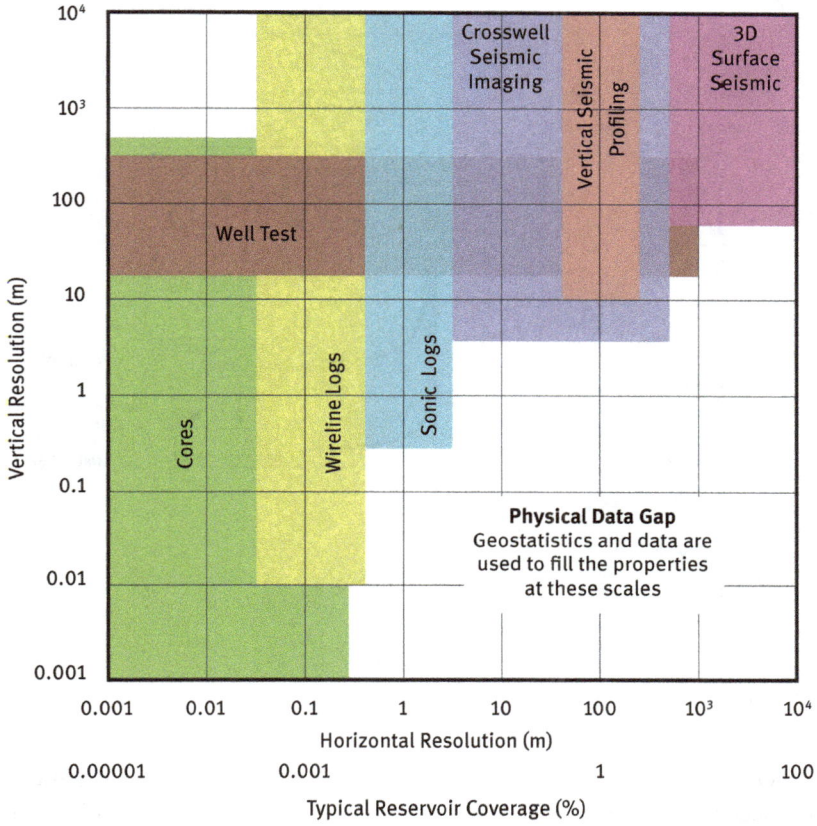

Fig. 3.4: Horizontal and vertical resolution for different reservoir characterization techniques, in which a physical data gap is clearly noticeable.

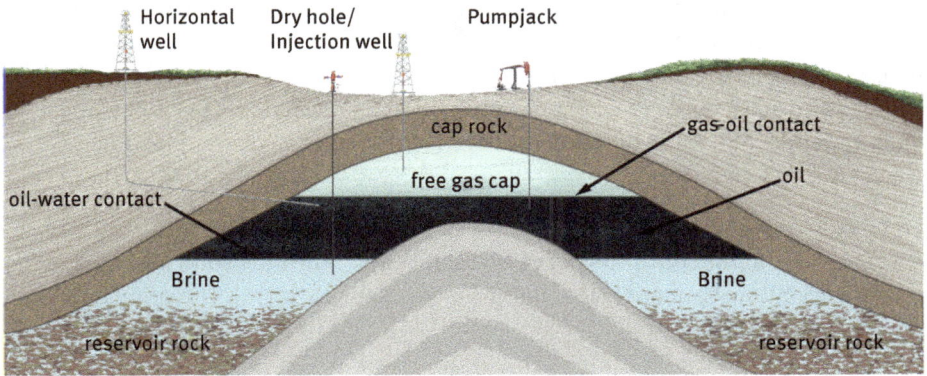

Fig. 3.5: Anticlinal type petroleum trap [9].

Lignin | Carbohydrates | Proteins | Lipids

Living Organisms

Microbial Degradation
Polymerization
Condensation

Minor or no Alteration
Retaining Carbon Skeleton

Geochemical Fossils

Zone	T_{max} [°C]	Temp [°C]	Depth [km]

DIAGENESIS — immature

415 — 50

430 — 100

CATAGENESIS — oil window / wet gas / dry gas

448

oil

455 — 150

475

METAGEN. — destruction

— 500 —

200

250

0   100
rel. abundance (%)

CH_4 aq   CO vap
CH_4 vap   CO aq

0
2
4
6
8

Fulvic Acids
Humic Acids
Humin

Release of Trapped Molecules

Kerogen

Thermal Degradation

Cracking

Carbon Residue

Small to Medium Hydrocarbons

Cracking

Methane and Light Hydrocarbons

Crude Oil

Heavy Hydrocarbons (and Bitumen)

Natural Gas

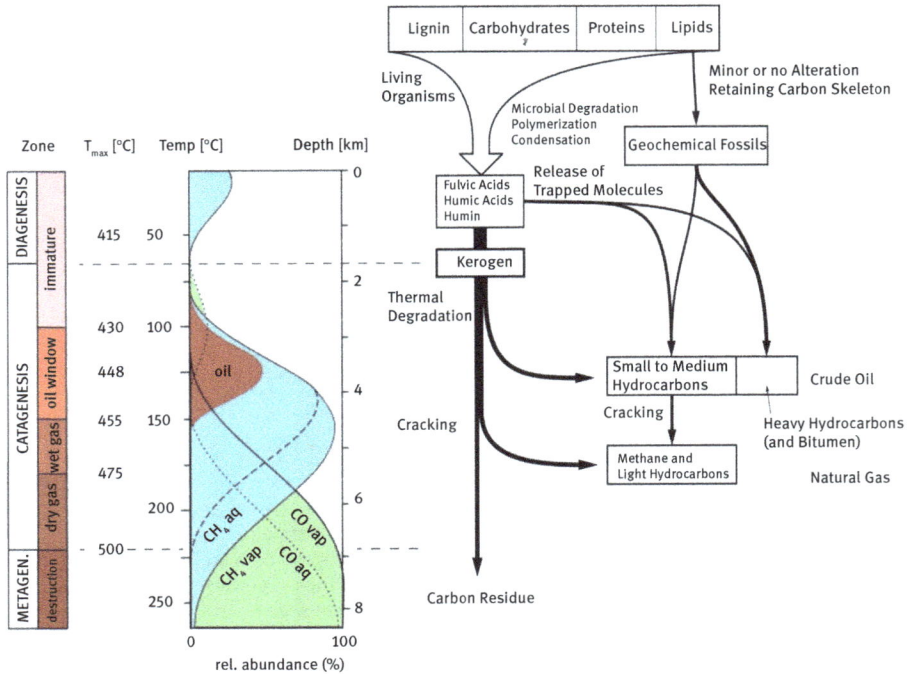

Fig. 3.6: Origin of oil [9].

tions, at low depths, the action of bacteria produce methane, water and CO_2, leaving as reaction products, kerogen substances (cyclic and large hydrocarbon molecules containing oxygen, nitrogen and sulfur). The second process, called catagenesis, occurs at greater depths and temperatures. In this region, known as the oil window zone, the kerogen molecules break into smaller, heavy hydrocarbons forming the oil phase. At higher temperatures the cracking of hydrocarbons continues creating lighter compounds, first wet gas and subsequently dry gas. Finally, at even greater depths, the last process takes place (called metagenesis), where the remaining kerogen is out of hydrocarbons and subsequent cracking processes terminate when there is no hydrogen in the compound. The result of these reactions is the formation of graphite (Fig. 3.6).

Porosity and permeability

A porous rock formation is composed of a solid part, called the solid matrix, and the remaining void space or microchannels whereto oil migrates [34]. The volumetric fraction of these channels is denominated the porosity of the porous media. The latter depends on the fluid pressure, if the rock is compressible, or in some other phenomena which may take place (e.g., adsorption of chemical components during EOR processes). The following list provides typical values of porosity according to the origin of the rock formation [35, 36]: Consolidated sandstones 0.05 to 0.3; limestones 0.1 to 0.4; uniform spheres with minimal porosity packing 0.25; uniform spheres with normal packing 0.35; unconsolidated sands with normal packing 0.40 and unconsolidated clays 0.60.

The permeability of the formation is a property that characterizes the ease with which fluids can flow when a pressure gradient is applied between two points. Nonetheless, reservoir rocks usually have no uniformity in their properties because of the mechanisms involved in its formation, thus the permeability will have a large dispersity in its values [37, 38]. The fluids commonly used in EOR operations, are more mobile than oil, occupy high-permeability zones (e.g., faults or fractures), with the result that large areas of oil will be bypassed, reducing the efficiency of the process. Mathematically, the permeability can be expressed as a diagonal tensor ($\underline{\underline{K}}$). When the medium is isotropic then the permeability can be represented as a scalar function. Due to transitions between different rock types, the permeability may vary swiftly throughout the reservoir, going from extremely low permeability areas of 1 mD to areas with permeabilities exceeding 10 D.

In order to develop a mathematical model of the porous medium, Corey [35] established several restrictions: the whole void space of the porous medium is interconnected; the mean free path length of the fluid molecules or molecules contained in the fluid must be negligible when compared to the dimensions of the pore channels, and the dimensions of the void space must be small enough so that the fluid flow is controlled by adhesive forces at fluid-solid interfaces and cohesive forces at fluid-fluid interfaces (in multiphase systems) [34]. These assumptions allow excluding of any disconnected channels in which there can be no fluid flow, eliminating the difference between the concepts of *total* and *effective* porosities. Furthermore, since the dimensions of the molecules or particles in the fluid are negligible with respect to the microchannels, a suitable replacement can be performed for a hypothetical continuous medium. Finally, considering the microscopic size of the channels allows taking into account physical phenomena that in other cases would be negligible. In order to derive a mathematical model at the macroscopic level, each point in the continuum is the assigned average values over the representative elementary volumes (REV's) of the quantities at the microscopic level. The advantage of this technique is that it leads to a set of macroscopic equations that do not need an exact description of the microscopic configuration, as would be the case with the Navier–Stokes equations [34].

Representative elementary volume

The flow of reservoir fluids in porous media can be described on several different scales, from a microscopic scale to a macroscopic/formation scale. In order to perform large-scale reservoir simulations, a microscopic description of the flow channels would be too demanding for the computational power available and besides, to characterize a reservoir rock so accurately to determine the geometry of the pore network is beyond the scope of modern techniques and equipment. A continuum scale description is then utilized, and its behavior is governed by forces acting between the different fluids and the rock formation. The goal of a reservoir continuum model is then to average both the fluids and reservoir rock [33, 39–42]. In order to develop the mathematical model based on a continuum, the concept of representative elementary volume (REV) (Fig. 3.4) is introduced. This is based on the hypothesis that certain properties of both the fluid and the rock may be considered constant along a certain range of scale and thus it establishes limits for the physical scales in the numerical models. If a REV cannot be identified for a specific porous medium then this concept cannot be applied and the macroscopic approach should be discarded [43].

The procedure for estimating REV dimensions and for establishing boundaries between microscopic and macroscopic scales is explained below using the porosity in Fig. 3.4 as an example [34]. A porous medium is then considered as occupying the domain Ω, with a volume $V(\Omega)$. A subdomain $\Omega_0(d) \subset \Omega$ with a characteristic dimension d is also determined. Furthermore, the porosity piece-wise function is defined in Eq. (3.1) as follows,

$$f(\overline{x}) = \begin{cases} 1 & \text{if } \overline{x} \in \text{void space} \\ 0 & \text{if } \overline{x} \in \text{solid matrix} \end{cases} \quad \forall \overline{x} \in \Omega \tag{3.1}$$

Then, the porosity of an element with characteristic dimension d is defined by Eq. (3.2) as,

$$\phi\,[\Omega_0(d)] = \frac{1}{V(\Omega_0)} \cdot \int_{\Omega_0} f(\overline{x})d\Omega_0 \tag{3.2}$$

This relationship allows explaining the evaluation of the porosity as a function of the dimension d. For sizes smaller then a value d_m the porosity varies significantly, with no particular pattern or trend; then, if the value of said dimension is between d_m and d_M, the value of the porosity plateaus and remains constant for the entire considered range. Finally, for values greater than d_M, the porosity may remain constant in the case of a homogeneous medium, while in the case of an heterogeneous one, the function again becomes chaotic [44, 45]. The volume with dimensions between d_m and d_M is called a representative elementary volume (REV) (Fig. 3.7).

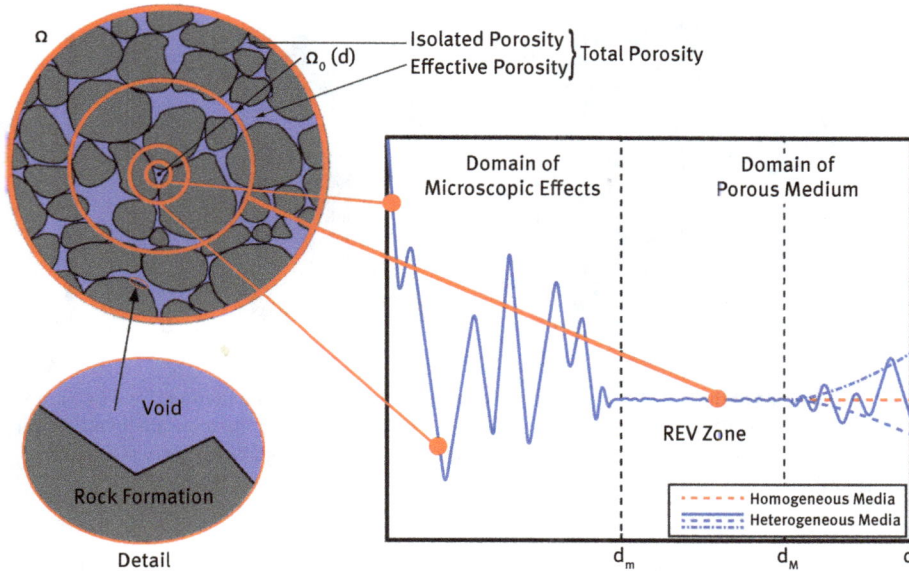

Ω

$\Omega_0\,(d)$

Isolated Porosity ⎫
Effective Porosity ⎭ Total Porosity

Void

Rock Formation

Detail

Domain of
Microscopic Effects

Domain of
Porous Medium

REV Zone

Homogeneous Media
Heterogeneous Media

d_m d_M d

Fig. 3.7: Scheme showing the boundaries to determine the REV [46].

3.1.3 Fluid flow models

In underground processes in porous media, fluid flow involves mainly convection (advection and diffusion) of the different phases and components through a heterogeneous medium. Generally speaking there are two approaches to calculate the flow in porous media: *direct* and *continuum* modeling. The former is also known as pore-scale approach, because the equations are solved without making any assumptions regarding the pore geometry. The equations used to describe the flow at a microscopic level or pore scale are variations of Navier–Stokes (creeping flow) and the mass conservation, and are used in limited cases in porous media flow. The second approach considers the quantities or properties involved averaged over control volumes. At this macroscopic level, Darcy's law [47] was derived and it is used to describe flow behavior (Fig. 3.8). Also, the effects of the displacing process on the rock formation will be considered negligible even though these mechanisms are sometimes necessary to represent first-order effects (e.g., adsorption) [48, 49]. When dealing with multiphase/multicomponent flows, some of the processes therein involved are characterized by the chemical and physical interaction among the components present in the

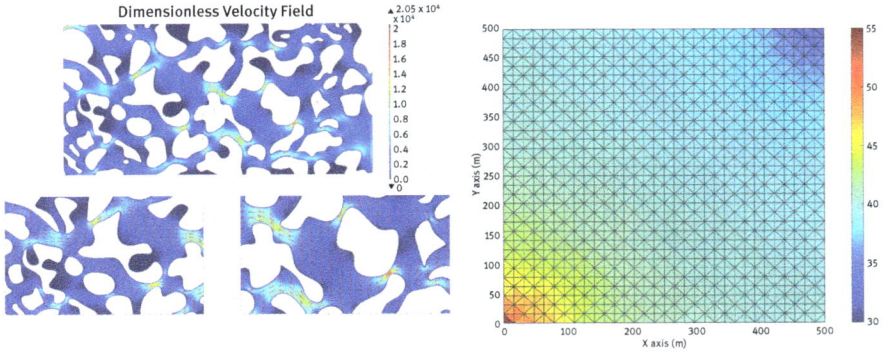

Fig. 3.8: Examples of the direct approach (left) solving the Navier–Stokes equations for a 640 μm × 320 μm porous domain, and of the continuum approach (right) depicting the pressure [in MPa] over a square reservoir, using the Darcy equation, with the properties averaged over every elementary volume [46].

fluids. Therefore, diffusive and/or dispersive mixing of these components is often critical and must be correctly understood and modeled in order to get accurate results. Molecular diffusion is typically quite small in porous media processes. Nevertheless, hydrodynamic dispersion may be important and therefore it should be incorporated in the flow equations. This can be done by means of the standard diffusion/dispersion tensor [50, 51], provided that the degree of the heterogeneity is not too large (Dykstra–Parsons coefficient < 0.50) resulting in a Fickian behavior. However, when the permeability variations are large a non-Fickian behavior was reported (anomalous dispersion) [52–56].

Generally, three fluid phases may exist inside a reservoir (Fig. 3.5): aqueous/brine (the phase containing predominantly water), hydrocarbon/organic (the phase containing liquid hydrocarbons) and gas phase, which contains lighter gaseous hydrocarbons. In the case of single phase systems, the void space of the porous medium is filled by a single fluid or by several fluids completely miscible with each other. In multiphase systems the void space is filled by two or more fluids that are immiscible with each other, thus maintaining a distinct boundary between them. Formally speaking, the solid matrix of the porous medium can also be considered as a phase called the solid phase. Each phase may also be composed by many chemical components. For example, oil is a very complex mixture of hundreds of hydrocarbons with different chemical properties. In order to derive a set of equations for the fluid flow some terms should be defined beforehand. Firstly, the term *phase* stands for matter that has a homogeneous chemical composition and physical state of a system under consideration that is separated from other such portions by a definite physical boundary. Secondly, it is defined as *component* present in a phase to the matter that is composed of an identifiable homogeneous unique chemical species or an assembly thereof [44].

The number of components needed to describe a phase is given by the conceptual model, i.e., it depends on the physical processes to be modeled and the desired accuracy. In many oil reservoirs (above the bubble point pressure) crude oil contains some amount of dissolved gas. It is acceptable to assume that the oil and gas compositions are fixed [20, 24, 31], and that the solubility of the gas in the oil depends on pressure only. Provided these conditions are met, then it is possible to consider the pseudo-components oil and gas.

Both microscopic and macroscopic effects control the movement of fluids in the reservoir. At the pore scale, interfacial tension (IFT) and capillary effects control the fluid behavior. Macroscopically, fluid flow is controlled by reservoir heterogeneity and mobility differences between the fluids. Viscosities, capillary pressure, IFT and mobility differences vary throughout the reservoir and depend mainly on phase saturations, their interactions and molecular compositions. Chemical components may transfer between contacting phases, altering the fluid properties of both. Interactions between the fluids or their components and the reservoir rock may also impact performance (e.g., adsorption of chemical components onto the surface of the rock altering the wettability). Thermal effects are generally very small due to the large heat capacity of the rock. However, in EOR thermal processes (steam flooding or in-situ combustion), the conservation of energy in the REV should be considered.

Single phase flow

The governing equations for single phase flow in porous media are the conservation of mass, Darcy's Law and an equation of state (EOS). Considering the flow of a single fluid with density ρ through a REV of a porous medium the differential form of the continuity Eq. (3.3) can be expressed as [20, 23, 24, 37, 39, 57–59],

$$\frac{\partial(\rho\phi)}{\partial t} + \nabla \cdot (\rho\overline{u}) = q \tag{3.3}$$

Where ϕ is the porosity of the rock formation, q represents the fluid source/sink term and $\overline{u}$ is the velocity vector. The fundamental law of fluid flow in a porous medium is Darcy's law [47]. It gives the effective flow velocity across a REV of porous media and thus does not analyze the flow at a microscopic scale. In differential form, the relationship between velocity and pressure drop is given by Eq. (3.4),

$$\overline{u} = -\frac{1}{\mu}\underline{\underline{K}}\left(\nabla p - \rho\mathbf{g}\nabla z\right) \tag{3.4}$$

Where $\underline{\underline{K}}$ is the absolute permeability tensor of the porous medium, μ the fluid viscosity, $\mathbf{g}$ the acceleration field, and z represents physical dimensions. In most of the cases, it is possible to assume that $\underline{\underline{K}}$ is a diagonal tensor as presented in Eq. (3.5),

$$\underline{\underline{K}} = \begin{pmatrix} k_{11} & 0 & 0 \\ 0 & k_{22} & 0 \\ 0 & 0 & k_{33} \end{pmatrix} \tag{3.5}$$

When $k_{11} = k_{22} = k_{33}$, the porous medium is called isotropic; otherwise, it is anisotropic. Generally in porous media it can be considered that both lateral permeabilities are of the same order of magnitude while the vertical permeability is considerably lower (at least one order of magnitude) than the other two components. Originally, Darcy's law was derived experimentally and was thus considered an empirical law [33]. Several authors have reported the derivation of Darcy's law based on volume averaging of the Navier–Stokes momentum equations [41, 44, 50, 60–65]. The assumptions needed for the derivation of Darcy's law include low flow speeds, Newtonian behavior and that the pore/fluid friction is the dominating force acting on the fluid. Also, the porous medium is assumed to be rigid and not compacted due to fluid flow. By introducing rock and fluid compressibilities in Eq. (3.6), c_r and c_f respectively, both equations can be coupled in the parabolic Eq. (3.7) for the fluid pressure,

$$c_r = \frac{1}{\phi}\frac{d\phi}{dp} \quad \text{and} \quad c_f = \frac{1}{\rho}\frac{d\rho}{dp} \tag{3.6}$$

$$\phi\rho(c_r + c_f)\frac{\partial p}{\partial t} - \nabla \cdot \left[\frac{\rho}{\mu}\underline{\underline{K}}(\nabla p - \rho\mathbf{g}\nabla z)\right] = q \tag{3.7}$$

In the special case of incompressible rock and fluid (generally acceptable for liquid systems) the PDE simplifies to a Poisson elliptic equation.

Two phase flow

The space in a reservoir is generally filled by both the hydrocarbon phase and brine. In addition, during secondary recovery processes, water is frequently injected in order to improve oil recovery. If the fluids are immiscible, they are referred to as phases. A two-phase system is commonly divided into a wetting and a non-wetting phase, given by the contact angle between the solid surface and the fluid-fluid interface on the microscale. For each pair of phases, one phase will wet the rock more than the other phase, and that phase will be referred to as the wetting phase ($j = w$). The other phase is then the non-wetting phase ($j = nw$). Normally, water is the wetting phase in a water-oil system, and oil is the wetting phase in an oil-gas system. In the absence of phase transitions, the saturations change when one phase displaces the other. During the displacement, the ability of one phase to move is affected by the interaction with the other phase at the pore scale. In the mathematical model at a macroscopic scale this effect is represented by the relative permeabilities k_r^j ($j = w, nw$), that are dimensionless scaling factors that depend on the saturation and modify the absolute permeability to account for the rock's reduced capability to make one phase flow in the presence of the other. Then, the mass conservation Eq. (3.8) for each phase yields [20],

$$\frac{\partial(\rho^j\phi S^j)}{\partial t} + \nabla \cdot (\rho^j\overline{u}^j) = q^j \tag{3.8}$$

and the multiphase extension of Darcy's law reads,

$$\overline{u}^j = -\frac{k_r^j}{\mu^j}\underline{\underline{K}}\left(\nabla p^j - \rho^j\mathbf{g}\nabla z\right) \tag{3.9}$$

Together, they form the basic system of equations. Because of the interfacial tension (IFT), the pressure in the two phases will differ. This difference is called capillary pressure ($p_c = p_{nw} - p_w$), which is usually assumed on the macroscale to be a function of saturation. From the formulation exposed previously, the following relationships can be derived,

$$\overline{u}_T = \sum_{i=1}^{N_p} \overline{u}_i \tag{3.10}$$

$$f_w(S_w) = \frac{\frac{k_{rw}}{\mu_w}}{\frac{k_{rw}}{\mu_w} + \frac{k_{rnw}}{\mu_{nw}}} \tag{3.11}$$

Where $\overline{u}_T$ is the total Darcy velocity, which is useful in schemes employed to solve the system of equations, and $f_w(S_w)$ is the fractional flow of the wetting phase. The system of equations derived from the formulation of two phases can be solved using various numerical techniques. The most used are: formulations using the pressure of both phases, known as simultaneous solutions (SS); formulations using a pressure and saturation phase, known as IMPES or IMPSAT, depending on the numerical treatment of the saturation equation (explicit or implicit, respectively); and the global pressure formulation [66].

Saturation

The volume fraction occupied by each phase is defined as the saturation of that phase. Thus, for a two-phase system, and considering no phase transitions, the sum of the saturation of both the wetting and non-wetting phases is equal to unity (Eq. (3.12)). Similar to what was done for the void space, the phase indicator piece-wise function is defined by Eq. (3.13),

$$\sum_{i=1}^{N_p} S_i(\overline{x}, t) = 1 \tag{3.12}$$

$$f^j(\overline{x}, t) = \begin{cases} 1 & \text{if } \overline{x} \in \text{phase } j \text{ at time } t \\ 0 & \text{if } \overline{x} \notin \text{phase } j \text{ at time } t \end{cases} \quad \forall \overline{x} \in \Omega \tag{3.13}$$

Then, the saturation of the phase j in an REV Ω_0 element with characteristic dimension x_0 will be defined by Eq. (3.14),

$$S^j(\overline{x}, t) = \frac{\text{Volume of phase } j \text{ in REV at time } t}{\text{Volume of void space in REV}} = \frac{\int_{\Omega_0} f^j(\overline{x}, t) \, d\Omega_0}{\int_{\Omega_0} f(\overline{x}) \, d\Omega_0} \tag{3.14}$$

Relative permeabilities

The relative permeability of each phase depends on the phase saturations but does not depend directly on fluid flow properties [67]. If only a single phase is present the

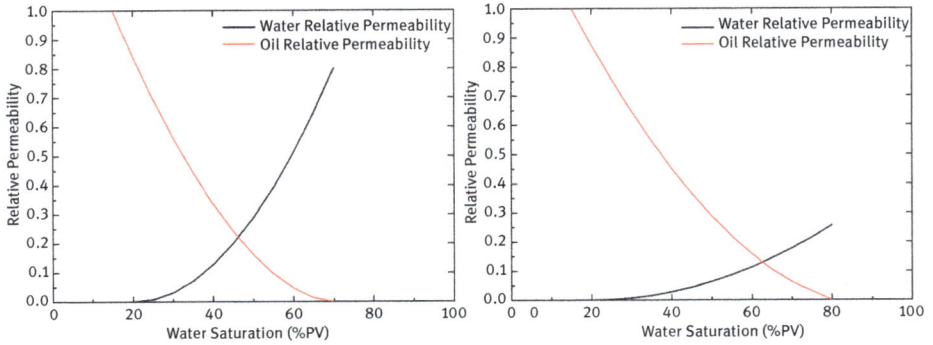

Fig. 3.9: Relative permeability for oil-wet (left) and water-wet (right) formation rocks [46].

relative permeability has no physical meaning, but in a multiphase system, the flow of one phase interferes with the others, hence this influence is taken into account in the Darcy equation ($k_r^j \leq 1$). It is usual in multiphase systems to use correlations of the relative permeabilities expressed as functions of the wetting phase saturation (Eq. (3.15) and Fig. 3.9),

$$k_{rw} = k_{rw}(S_w), \quad k_{rnw} = k_{rnw}(S_w) \tag{3.15}$$

In addition to relative permeability correlations, also analytical capillary pressure functions are needed. In two phase simulations it is standard to use the relations provided by either Brooks–Corey or Van Genuchten [39]. As for the relative permeability, these depend on empirical constants (e.g., if the system is oil-wet or water-wet), so several models have been developed through the years [68–74].

Compositional models

In the two phase models a no mass transfer condition between the phases is assumed. This assumption is valid for two phase flows of water/brine and oil, which is often the case in primary and secondary recovery mechanisms. In EOR processes, mass transfer and compositional effects are deemed essential to accurately model as they may become the driving mechanisms for the displacing process. A typical reservoir fluid may consist of several different chemical pseudo-components and fully compositional models must be used when the fluid flow depends strongly on component transfer between phases. In fact, many EOR techniques, mainly chemical and miscible gas processes, are specifically designed to take advantage of the phase behavior of multicomponent fluid systems. Because these components may be transferred between phases (and change their composition), the basic conservation laws must be expressed for each component instead for each phase. For a chemical flooding compositional model, the governing differential equations consist of a mass conservation equation for each component and Darcy's law for each phase [24, 75, 76].

$$\frac{\partial}{\partial t}\left(\rho^j \phi z_i\right) = -\nabla \cdot \left[\sum_{j=1}^{N_p} \rho^j \left(c_i^j \overline{u}^j - \underline{\underline{D}}_i^j \nabla c_i^j\right)\right] + q_i - \frac{\partial}{\partial t}\left(\rho^j \phi \mathrm{Ad}_i\right), \quad i = 1, 2, \ldots, N_{\mathrm{comp}}$$

(3.16)

$$z_i = \sum_{j=1}^{N_p} S^j c_i^j, \quad \sum_{i=1}^{N_{\mathrm{comp}}} c_i^j = 1$$

(3.17)

Here z_i is the overall concentration of component i calculated by Eq. (3.17), N_{comp} is the number of components in the system, c_i^j is the volumetric concentration of the component i in the phase j, $\underline{\underline{D}}_i^j$ is the diffusion-dispersion tensor of the component i in the phase j, and Ad_i is the amount of component i adsorbed by the rock formation. In this formulation the reduction of pore volume due to adsorption of chemical components onto the rock surface is neglected. As in the previous models, the velocities are modeled using the multiphase extension of Darcy's law. This system is just the starting point of modeling and must be further manipulated and supplied with extra equations (PVT models, phase equilibrium conditions, see Phase Partitioning) for specific fluid systems. This model is developed under the following assumptions: local thermodynamic equilibrium, immobile solid phase, Fickian dispersion, ideal mixing and Darcy's law, though some of these fluids may not have a Newtonian behavior. Sources and sinks of a component can result from injection and production of the latter by external means. The advantage of the compositional approach is the ability to handle various processes within the fluid phase, such as chemical reactions among components, radioactive decay, any kind of degradation and growth due to bacterial activities that cause the quantity of this component and/or its properties to increase or decrease. When miscibility develops, relative permeabilities vary with IFT due to the influence of the latter in the capillary desaturation curves. Phase viscosities are generally given by empirical correlations that consider the viscosity a function of the mole fractions and molar density for the phase. However, the non-Newtonian behavior of some oils and solutions used in chemical EOR should be taken into account by means of proper rheological correlations (Power-Law, Carreau–Yasuda, unified viscosity model [77]).

Dispersion

In addition to the advective, directional movement of a component described by the Darcy phase velocity, components may also move due to dispersive forces. The simplest movement is molecular diffusion described by the random Brownian motion of molecules. Such motion in reservoir simulation, is usually considered to be of negligible importance compared to other forces acting on the fluid. A more substantial phenomenon is mechanical dispersion. Narrow channel flows experience parabolic diffusion along the fronts (Taylor dispersion) and the irregular pore networks disperse

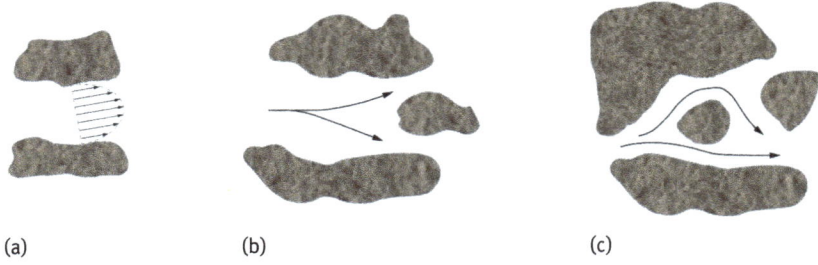

(a)   (b)   (c)

Fig. 3.10: Different types of mechanical dispersion phenomena: the velocity profile developed due to the no-slip boundary condition (a) leads to a longitudinal spreading of the component (Taylor diffusion); the stream splitting in (b) leads to a transversal spreading, and the tortuosity effect in (c) leads also to a longitudinal spreading [46].

the mass at a microscale (Fig. 3.10). The tensor of hydrodynamic dispersion taking into account the mentioned effects is expressed in Eq. (3.18) [24, 42, 50, 51, 78],

$$\underline{\underline{D}}^j_i = \phi S^j \mathrm{dm}^j_i \underline{\underline{I}} + \|\overline{u}^j\|_2 \left[\mathrm{dl}^j \underline{\underline{E}} \left(\overline{u}^j \right) + \mathrm{dt}^j \underline{\underline{E}}^\perp \left(\overline{u}^j \right) \right] \tag{3.18}$$

$$\underline{\underline{E}} \left(\overline{u}^j \right) = \frac{1}{\left(\|\overline{u}^j\|_2 \right)^2} \begin{pmatrix} \left(u^j_x \right)^2 & u^j_x u^j_y & u^j_x u^j_z \\ u^j_y u^j_x & \left(u^j_y \right)^2 & u^j_y u^j_z \\ u^j_z u^j_x & u^j_z u^j_y & \left(u^j_z \right)^2 \end{pmatrix} \tag{3.19}$$

$$\underline{\underline{E}}^\perp \left(\overline{u}^j \right) = \underline{\underline{I}} - \underline{\underline{E}} \left(\overline{u}^j \right) \tag{3.20}$$

where dm^j_i denotes the molecular diffusion constant of the component i in the phase j, the factors dl^j and dt^j are the parameters of longitudinal and transversal dispersivity of the phase j, $\|\overline{u}^j\|_2$ is the Euclidean norm of the phase velocity, and $\underline{\underline{E}}(\overline{u}^j)$ is the orthogonal projection along the velocity field as expressed in Eqs. (3.19) and (3.20). Mechanical dispersion models the spreading of the component on the macroscopic level due to the random structure of the porous medium and depends on the size and direction of the flow velocity.

The dispersion part of the tensor is significantly larger than the molecular diffusion; also, dl^j is usually considerably larger than dt^j and their relationship can be expressed as a function of the Peclet number [24, 50]. Nevertheless, at low Peclet numbers (Pe < 10) both dispersivities are of the same order of magnitude [79–81]. Dispersion can represent small scale movements not captured by the REV used in the mathematical model, but according to Heimsund [33], taking this into account in the numerical model may be troublesome [82, 83]. It is worth mentioning that some numerical schemes, especially first order methods (see Point 3.2.2), add artificial diffusion which is most of the times far greater than the physical dispersion discussed here. It is then advisable that when using numerical dissipative schemes the physical dispersion should be neglected [33, 58, 84].

Phase partitioning

If a compositional model is formed by a system of N_{comp} components and N_p phases, there are a total of $N_p \cdot (N_{comp} + 3)$ unknowns. These in each grid block are: $N_p \cdot N_{comp}$ concentrations, for each component on each phase, and $3N_p$ values for saturation, pressure and Darcy velocities. The governing system described previously include N_{comp} conservation of mass equations, $N_p - 1$ capillary pressure relationships, N_p Darcy equations, N_p phase constraints, and one saturation constraint. To define the system of equations a number of $N_{comp} \cdot (N_p - 1)$ extra relationships shall be defined so as to define the system considering an isothermal medium and local thermodynamic equilibrium [24, 45, 85–87]. These relationships take the form of Eq. (3.21) relating components i_1 and i_2 of phase j,

$$\frac{c_{i_1}^j}{c_{i_2}^j} = K_{i_1,i_2}^j \left(T, p_{i_1}, p_{i_2}, c_{i_1}^j, c_{i_2}^j \right) \tag{3.21}$$

Energy equation

In case the flow cannot be considered isothermal, or the recovery process involves the addition of considerable amounts of energy to the reservoir, an extra condition and variable must be introduced to the system. The energy conservation (Eq. (3.22)) and its dependent variable, the temperature, are then added to the system. The major difference with respect to the other equations is that the energy is also conducted by the rock formation, and not only between the phases. If the local thermal equilibrium concept is applied, the temperature in the REV for all the phases and the porous medium is considered to be the same and the energy equation is as follows [24],

$$\frac{\partial}{\partial t} \left[\phi \sum_{j=1}^{N_p} \left(\rho^j S^j U^j \right) + (1 - \phi) \rho_s C_s T \right] + \nabla \cdot \left[\sum_{j=1}^{N_p} \left(\rho^j \overline{u}^j H^j \right) - \underline{\underline{k}}_t \nabla T \right] = q_H - q_L \tag{3.22}$$

where U^j is the specific internal energy, C_s the specific heat capacity of the rock, H^j the enthalpy of the phase, $\underline{\underline{k}}_t$ the thermal conductivity tensor, q_H the enthalpy source term, and q_L the heat loss. In most of the chemical EOR operations, the heat transfer is considered negligible and therefore an isothermal assumption is valid.

Well models

A production/injection well is a vertical (or vertical/horizontal in case of horizontal wells), open hole through which fluid can flow in and out of the reservoir, according to the strategies or its degree of maturity. These are cemented and then perforated along specific intervals (multi-zone wells). The primary function of production wells is to extract hydrocarbons and later on, the water/chemical products injected as part of EOR processes. Injection wells can also be used for disposal of certain fluids (e.g., CO_2 storage) as well as to inject chemical solutions so as to increase the recovery efficiency, sweeping the oil towards production wells. These wells are controlled through surface facilities (e.g., choke valves, Christmas trees) (Fig. 3.11).

The main purpose of a well model is to represent the flow in the wellbore and provide equations that serve as input for the mass conservation and Darcy equations, to calculate the flow rate of each component being injected or produced. Generally, the bottomhole pressure is significantly different from the average one in the perforated grid blocks. Modeling injection and production of fluids using point sources causes numerical problems in the flow field, so the concept of a productivity/well index (PI) was introduced in the form $-q = \text{PI}(p - p_{wf})$, to relate the bottomhole pressure p_{wf} to the numerically computed pressure p inside the model [23]. Here, the well index PI takes into account the geometric characteristics of the well and the properties of the surrounding rock, as is indicated in Eq. (3.23). The most used model was developed by Peaceman [88]. Assuming steady-state radial flow, the well index for an anisotropic medium represented on a Cartesian grid in three dimensions yields,

$$\text{PI}^j = \frac{2\pi \sqrt{K_x K_y} \Delta z}{\left[\ln\left(\frac{r_0}{r_w}\right) + s\right]} \cdot \frac{k_r^j}{\mu^j}, \quad r_0 = 0.28 \frac{\left[\left(\frac{K_x}{K_y}\right)^{1/2} \Delta y^2 + \left(\frac{K_y}{K_x}\right)^{1/2} \Delta x^2\right]^{1/2}}{\left(\frac{K_x}{K_y}\right)^{1/4} + \left(\frac{K_y}{K_x}\right)^{1/4}} \tag{3.23}$$

where s is the skin factor, r_w is the well radius and r_0 is the effective block radius at which the steady-state pressure equals the computed block one. The Peaceman model has been also extended to horizontal wells and modified to take into account non-square grids, boundary blocks, and non-Darcy effects [49, 89–92].

3.2 Numerical techniques for fluid flow in porous media

Reservoir flow problems can be highly complex, consisting of many different physical effects when it comes to EOR processes. The analysis of all these phenomena can be achieved, up to some extent, by laboratory experiments or field tests at small scale, but these tend to be expensive to conduct and may not be extrapolated to the whole reservoir. In order to solve this problem, mathematical models became progressively more important. Using these along with analytical solutions, engineers provided basic performance predictions so as to modify production strategies.

Several numerical formulations are employed to solve the non-linear systems of equations. The most stable approach is a fully implicit solution technique in pressure and saturation/concentration, but this generally leads to large, ill-conditioned matrices. Another scheme broadly utilized in compositional formulation consists, in order to reduce the level of implicitness, to solve the pressure equation system implicitly (which can be viewed as an overall volume balance) plus a sequence of ($N_{comp} - 1$) components conservation equations [24, 26, 93–97]. This equation has a strongly hyperbolic character due to the advective term. The overall technique chosen to solve

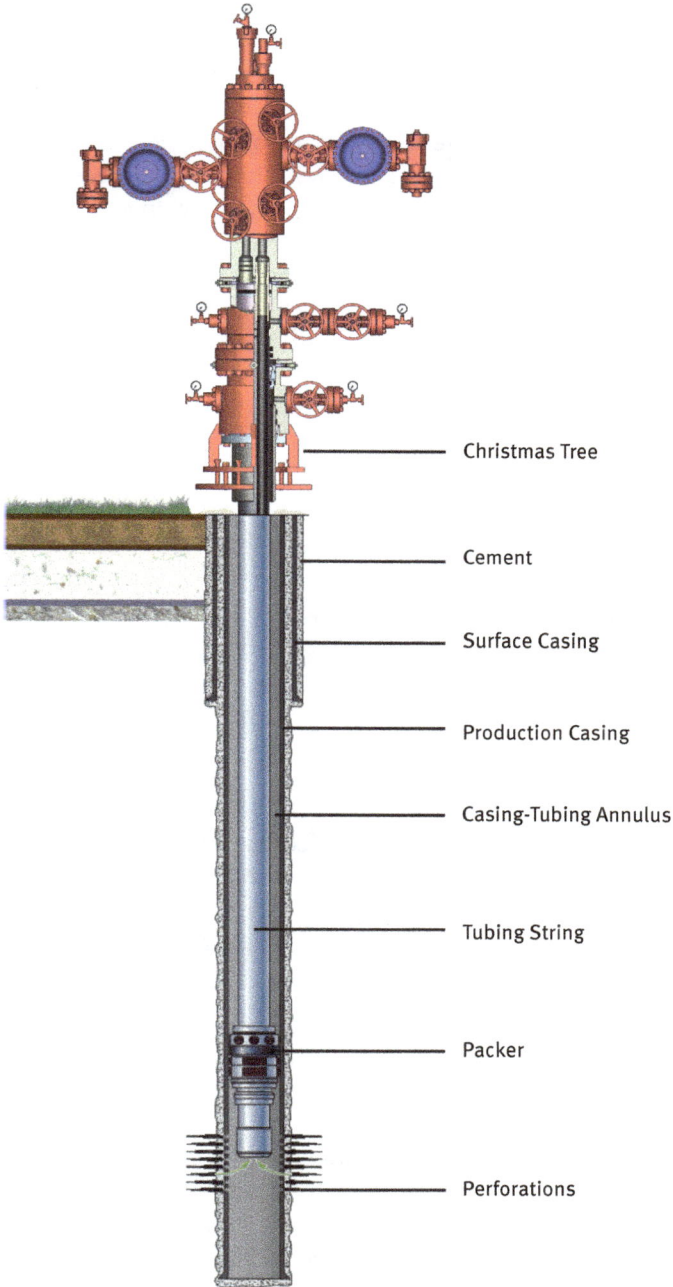

— Christmas Tree

— Cement

— Surface Casing

— Production Casing

— Casing-Tubing Annulus

— Tubing String

— Packer

— Perforations

Fig. 3.11: Schematic representation (out of scale) of a single-zone, conventional oil well [9].

the system during this chapter is called IMPEC (implicit in pressure, explicit in concentration). IMPEC methods are often limited by stability restrictions on the time step size due to their explicit scheme, but solutions do not suffer less smoothing than fully implicit methods, which strongly affect the performance prediction of compositional models [76]. Different implicit techniques are also used and often provide enhanced efficiency, allowing bigger time steps than those employed with IMPEC. One of these methods is the implicit pressure and saturation (IMPSAT) procedure, in which pressures and $(N_p - 1)$ saturations (but not compositions) are determined implicitly [86]. Different techniques worth of mentioning are: Adaptive implicit (AIM) [86, 98], bilinear approximation techniques [99], preconditioning schemes [100, 101], parallel computing and adaptive mesh refinement (AMR) in compositional simulation [76, 102, 103].

3.2.1 Numerical schemes

The aim of this section is the derivation and explanation of the numerical schemes to be used for solving differential equations presented above, as well as to also explain the reasons for the occurrence and possible numerical solutions of certain phenomena that affect simulation results [104]. The equation to be used as a model is the advection-diffusion equation in 1D, which is a simplification of the continuity equation presented for the compositional model.

$$\frac{\partial u}{\partial t} + v \frac{\partial u}{\partial x} = D \frac{\partial^2 u}{\partial x^2} \tag{3.24}$$

By means of finite-difference techniques a continuous medium is transformed into a discrete representation with a finite number of points in a spatial (i) and temporal ($\langle n \rangle$) grids. Then, the time derivative in previous equation is expressed using a Taylor series expansion around the point $u_i^{\langle n \rangle}$ yielding Eqs. (3.25) and (3.26),

$$u_i^{\langle n+1 \rangle} = u_i^{\langle n \rangle} + \Delta t \frac{\partial u}{\partial t}\Big|_i^{\langle n \rangle} + \frac{\Delta t^2}{2} \frac{\partial^2 u}{\partial t^2}\Big|_i^{\langle n \rangle} + \mathcal{O}\left(\Delta t^3\right) \tag{3.25}$$

$$\frac{\partial u}{\partial t}\Big|_i^{\langle n \rangle} = \frac{u_i^{\langle n+1 \rangle} - u_i^{\langle n \rangle}}{\Delta t} + \mathcal{O}\left(\Delta t\right) \tag{3.26}$$

where $\mathcal{O}(\Delta t)$ is part of the Bachmann–Landau notation, used to describe the error term in an approximation to a mathematical function (of one or several variables). Therefore, the significant terms in the approximation are written explicitly, whilst the least-important terms are summarized using this notation. In numerical analysis, the big-$\mathcal{O}$ is used to describe how closely a finite series using Taylor expansion approximates the derivative being discretized. Time and spatial operators may have finite-difference schemes with different orders of accuracy and in this case the overall order of the equation is determined by the differential operator with the largest truncation error. Noteworthy is that while the latter is expressed for the differential operator, the numerical algorithms will not be expressed in terms of the differential operators and will therefore have different truncation errors. Following a similar procedure, finite-difference approximations can be obtained (Eqs. (3.27) and (3.28)) for the space derivative in backwards and centered forms as,

$$\frac{\partial u}{\partial x}\bigg|_i^{\langle n\rangle} = \frac{u_i^{\langle n\rangle} - u_{i-1}^{\langle n\rangle}}{\Delta x} + \mathcal{O}\left(\Delta x\right) \tag{3.27}$$

$$\frac{\partial u}{\partial x}\bigg|_i^{\langle n\rangle} = \frac{u_{i+1}^{\langle n\rangle} - u_{i-1}^{\langle n\rangle}}{2\Delta x} + \mathcal{O}\left(\Delta x^2\right) \tag{3.28}$$

Even though the centered scheme has a higher order of precision, it generates unstable results when applied to the advection equation (wave transport equation). Hence, numerical methods are employed which only use points located "upwind" of the wave-front [105].

The inclusion of diffusion phenomena in the description of a fluid flow leads to non-trivial complications in the numerical solution of the mass conservation equations. From an analytical point of view, the resulting equations are no longer purely hyperbolic partial differential equations (PDE's) but rather mixed hyperbolic-parabolic PDE's. This means that the numerical method used to solve them must necessarily be able to cope with the parabolic part of the equations. For the diffusive term, the second order derivative is usually discretized using a centered scheme yielding,

$$\frac{\partial^2 u}{\partial x^2}\bigg|_i^{\langle n\rangle} = \frac{u_{i+1}^{\langle n\rangle} - 2u_i^{\langle n\rangle} + u_{i-1}^{\langle n\rangle}}{\Delta x^2} + \mathcal{O}\left(\Delta x^2\right) \tag{3.29}$$

The final upwind discretized equation (FTUS – forward in time, upwind in space) and its matrix form for the advective-diffusive system are then presented in Eqs. (3.30) and (3.31), respectively.

$$\frac{u_i^{\langle n+1\rangle} - u_i^{\langle n\rangle}}{\Delta t} + v\frac{u_i^{\langle n\rangle} - u_{i-1}^{\langle n\rangle}}{\Delta x} = D\frac{u_{i+1}^{\langle n\rangle} - 2u_i^{\langle n\rangle} + u_{i-1}^{\langle n\rangle}}{\Delta x^2} + \mathcal{O}\left(\Delta x, \Delta t\right) \tag{3.30}$$

$$\underline{\underline{A}}\overline{u}^{\langle n+1\rangle} = \underline{\underline{B}}\overline{u}^{\langle n\rangle} + \overline{C} \tag{3.31}$$

In Eq. (3.30) the order of accuracy is expressed as a function of both independent variables $\mathcal{O}(\Delta x, \Delta t)$, inferring that both discretization schemes for spatial and temporal grids have influence in the total error of the numerical model. The solution at the new time-step $\langle n+1 \rangle$ can be calculated explicitly from the quantities that are already known at the previous step $\langle n \rangle$. This differs with an implicit scheme in which the finite-difference representations of the differential equation are expressed in terms of the new time-level $\langle n+1 \rangle$. These methods require solving a number of coupled algebraic equations. In Tab. 3.1 several schemes of different finite-difference operators are summarized indicating the truncation errors and the their representation in a purely advective or diffusive 1D equation.

Tab. 3.1: Most common numerical schemes used in reservoir simulation.

Method	Order	Finite-Difference Form
Purely Advective		
Upwind (FTUS)	$\mathcal{O}(\Delta x, \Delta t)$	$\dfrac{u_j^{\langle n+1 \rangle} - u_j^{\langle n \rangle}}{\Delta t} + v\dfrac{u_j^{\langle n \rangle} - u_{j-1}^{\langle n \rangle}}{\Delta x} = 0$
Centered (FTCS)	$\mathcal{O}(\Delta x^2, \Delta t)$	$\dfrac{u_j^{\langle n+1 \rangle} - u_j^{\langle n \rangle}}{\Delta t} + v\dfrac{u_{j+1}^{\langle n \rangle} - u_{j-1}^{\langle n \rangle}}{2\Delta x} = 0$
Lax–Friedrichs	$\mathcal{O}(\Delta x^2, \Delta t)$	$\dfrac{u_j^{\langle n+1 \rangle} - {}^{1}\!/\!{}_{2}\left(u_{j-1}^{\langle n \rangle} + u_{j+1}^{\langle n \rangle}\right)}{\Delta t} + v\dfrac{u_{j+1}^{\langle n \rangle} - u_{j-1}^{\langle n \rangle}}{2\Delta x} = 0$
Lax–Wendroff	$\mathcal{O}(\Delta x^2, \Delta t^2)$	$\dfrac{u_j^{\langle n+1 \rangle} - u_j^{\langle n \rangle}}{\Delta t} + v\dfrac{u_{j+1}^{\langle n \rangle} - u_{j-1}^{\langle n \rangle}}{2\Delta x} = \dfrac{v^2 \Delta t}{2}\left(\dfrac{u_{j+1}^{\langle n \rangle} - 2u_j^{\langle n \rangle} + u_{j-1}^{\langle n \rangle}}{\Delta x^2}\right)$
Beam–Warming	$\mathcal{O}(\Delta x^2, \Delta t^2)$	$\dfrac{u_j^{\langle n+1 \rangle} - u_j^{\langle n \rangle}}{\Delta t} + v\dfrac{3u_j^{\langle n \rangle} - 4u_{j-1}^{\langle n \rangle} + u_{j-2}^{\langle n \rangle}}{2\Delta x} = \dfrac{v^2 \Delta t}{2}\left(\dfrac{u_{j+1}^{\langle n \rangle} - 2u_j^{\langle n \rangle} + u_{j-1}^{\langle n \rangle}}{\Delta x^2}\right)$
Purely Diffusive		
Leapfrog	$\mathcal{O}(\Delta x^2, \Delta t^2)$	$\dfrac{u_j^{\langle n+1 \rangle} - u_j^{\langle n-1 \rangle}}{2\Delta t} = D\dfrac{u_{j+1}^{\langle n \rangle} - 2u_j^{\langle n \rangle} + u_{j-1}^{\langle n \rangle}}{\Delta x^2}$
Crank–Nicholson	$\mathcal{O}(\Delta x^2, \Delta t^2)$	$\dfrac{u_j^{\langle n+1 \rangle} - u_j^{\langle n \rangle}}{\Delta t} = \dfrac{D}{2}\left(\dfrac{u_{j+1}^{\langle n \rangle} - 2u_j^{\langle n \rangle} + u_{j-1}^{\langle n \rangle}}{\Delta x^2} + \dfrac{u_{j+1}^{\langle n+1 \rangle} - 2u_j^{\langle n+1 \rangle} + u_{j-1}^{\langle n+1 \rangle}}{\Delta x^2}\right)$
DuFort–Frankel	$\mathcal{O}(\Delta x^2, \Delta t^2, {}^{\Delta t^2}\!/\!{}_{\Delta x^2})$	$\dfrac{u_j^{\langle n+1 \rangle} - u_j^{\langle n-1 \rangle}}{2\Delta t} = D\dfrac{u_{j+1}^{\langle n \rangle} - u_j^{\langle n+1 \rangle} - u_j^{\langle n-1 \rangle} + u_{j-1}^{\langle n \rangle}}{\Delta x^2}$

Moreover, Tab. 3.2 summarizes the most common discretization stencils and their orders of accuracy [104]. Higher-order schemes allow increasing the latter at the cost of requiring a higher number of points to make the evaluation of the derivatives. This increases both the computational cost and the difficulty of evaluating derivatives near the boundaries.

A key factor in all numerical schemes is the issue of how treating the solution on the boundaries of the spatial grid as the time evolution proceeds. Two types of conditions are generally used in reservoir simulation to describe whether the Darcy velocity or the pressure of a phase at the boundaries. These are: Dirichlet type conditions,

Tab. 3.2: Higher order schemes for first and second order derivatives.

Type	Difference Stencil	Order
$\partial u/\partial x$		
Backward	$\dfrac{u_j^{\langle n\rangle} - u_{j-1}^{\langle n\rangle}}{\Delta x}$	$\mathcal{O}(\Delta x)$
Backward	$\dfrac{3u_j^{\langle n\rangle} - 4u_{j-1}^{\langle n\rangle} + u_{j-2}^{\langle n\rangle}}{2\Delta x}$	$\mathcal{O}(\Delta x^2)$
Centered	$\dfrac{u_{j+1}^{\langle n\rangle} - u_{j-1}^{\langle n\rangle}}{2\Delta x}$	$\mathcal{O}(\Delta x^2)$
Backward	$\dfrac{2u_{j+1}^{\langle n\rangle} + 3u_j^{\langle n\rangle} - 6u_{j-1}^{\langle n\rangle} + u_{j-2}^{\langle n\rangle}}{6\Delta x}$	$\mathcal{O}(\Delta x^3)$
Backward	$\dfrac{25u_j^{\langle n\rangle} - 48u_{j-1}^{\langle n\rangle} + 36u_{j-2}^{\langle n\rangle} - 16u_{j-3}^{\langle n\rangle} + 3u_{j-4}^{\langle n\rangle}}{12\Delta x}$	$\mathcal{O}(\Delta x^4)$
$\partial^2 u/\partial x^2$		
Centered	$\dfrac{u_{j+1}^{\langle n\rangle} - 2u_j^{\langle n\rangle} + u_{j-1}^{\langle n\rangle}}{\Delta x^2}$	$\mathcal{O}(\Delta x^2)$
Centered	$\dfrac{-u_{j+2}^{\langle n\rangle} + 16u_{j+1}^{\langle n\rangle} - 30u_j^{\langle n\rangle} + 16u_{j-1}^{\langle n\rangle} - u_{j-2}^{\langle n\rangle}}{12\Delta x^2}$	$\mathcal{O}(\Delta x^4)$

when the values of the relevant quantity are imposed at the boundaries of the grid (these values can be either functions of time or be held constant), and Neumann type conditions, when the values of the derivatives of the relevant quantity are imposed.

3.2.2 Numerical dissipation and dispersion

The exact solution of the discretized equation satisfies a PDE different from the one being solved. This difference is represented by the local truncation error (LTE) of the numerical scheme. The LTE can be expressed as a function of higher order derivatives [106, 107],

Original PDE	Modified Equation
$\dfrac{\partial u}{\partial t} + v\dfrac{\partial u}{\partial x} - D\dfrac{\partial^2 u}{\partial x^2} = 0$	$\dfrac{\partial u}{\partial t} + v\dfrac{\partial u}{\partial x} - D\dfrac{\partial^2 u}{\partial x^2} = \sum_{n=1}^{\infty} a_{2n}\dfrac{\partial^{2n} u}{\partial x^{2n}} + \sum_{n=1}^{\infty} a_{2n+1}\dfrac{\partial^{2n+1} u}{\partial x^{2n+1}}$

The procedure to calculate this error and assess its contribution to the numeric solution is straightforward. It consists of performing an expansion in a double Taylor series around a single point $u_i^{\langle n\rangle}$, both in spatial and temporal grids to obtain a modified PDE. Besides, the high-order time derivatives as well as mixed derivatives must be transformed in terms of space derivatives using this modified PDE. The analysis for the truncation error in the 1D advective-diffusive equation is presented using two numerical schemes: firstly the upwind scheme and subsequently the Lax–Wendroff. For the upwind scheme it yields Eq. (3.32),

$$\frac{u_i^{\langle n+1\rangle} - u_i^{\langle n\rangle}}{\Delta t} + v\frac{u_i^{\langle n\rangle} - u_{i-1}^{\langle n\rangle}}{\Delta x} = D\frac{u_{i+1}^{\langle n\rangle} - 2u_i^{\langle n\rangle} + u_{i-1}^{\langle n\rangle}}{\Delta x^2} + \mathcal{O}(\Delta x, \Delta t) \tag{3.32}$$

$$u_i^{\langle n+1\rangle} = u_i^{\langle n\rangle} + \Delta t\frac{\partial u}{\partial t}\Big|_i^{\langle n\rangle} + \frac{\Delta t^2}{2}\frac{\partial^2 u}{\partial t^2}\Big|_i^{\langle n\rangle} + \frac{\Delta t^3}{6}\frac{\partial^3 u}{\partial t^3}\Big|_i^{\langle n\rangle} + \mathcal{O}(\Delta t^4) \tag{3.33}$$

$$u_{i+1}^{\langle n\rangle} = u_i^{\langle n\rangle} + \Delta x\frac{\partial u}{\partial x}\Big|_i^{\langle n\rangle} + \frac{\Delta x^2}{2}\frac{\partial^2 u}{\partial x^2}\Big|_i^{\langle n\rangle} + \frac{\Delta x^3}{6}\frac{\partial^3 u}{\partial x^3}\Big|_i^{\langle n\rangle} + \frac{\Delta x^4}{24}\frac{\partial^4 u}{\partial x^4}\Big|_i^{\langle n\rangle} + \mathcal{O}(\Delta x^5) \tag{3.34}$$

$$u_{i-1}^{\langle n\rangle} = u_i^{\langle n\rangle} - \Delta x\frac{\partial u}{\partial x}\Big|_i^{\langle n\rangle} + \frac{\Delta x^2}{2}\frac{\partial^2 u}{\partial x^2}\Big|_i^{\langle n\rangle} - \frac{\Delta x^3}{6}\frac{\partial^3 u}{\partial x^3}\Big|_i^{\langle n\rangle} + \frac{\Delta x^4}{24}\frac{\partial^4 u}{\partial x^4}\Big|_i^{\langle n\rangle} + \mathcal{O}(\Delta x^5) \tag{3.35}$$

Introducing these terms from Eqs. (3.33), (3.34) and (3.35) in the numerical scheme presented in Eq. (3.32) yields

$$\left(\frac{\partial u}{\partial t} + \frac{\Delta t}{2}\frac{\partial^2 u}{\partial t^2} + \frac{\Delta t^2}{6}\frac{\partial^3 u}{\partial t^3}\right)\Big|_i^{\langle n\rangle} + \frac{v}{\Delta x}\left(\Delta x\frac{\partial u}{\partial x} - \frac{\Delta x^2}{2}\frac{\partial^2 u}{\partial x^2} + \frac{\Delta x^3}{6}\frac{\partial^3 u}{\partial x^3} - \frac{\Delta x^4}{24}\frac{\partial^4 u}{\partial x^4}\right)\Big|_i^{\langle n\rangle}$$

$$= \frac{D}{\Delta x^2}\left(\Delta x^2\frac{\partial^2 u}{\partial x^2} + \frac{\Delta x^2}{12}\frac{\partial^4 u}{\partial x^4}\right)\Big|_i^{\langle n\rangle} + \mathcal{O}(\Delta x^4, \Delta t^3) \tag{3.36}$$

Rearranging Eq. (3.36) to split the original PDE and the truncation error gives

$$\frac{\partial u}{\partial t}\Big|_i^{\langle n\rangle} + v\frac{\partial u}{\partial x}\Big|_i^{\langle n\rangle} - D\frac{\partial^2 u}{\partial x^2}\Big|_i^{\langle n\rangle} = -\frac{\Delta t}{2}\frac{\partial^2 u}{\partial t^2}\Big|_i^{\langle n\rangle} - \frac{\Delta t^2}{6}\frac{\partial^3 u}{\partial t^3}\Big|_i^{\langle n\rangle}$$

$$+ \frac{v\Delta x}{2}\frac{\partial^2 u}{\partial x^2}\Big|_i^{\langle n\rangle} - \frac{v\Delta x^2}{6}\frac{\partial^3 u}{\partial x^3}\Big|_i^{\langle n\rangle} + \frac{v\Delta x^3}{24}\frac{\partial^4 u}{\partial x^4}\Big|_i^{\langle n\rangle}$$

$$+ \frac{D\Delta x^2}{12}\frac{\partial^4 u}{\partial x^4}\Big|_i^{\langle n\rangle} + \mathcal{O}(\Delta x^3, \Delta t^3) \tag{3.37}$$

The temporal derivatives in Eq. (3.37) are transformed into space derivatives. Furthermore, using Courant and Peclet dimensionless groups, the LTE for the method is derived in Eq. (3.39).

$$Cr = \frac{v\Delta t}{\Delta x}, \quad Pe = \frac{\Delta x v}{D} \tag{3.38}$$

$$\frac{\partial u}{\partial t}\Big|_i^{\langle n\rangle} + v\frac{\partial u}{\partial x}\Big|_i^{\langle n\rangle} - D\frac{\partial^2 u}{\partial x^2}\Big|_i^{\langle n\rangle} = -\frac{v\Delta x}{2}(1 - Cr)\frac{\partial^2 u}{\partial x^2}\Big|_i^{\langle n\rangle}$$

$$+ \frac{v\Delta x^2}{6}\left(3Cr - 2Cr^2 - 1 + \frac{6Cr}{Pe}\right)\frac{\partial^3 u}{\partial x^3}\Big|_i^{\langle n\rangle}$$

$$+ \mathcal{O}(\Delta x^3, \Delta t^3) \tag{3.39}$$

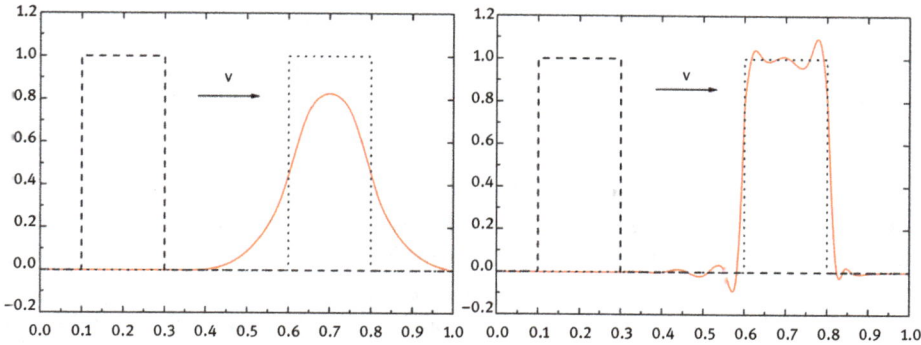

Fig. 3.12: Influence of even (left) and odd (right) higher order derivatives in the numerical solution [9].

The numerical scheme does not solve the original PDE, but a modified PDE with extra terms of higher order derivatives. The extra term containing the second order derivative is interpreted as a numerical diffusion, additional to the physical coefficient D. As long as the Cr < 1 condition is met, the numerical solution will produce an artificial smearing given by the term $(v\Delta x/2)(1 - \text{Cr})$. The term containing the third order derivative is interpreted as a numerical dispersion, which causes phase errors in the wave speed v. As this term is positive spurious oscillations occur ahead of steep wave fronts, and vice versa. To summarize this analysis, the terms of derivatives of even-order provoke a numerical diffusion which modifies the amplitude of the wave, while odd-terms cause numerical dispersion, which translates as oscillations in the wave front (Fig. 3.12).

One way to reduce these numerical errors is by means of additional, artificial factors which stabilize or decrease the previously seen effects. As an example, the streamline diffusion method consists of adding a term of artificial diffusion to counteract the added terms by the numerical scheme; the non-oscillatory shock-capturing methods, TVD (total variation diminishing) or flux-limiters [23] can also be listed as improved schemes to overcome these effects. To reduce the influence of these differences a possible solution is also to use schemes of superior orders (Tab. 3.2). As an example of these techniques, the same analysis done for the upwind scheme is performed using the Lax–Wendroff method. In the latter the time derivative is expressed as a second order Taylor series expansion [106, 107]. The numerical model to solve is presented in Eq. (3.40),

$$\frac{u_i^{\langle n+1\rangle} - u_i^{\langle n\rangle}}{\Delta t} + v\frac{u_{i+1}^{\langle n\rangle} - u_{i-1}^{\langle n\rangle}}{2\Delta x} = \left(D + \frac{v^2\Delta t}{2}\right)\frac{u_{i+1}^{\langle n\rangle} - 2u_i^{\langle n\rangle} + u_{i-1}^{\langle n\rangle}}{\Delta x^2} + \mathcal{O}\left(\Delta x^2, \Delta t^2\right) \quad (3.40)$$

Following a procedure similar to the previous scheme it renders Eq. (3.41),

$$\left.\frac{\partial u}{\partial t}\right|_i^{\langle n \rangle} + v \left.\frac{\partial u}{\partial x}\right|_i^{\langle n \rangle} - D \left.\frac{\partial^2 u}{\partial x^2}\right|_i^{\langle n \rangle} = -\frac{v \Delta x^2}{6}\left(1 - \text{Cr}^2 + \frac{6\text{Cr}}{\text{Pe}}\right)\left.\frac{\partial^3 u}{\partial x^3}\right|_i^{\langle n \rangle} + \mathcal{O}\left(\Delta x^3, \Delta t^3\right)$$

(3.41)

As shown, the first diffusive term in the LTE has disappeared, leaving the dispersive term as the main source of error. Since this term is negative, the spurious oscillations occur behind steep fronts. It is worth mentioning that a more accurate numerical scheme is not necessarily a preferable one. As an example, the upwind and the Lax–Friedrichs methods are both dissipative, although the latter is generically more despite being of higher order accuracy in space.

3.2.3 Flux limiters

In the previous section two of the most common numerical schemes utilized were introduced, and the advantages/disadvantages of each were studied and inferred. While the upwind scheme can handle steep gradients, it is very diffusive and moreover a first order scheme; on the other hand, Lax–Wendroff is a second order scheme, less diffusive but presents serious problems when sharp gradients are present in the system. Therefore, new numerical schemes were published coupling low- and high-resolution methods, taking advantage of the mentioned characteristics [106, 108–117]. For this analysis, the 1D advection-diffusion equation is considered in terms of fluxes ($F_i^{\langle n \rangle}$) (Eq. (3.42)), with no diffusive terms.

$$\frac{u_i^{\langle n+1 \rangle} - u_i^{\langle n \rangle}}{\Delta t} + \frac{1}{\Delta x}\left(F_{i+1/2}^{\langle n \rangle} - F_{i-1/2}^{\langle n \rangle}\right) = 0$$

(3.42)

The idea behind this concept is then to write the fluxes as a function of low- and high-resolution numerical schemes, using a proportionality factor.

$$F_{i\pm1/2}^{\langle n \rangle} = \psi_{i\pm1/2}\left(r_i\right) \cdot F_{i\pm1/2}^{\text{high}} + \left[1 - \psi_{i\pm1/2}\left(r_i\right)\right] \cdot F_{i\pm1/2}^{\text{low}}$$

(3.43)

The proportionality factor $\psi_{i\pm1/2}(r_i)$, also called the flux limiter function, depends on the ratio of consecutive gradients in the numerical mesh (Eq. (3.44)), this is,

$$r_i = \frac{u_i - u_{i-1}}{u_{i+1} - u_i}$$

(3.44)

Using the FTUS and Lax–Wendroff as low- and high-resolution schemes respectively, the flux is calculated according to Eq. (3.46).

$$F_{i+1/2}^{\text{low}} = v \cdot u_i \,, \quad F_{i+1/2}^{\text{high}} = v \cdot u_i + \frac{v\left(1 - \text{Cr}\right)}{2}\left(u_{i+1} - u_i\right)$$

(3.45)

$$F_{i+1/2}^{\langle n \rangle} = v \cdot u_i + \psi_{i+1/2}\left(r_i\right) \cdot \frac{v\left(1 - \text{Cr}\right)}{2}\left(u_{i+1} - u_i\right)$$

(3.46)

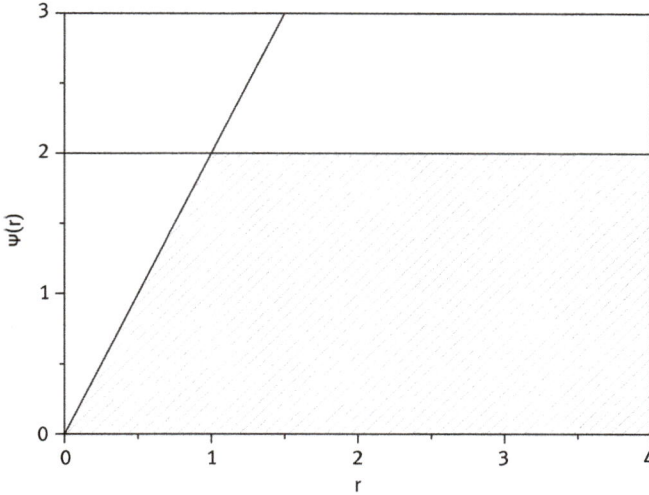

Fig. 3.13: Total variation diminishing (TVD) region for the flux limiter function [9].

Finally, the discretized advection Eq. (3.47) is written in terms of the flux limiter parameter.

$$u_i^{\langle n+1 \rangle} = u_i^{\langle n \rangle} - \left[Cr - \frac{Cr\,(1 - Cr)}{2} \psi_{i-1/2}\,(r_i) \right] \left(u_i^{\langle n \rangle} - u_{i-1}^{\langle n \rangle} \right)$$
$$- \frac{Cr\,(1 - Cr)}{2} \psi_{i+1/2}\,(r_i) \left(u_{i+1}^{\langle n \rangle} - u_i^{\langle n \rangle} \right) \tag{3.47}$$

$$u_i^{\langle n+1 \rangle} = u_i^{\langle n \rangle} - Cr \left[1 - \frac{(1 - Cr)}{2} \psi_{i-1/2}\,(r_i) \right.$$
$$\left. + \frac{(1 - Cr)}{2} \frac{\psi_{i+1/2}\,(r_i)}{r_{i+1/2}} \right] \left(u_i^{\langle n \rangle} - u_{i-1}^{\langle n \rangle} \right) \tag{3.48}$$

Equation (3.48) resembles the FTUS scheme with a modified Courant number [109, 112, 113]. The properties of stability and monotony of the FTUS are well-known. So for the new numerical scheme to meet these requirements the inequation (3.49) must be valid,

$$0 \leq Cr \left[1 - \frac{(1 - Cr)}{2} \psi_{i-1/2}\,(r_i) + \frac{(1 - Cr)}{2} \frac{\psi_{i+1/2}\,(r_i)}{r_{i+1/2}} \right] \leq 1 \tag{3.49}$$

This is valid for positives values of r, when the following two conditions in equation (3.50) are met (Fig. 3.13).

$$0 \leq \left(\frac{\psi(r)}{r}, \psi(r) \right) \leq 2 \tag{3.50}$$

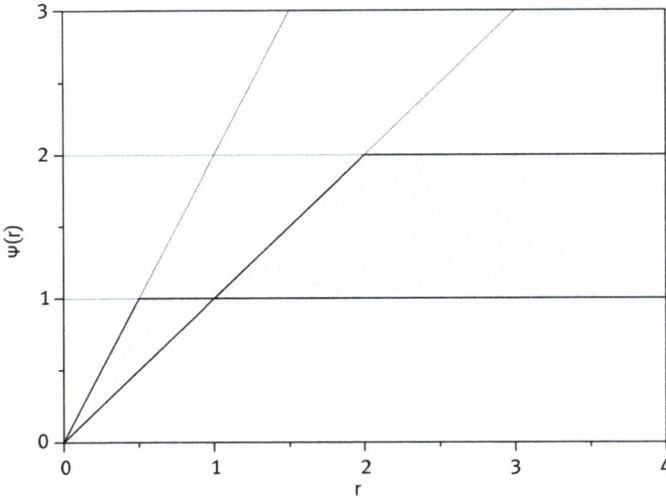

Fig. 3.14: Total variation diminishing, second-order accuracy region for the flux limiter function [9].

Further, more restrictive constraints are applied in order to make the scheme second order in accuracy (Fig. 3.14). Several high-order flux limiter functions were developed within this region. These are characterized by a low numerical dispersion in high gradient fields as well as being less diffusive than traditional schemes (e.g., FTUS) in low gradient advection phenomena (Tab. 3.3 and Fig. 3.15).

Tab. 3.3: Most commonly used flux limiter functions.

Type	Flux Limiter Function	Reference
Superbee	$\max[0, \min(2r, 1), \min(r, 2)]$	Roe [118]
Minmod	$\max[0, \min(r, 1)]$	Roe [119]
Van Leer	$\frac{r+|r|}{1+|r|}$	Van Leer [120]
Van Albada	$\frac{r+r^2}{1+r^2}$	Van Albada [121]
Koren	$\max[0, \min(2r, \frac{2+r}{3}, 2)]$	Koren [122]
CHARM	$\frac{r \cdot (3r+1)}{(r+1)^2}$	Zhou [123]
MUSCL	$\max[0, \min(2r, \frac{1+r}{2}, 2)]$	Van Leer [124]

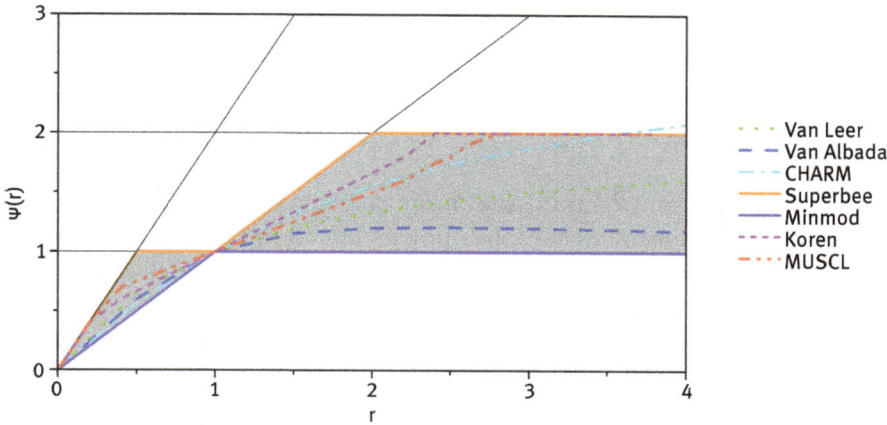

Fig. 3.15: Second-order TVD functions presented in Table 3.3 [9].

3.2.4 Consistency and stability

This chapter concludes with the study of concepts related to numerical simulation. The first to be defined is the consistency of a numerical scheme. Given a PDE in its operator form, $\mathcal{L}(u) - f = 0$, a finite difference scheme applied to this PDE, $\mathcal{L}_\Delta(u_i^{\langle n \rangle}) - f_i^{\langle n \rangle} = 0 + \mathcal{O}(\Delta x^p, \Delta t^p)$, and the LTE being expressed by means of a representative variable, $\delta(h) = Ch^p = \mathcal{O}(h^p), \forall C \in \mathbb{R}$, then the finite difference scheme is consistent with the PDE if and only if $\lim_{h \to 0} \delta(h) = 0$. This means that the operator of the finite difference scheme converges towards the continuous operator of the PDE as the increments in the independent variables $\Delta x, \Delta t$ vanish ($LTE \to 0$).

While the concept of consistency associates the original PDE with the discretized equation, it is necessary but not sufficient to ensure that the numerical results converge to the exact solution. It should also be ensured that the numerical results of the discretized equation converge to the exact results of the latter. This concept may seem trivial, but numerical errors introduced during simulation can grow boundlessly, thus amplifying the errors until the system eventually collapses. This is ensured by introducing the concept of numerical stability. To understand this, a system of equations is analyzed, and written in the form of the modified PDE (Eq. (3.31)). Perturbations at the baseline as well as at a generic time $\langle n \rangle = m\Delta t$ are introduced due to numerical errors during the simulation. This is expressed as $\tilde{u}_i^{\langle n \rangle} = u_i^{\langle n \rangle} + \epsilon_i^{\langle n \rangle}$ and $\tilde{u}_i^{\langle 0 \rangle} = u_i^{\langle 0 \rangle} + \epsilon_i^{\langle 0 \rangle}$. Replacing these values it renders Eq. (3.51),

$$\underline{\underline{A}}\,\overline{\tilde{u}}^{\langle n+1 \rangle} = \underline{\underline{B}}\,\overline{\tilde{u}}^{\langle n \rangle} \quad \rightarrow \quad \underline{\underline{A}}\,\overline{\epsilon}^{\langle n+1 \rangle} = \underline{\underline{B}}\,\overline{\epsilon}^{\langle n \rangle} \quad \rightarrow \quad \overline{\epsilon}^{\langle n+1 \rangle} = \underline{\underline{G}}\,\overline{\epsilon}^{\langle n \rangle} \qquad (3.51)$$

where $\underline{\underline{G}}$ is the amplification matrix of the numerical perturbations in the system. Using Eq. (3.51), a relationship is established between the perturbation at time n with the initial perturbation.

$$\overline{\epsilon}^{\langle n+1\rangle} = \underline{\underline{G}}\overline{\epsilon}^{\langle n\rangle} = \underline{\underline{G}}^2\overline{\epsilon}^{\langle n-1\rangle} = \cdots = \underline{\underline{G}}^n\overline{\epsilon}^{\langle 0\rangle} \tag{3.52}$$

Equation (3.52) exposes a central issue in the stability analysis: the question of whether amplification matrices have their powers uniformly bounded. In order for the numerical errors to not be amplified during the simulation, a stability restriction is then defined, $\|\underline{\underline{G}}^n\|_2 \leq K$, $K \in \mathbb{R}$. According to this condition, the only way to keep the errors limited and prevent their propagation and amplification is fulfilling the condition established in Eq. (3.53), ensuring the existence of K. Therefore, a numerical scheme applied to a differential equation is stable if the error caused by a small perturbation in the numerical solution remains bound. Thus, stability means that the numerical solution of the discretized equation converges to the exact solution of the latter.

$$\left\|\underline{\underline{G}}^n\right\|_2 \leq K, \quad n \to \infty \Leftrightarrow \rho\left(\underline{\underline{G}}\right) \leq 1 \tag{3.53}$$

Where $\rho(\underline{\underline{G}})$ is the spectral radius of the amplification matrix. Hence, a one-step finite difference scheme approximating a PDE is convergent *if and only if* it renders the exact

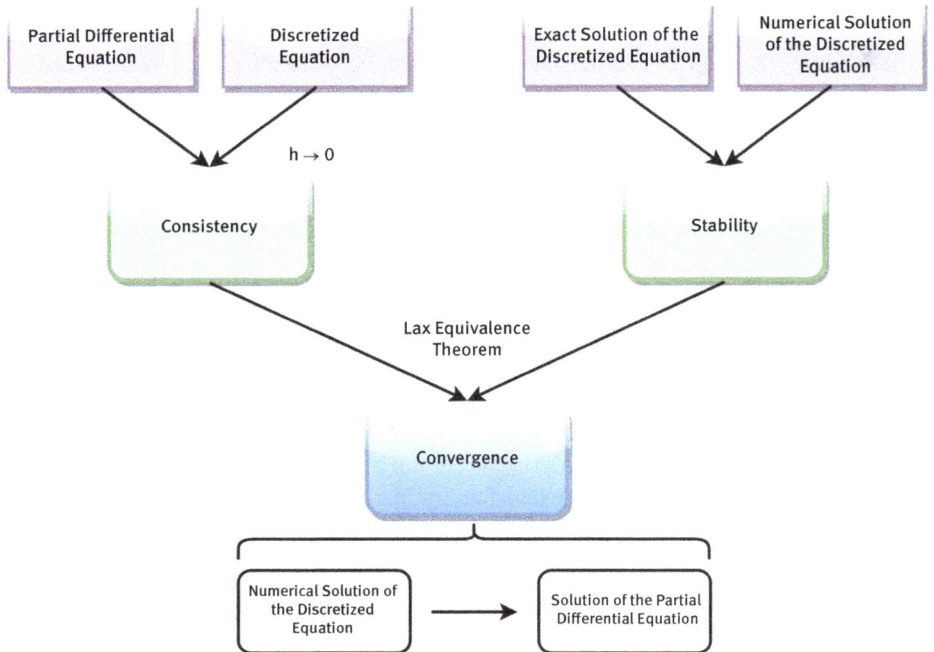

Fig. 3.16: Scheme of the Lax–Richtmyer equivalence theorem [46].

solution of the original PDE, expressed mathematically in Eq. (3.54),

$$\lim_{\substack{\Delta x \to 0 \\ \Delta t \to 0}} \left[u(x, t) - u_{n\Delta x}^{\langle m\Delta t \rangle} \right] = 0 \, , \quad \forall m, n \in \mathbb{N}_0^+ \wedge x = n\Delta x, \, t = m\Delta t \quad (3.54)$$

These three concepts can be summarized by the following theorem: Given a properly posed initial-value problem and a finite difference approximation to it, that satisfies the consistency criterion, stability is the necessary and sufficient condition for convergence. This theorem, known as the "Lax–Richtmyer equivalence theorem" or the fundamental theorem of numerical analysis (Fig. 3.16), is important since it shows that consistency, stability and convergence are strictly related. In general, proving that the numerical scheme adopted is stable will validate that the discretized equations represent the PDE as well as the numerical errors in the simulation are bounded at all times [104].

3.3 Conclusions

Reservoir simulation is a branch of engineering that emerged in recent years, since oil companies needed to justify and evaluate E&P investments. The former is not related to only one discipline, but includes the assistance and collaboration of various specialists so as to characterize a reservoir, estimate its profitability and give the "green light" to the project development phase. Oil reservoirs are geological traps where oil migrated and remained for long periods of time. An accurate determination of their physical characteristics has not yet been developed and statistical tools are used along with complex field tests to extrapolate these properties. In addition, oil is not a homogeneous, pure fluid but is composed by a large group of different components which may alter its properties. Hence, a set of variables must be previously evaluated and studied in order to get feasible results. The research and development of new exploration technologies to reduce the model uncertainties is deemed essential. These, along with production studies at reservoir scale on pilot wells will allow the performing of accurate history matching analysis. Due to the current oil reserve conditions and future production estimates, these new technologies should also consider their applicability also in non-conventional oil reservoirs (e.g., oil sands, tight oil, shale oil), or in geographical areas with harsh conditions (e.g., off-shore platforms).

Fluid flow simulations are performed using two different approaches: the Navier–Stokes equations throughout a complex network of microchannels in the porous medium; or the assumption of a continuum with averaged properties using Darcy's equation, rendering a system independent of the geometry at a microscopic scale. The first is only circumscribed to specific laboratory tests and has limited application in field studies. This is due to several factors, among them the uncertainties associated with the poral geometry in the field as well as high computational costs required to solve the system of equations. However, this approach might be useful in the design of new chemicals while being evaluated at a microscopic level. These studies should then be supplemented with scale reservoir simulations and field tests using Darcy's equation.

Subsequently, the mathematical tools for reservoir simulation were presented using finite difference methods (FDM). The errors introduced by these schemes have been addressed, as well as possible solutions to tackle these problems. The numerical convergence of a system of PDE's is a critical aspect that must be taken into account in order to limit numerical errors. In addition, the continuous increase in the complexity of numerical models has demanded a proportional increase of computational power to obtain results in a reasonable time frame. The development of new schemes of higher orders of accuracy as well as models dealing with the non-linearities present in the simulation could reduce either the computational requirements or the numerical errors produced. Besides the FDM's discussed in this chapter, other numerical schemes of higher complexity are used and offer certain advantages, such as the capability of dealing with complex geometries or geological faults. These techniques are, among others: finite element methods (FEM), finite volume methods (FVM), immersed boundary (IB), hybrid methods, etc. However, these advantages are related to the degree of certainty in the definition of the physical boundaries and properties of the reservoir. Then, the development of the aforementioned technologies and the application of these methods are strongly connected and will allow for increasing computational efficiency and reliability of reservoir simulations.

References

[1] Owen, N. A., Inderwildi, O. R., and King, D. A. The status of conventional world oil reserves-hype or cause for concern? *Energy Policy*, 38:4743–4749, 2010.

[2] Maugeri, L. Oil: The next revolution discussion paper. Tech. Rep. 2012-10, Belfer Center for Science and International Affairs, 2012.

[3] Maggio, G. and Cacciola, G. When will oil, natural gas, and coal peak? *Fuel*, 98:111–123, 2012.

[4] Chapman, I. The end of peak oil? Why this topic is still relevant despite recent denials. *Energy Policy*, 64:93–101, 2014.

[5] Hughes, L. and Rudolph, J. Future world oil production: growth, plateau, or peak? *Current Opinion in Environmental Sustainability*, 3:225–234, 2011.

[6] Pinto, A. C. C. et al. Marlim complex development: A reservoir engineering overview. In *SPE Latin American and Caribbean Petroleum Engineering Conference*. Society of Petroleum Engineers, Buenos Aires, Argentina, 2001.

[7] Asrilhant, B. A program for excellence in the management of exploration and production projects. In *Offshore Technology Conference*. Society of Petroleum Engineers, Houston, USA, 2005.

[8] Suslick, S. B., Schiozer, D., and Rodriguez, M. R. Uncertainty and risk analysis in petroleum exploration and production. *Terrae*, 6:30–41, 2009.

[9] Druetta, P., Tesi, P., de Persis, C., and Picchioni, F. Methods in Oil Recovery Processes and Reservoir Simulation. *Advances in Chemical Engineering and Science*, 6:399–435, 2016.

[10]   Jürgenson, G. A., Bittner, C., Stein, S., and Büschel, M. Chemical EOR – a multidisciplinary effort to maximize value. *Journal of Petroleum Technology*, 69:52–53, 2017. J2: SPE-0617-0052-JPT.

[11]   Muggeridge, A. et al. Recovery rates, enhanced oil recovery and technological limits. *Philosophical Transactions of the Royal Society A-Mathematical Physical and Engineering Sciences*, 372:20120320, 2014.

[12]   Yilmaz, O. Earthquake seismology, exploration seismology, and engineering seismology: How sweet it is – listening to the earth. In *2007 SEG Annual Meeting*. Society of Petroleum Engineers, San Antonio, USA, 2007.

[13]   Yilmaz, O. and Doherty, S. M. *Seismic Data Analysis: Processing, Inversion and Interpretation of Seismic Data*. Society of Exploration Geophysicists, Tulsa, USA, 2000.

[14]   Lumley, D. E. Time-lapse seismic reservoir monitoring. *Geophysics*, 66:50–53, 2001.

[15]   Lumley, D. Business and technology challenges for 4D seismic reservoir monitoring. *The Leading Edge*, 23:1166–1168, 2004.

[16]   Goloshubin, G., Schuyver, C. V., Korneev, V., Silin, D., and Vingalov, V. Reservoir imaging using low frequencies of seismic reflections. *The Leading Edge*, 25:527–531, 2006.

[17]   Alford, J. et al. Sonic logging while drilling-shear answers. *Oilfield Review*, 24:4–15, 2012.

[18]   Farfour, M. Seismic Attributes in Hydrocarbon Reservoirs Detection and Characterization. In *Advances in Data, Methods, Models and Their Applications in Oil/Gas Exploration*, pp. 151–196. Science Publishing Group, New York, USA, 2014.

[19]   Fanchi, J. R. *Principles of Applied Reservoir Simulation*. Gulf Professional Publishing, Burlington, USA, 2005.

[20]   Aziz, K., Aziz, K., and Settari, A. *Petroleum reservoir simulation*. Springer, Amsterdam, the Netherlands, 1979.

[21]   Cancelliere, M., Viberti, D., and Verga, F. A step forward to closing the loop between static and dynamic reservoir modeling. *Oil & Gas Science and Technology-Revue D Ifp Energies Nouvelles*, 69:1201–1225, 2014.

[22]   Seiler, A., Rivenaes, J. C., Aanonsen, S. I., and Evensen, G. Structural uncertainty modelling and updating by production data integration. In *SPE/EAGE Reservoir Characterization and Simulation Conference*. Society of Petroleum Engineers, Abu Dhabi, UAE, 2009.

[23]   Lie, K. A. and Mallison, B. T. Mathematical models for oil reservoir simulation. *Encyclopedia of Applied and Computational Mathematics*, pp. 1–8, 2013.

[24]   Chen, Z., Huan, G., and Ma, Y. *Computational Methods for Multiphase Flows in Porous Media*. Society for Industrial and Applied Mathematics, Philadelphia, USA, 2006.

[25]   Horgue, P., Soulaine, C., Franc, J., Guibert, R., and Debenest, G. An open-source toolbox for multiphase flow in porous media. *Computer Physics Communications*, 187:217–226, 2015.

[26]   Edwards, D. et al. Reservoir simulation: Keeping pace with oilfield complexity. *Oilfield Review*, 23:4–15, 2011.

[27]   Zafari, M. and Reynolds, A. C. Assessing the uncertainty in reservoir description and performance predictions with the ensemble kalman filter. In *SPE Annual Technical Conference and Exhibition*. Society of Petroleum Engineers, Dallas, USA, 2005.

[28]   Zhang, Y. and Oliver, D. S. History matching using a hierarchical stochastic model with the ensemble kalman filter: A field case study. In *SPE Reservoir Simulation Symposium*. Society of Petroleum Engineers, The Woodlands, USA, 2009.

[29]   Aarnes, J. E., Gimse, T., and Lie, K. A. An introduction to the numerics of flow in porous media using Matlab. In *Geometric modelling, numerical simulation, and optimization*, pp. 265–306. Springer-Verlag Berlin Heidelberg, Berlin, Germany, 2007.

[30]   Huc, A. Y. *Heavy crude oils: from geology to upgrading: an overview*. Editions Technip, Paris, France, 2010.

[31] Bidner, M. S. *Propiedades de la Roca y los Fluidos en Reservorios de Petroleo*. EUDEBA, Buenos Aires, Argentina, 2001.

[32] McCarthy, K. et al. Basic petroleum geochemistry for source rock evaluation. *Oilfield Review*, 23:32–43, 2011.

[33] Heimsund, B. O. *Mathematical and Numerical Methods for Reservoir Fluid Flow Simulation*. PhD thesis, University of Bergen, Bergen, Norway, 2005.

[34] Bastian, P. *Numerical Computation of Multiphase Flows in Porous Media*. PhD thesis, Christian-Albrechts-Universität zu Kiel, Kiel, Germany, 1999.

[35] Corey, A. T. *Mechanics of Immiscible Fluids in Porous Media*. Water Resources Publications, Highlands Ranch, USA, 1994.

[36] Dietrich, P. et al. *Flow and Transport in Fractured Porous Media*. Springer Berlin Heidelberg, Berlin, Germany, 2005.

[37] Dake, L. P. *Fundamentals of reservoir engineering*. Elsevier, Amsterdam, the Netherlands, 1978.

[38] Sahimi, M. *Flow and transport in porous media and fractured rock: from classical methods to modern approaches*. Wiley, Morlenbach, Germany, 2011.

[39] Szymkiewicz, A. Upscaling from Darcy Scale to Field Scale. In *Modelling Water Flow in Unsaturated Porous Media: Accounting for Nonlinear Permeability and Material Heterogeneity*, pp. 139–175. Springer Berlin Heidelberg, Berlin, Germany, 2013.

[40] Costanza-Robinson, M. S., Estabrook, B. D., and Fouhey, D. F. Representative elementary volume estimation for porosity, moisture saturation, and air-water interfacial areas in unsaturated porous media: Data quality implications. *Water Resources Research*, 47:W07513, 2011.

[41] Bear, J. and Bachmat, Y. Heat and Mass Transport. In *Introduction to Modeling of Transport Phenomena in Porous Media*, pp. 449–480. Springer Netherlands, Dordrecht, the Netherlands, 1990.

[42] Bear, J. Modelling Transport Phenomena in Porous Media. In *Convective Heat and Mass Transfer in Porous Media*, pp. 7–69. Springer Netherlands, Dordrecht, the Netherlands, 1991.

[43] Hassanizadeh, M. and Gray, W. G. General conservation equations for multi-phase systems: 1. averaging procedure. *Advances in Water Resources*, 2:131–144, 1979.

[44] Bear, J. and Bachmat, Y. Macroscopic modeling of transport phenomena in porous-media .2. applications to mass, momentum and energy-transport. *Transport in Porous Media*, 1:241–269, 1986.

[45] Helmig, R. *Multiphase flow and transport processes in the subsurface: a contribution to the modeling of hydrosystems*. Springer-Verlag Berlin Heidelberg, Berlin, Germany, 1997.

[46] Druetta, P. *Numerical Simulation of Chemical EOR Processes*. PhD thesis, University of Groningen, Groningen, the Netherlands, 2018.

[47] Darcy, H. *Les fontaines publiques de la ville de Dijon: exposition et application des principes a suivre et des formules a employer dans les questions de distribution d'eau*. Victor Dalmont, editeur, Paris, 1856.

[48] Ewing, R. E. Mathematical Modeling and Simulation for Applications of Fluid Flow in Porous Media. In *Current and Future Directions in Applied Mathematics*, pp. 161–182. Birkhäuser Boston, Boston, USA, 1997.

[49] Ewing, R. and Lin, Y. A mathematical analysis for numerical well models for non-Darcy flows. *Applied Numerical Mathematics*, 39:17–30, 2001.

[50] Bear, J. *Dynamics of Fluids In Porous Media*. American Elsevier Publishing Company, New York, USA, 1972.

[51] Jacob, B. *Hydraulics of groundwater*. MacGraw-Hill, New York, USA, 1979.

[52] Glimm, J. and Sharp, D. H. A random field model for anomalous diffusion in heterogeneous porous media. *Journal of Statistical Physics*, 62:415–424, 1991.

[53] Glimm, J. and Lindquist, W. Scaling laws for macrodispersion. In *Proceedings of the Ninth Int. Conf. on Comp. Meth. in Water Resources*, vol. 2, pp. 35–49, 1992.

[54] de Anna, P. et al. Flow intermittent, dispersion, and correlated continuous time random walks in porous media. *Physical Review Letters*, 110:184502, 2013.

[55] Berkowitz, B. and Scher, H. On characterization of anomalous-dispersion in porous and fractured media. *Water Resources Research*, 31:1461–1466, 1995.

[56] Zhang, X. and Lv, M. Persistence of anomalous dispersion in uniform porous media demonstrated by pore-scale simulations. *Water Resources Research*, 43:W07437, 2007.

[57] Chavent, G. and Salzano, G. A finite-element method for the 1-D water flooding problem with gravity. *Journal of Computational Physics*, 45:307–344, 1982.

[58] Ewing, R. E. *The mathematics of reservoir simulation*. SIAM, Philadelphia, USA, 1983.

[59] Peaceman, D. W. *Fundamentals of Numerical Reservoir Simulation*. Elsevier Science, Amsterdam, the Netherlands, 2000.

[60] Bachmat, Y. and Bear, J. Macroscopic modeling of transport phenomena in porous-media: 1. the continuum approach. *Transport in Porous Media*, 1:213–240, 1986.

[61] Hassanizadeh, M. and Gray, W. G. General conservation equations for multi-phase systems: 3. constitutive theory for porous media flow. *Advances in Water Resources*, 3:25–40, 1980.

[62] Hassanizadeh, M. and Gray, W. G. General conservation equations for multi-phase systems: 2. mass, momenta, energy, and entropy equations. *Advances in Water Resources*, 2:191–203, 1979.

[63] Whitaker, S. Flow in porous-media: 1. a theoretical derivation of Darcy's-Law. *Transport in Porous Media*, 1:3–25, 1986.

[64] Whitaker, S. Flow in porous-media: 2. the governing equations for immiscible, 2-phase flow. *Transport in Porous Media*, 1:105–125, 1986.

[65] Whitaker, S. Flow in porous-media: 3. deformable media. *Transport in Porous Media*, 1:127–154, 1986.

[66] Stevenson, M., Kagan, M., and Pinczewski, W. Computational methods in petroleum reservoir simulation. *Computers & Fluids*, 19:1–19, 1991.

[67] Lake, L. W. *Enhanced Oil Recovery*. Prentice-Hall Inc., Englewood Cliffs, USA, 1989.

[68] Brooks, R. H. and Corey, A. T. Properties of porous media affecting fluid flow. *Journal of the Irrigation and Drainage Division*, 92:61–90, 1966.

[69] Camilleri, D. et al. Description of an improved compositional micellar/polymer simulator. *SPE Reservoir Engineering*, 2:427–432, 1987.

[70] Lassabatere, L. et al. Beerkan estimation of soil transfer parameters through infiltration experiments – best. *Soil Science Society of America Journal*, 70:521–532, 2006.

[71] Pinder, G. F. and Gray, W. G. *Essentials of multiphase flow in porous media*. Wiley, Hoboken, USA, 2008.

[72] Burdine, N. Relative permeability calculations from pore size distribution data. *Transactions of the American Institute of Mining and Metallurgical Engineers*, 198:71–78, 1953.

[73] Mualem, Y. New model for predicting hydraulic conductivity of unsaturated porous-media. *Water Resources Research*, 12:513–522, 1976.

[74] Delshad, M. and Pope, G. A. Comparison of the three-phase oil relative permeability models. *Transport in Porous Media*, 4:59–83, 1989.

[75] Acs, G., Doleschall, S., and Farkas, E. General-purpose compositional model. *Society of Petroleum Engineers Journal*, 25:543–553, 1985.

[76] Gerritsen, M. and Durlofsky, L. Modeling fluid flow in oil reservoirs. *Annual Review of Fluid Mechanics*, 37:211–238, 2005.

[77] Delshad, M. et al. Mechanistic interpretation and utilization of viscoelastic behavior of polymer solutions for improved polymer-flood efficiency. In *SPE Symposium on Improved Oil Recovery*. Society of Petroleum Engineers, Tulsa, USA, 2008.

[78] Scheidegger, A. General theory of dispersion in porous media. *Journal of Geophysical Research*, 66:3273–&, 1961.

[79] Delgado, J. M. P. Q. Longitudinal and transverse dispersion in porous media. *Chemical Engineering Research & Design*, 85:1245–1252, 2007.

[80] Bijeljic, B. and Blunt, M. J. Pore-scale modeling of transverse dispersion in porous media. *Water Resources Research*, 43:W12S11, 2007.

[81] Bruining, H., Darwish, M., and Rijnks, A. Computation of the longitudinal and transverse dispersion coefficient in an adsorbing porous medium using homogenization. *Transport in Porous Media*, 91:833–859, 2012.

[82] Espedal, M. S., Langlo, P., Saevareid, O., Gislefoss, E., and Hansen, R. Heterogeneous reservoir models: Local refinement and effective parameters. In *SPE Symposium on Reservoir Simulation*. Society of Petroleum Engineers, Anaheim, USA, 1991.

[83] Langlo, P. *Macrodispersion for 2-phase, immiscible flow in heterogeneous media*. PhD thesis, University of Bergen, Bergen, Norway, 1992.

[84] Bidner, M. S. and Savioli, G. B. On the numerical modeling for surfactant flooding of oil reservoirs. *Mecanica Computacional*, XXI:566–585, 2002.

[85] Allen, M. B. I. I. I., Behie, G. A., and Trangenstein, J. A. *Multiphase Flow in Porous Media: Mechanics, Mathematics, and Numerics*. Springer New York, New York, USA, 2013.

[86] Cao, H. and Aziz, K. Performance of impsat and IMPSAT-AIM models in compositional simulation. In *SPE Annual Technical Conference and Exhibition*. Society of Petroleum Engineers, San Antonio, USA, 2008.

[87] Chen, Z. Homogenization and simulation for compositional flow in naturally fractured reservoirs. *Journal of Mathematical Analysis and Applications*, 326:12–32, 2007.

[88] Peaceman, D. Interpretation of well-block pressures in numerical reservoir simulation with non-square grid blocks and anisotropic permeability. *Society of Petroleum Engineers Journal*, 23:531–543, 1983.

[89] Babu, D. K., Odeh, A. S., Al-Khalifa, A., and McCann, R. C. The relation between wellblock and wellbore pressures in numerical simulation of horizontal wells. *SPE Reservoir Engineering*, 6:324–328, 1991.

[90] Chen, Z. and Zhang, Y. Well flow models for various numerical methods. *International Journal of Numerical Analysis and Modeling*, 6:375–388, 2009.

[91] Ewing, R. et al. Numerical well model for non-Darcy flow through isotropic porous media. *Computational Geosciences*, 3:185–204, 1999.

[92] Wolfsteiner, C., Durlofsky, L., and Aziz, K. Calculation of well index for nonconventional wells on arbitrary grids. *Computational Geosciences*, 7:61–82, 2003.

[93] Coats, K. H. A note on IMPES and some IMPES-based simulation models. *SPE Journal*, 5:245–251, 2000.

[94] Trangenstein, J. and Bell, J. Mathematical structure of compositional reservoir simulation. *SIAM Journal on Scientific and Statistical Computing*, 10:817–845, 1989.

[95] Wong, T. W., Firoozabadi, A., and Aziz, K. Relationship of the volume-balance method of compositional simulation to the Newton-Raphson method. *SPE Reservoir Engineering*, 5:415–422, 1990.

[96] Zaydullin, R., Voskov, D. V., James, S. C., Henley, H., and Lucia, A. Fully compositional and thermal reservoir simulation. *Computers & Chemical Engineering*, 63:51–65, 2014.

[97] Moortgat, J., Sun, S., and Firoozabadi, A. Compositional modeling of three-phase flow with gravity using higher-order finite element methods. *Water Resources Research*, 47:W05511, 2011.

[98] Thomas, G. and Thurnau, D. Reservoir simulation using an adaptive implicit method. *Society of Petroleum Engineers Journal*, 23:759–768, 1983.

[99] Ghasemi, M., Ibrahim, A., and Gildin, E. Reduced order modeling in reservoir simulation using the bilinear approximation techniques. In *SPE Latin America and Caribbean Petroleum Engineering*. Society of Petroleum Engineers, Maracaibo, Venezuela, 2014.

[100] Hu, X. et al. Combined preconditioning with applications in reservoir simulation. *Multiscale Modeling & Simulation*, 11:507–521, 2013.

[101] Liu, H., Wang, K., Chen, Z., and Jordan, K. E. Efficient multi-stage preconditioners for highly heterogeneous reservoir simulations on parallel distributed systems. In *SPE Reservoir Simulation Symposium*. Society of Petroleum Engineers, Houston, USA, 2015.

[102] Sammon, P. H. Dynamic grid refinement and amalgamation for compositional simulation. In *SPE Reservoir Simulation Symposium*. Society of Petroleum Engineers, Houston, USA, 2003.

[103] Jackson, M. D. et al. Reservoir modeling for flow simulation using surfaces, adaptive unstructured meshes, and control-volume-finite-element methods. In *SPE Reservoir Simulation Symposium*. Society of Petroleum Engineers, The Woodlands, USA, 2013.

[104] Rezzolla, L. Numerical methods for the solution of partial differential equations. *Lecture Notes for the COMPSTAR School on Computational Astrophysics*, pp. 8–13, 2011.

[105] Faires, J. D. and Burden, R. L. *Numerical Methods*. Cengage Learning, Boston, USA, 2012.

[106] Kuzmin, D. and Turek, S. Flux correction tools for finite elements. *Journal of Computational Physics*, 175:525–558, 2002.

[107] Kuzmin, D. and Turek, S. High-resolution FEM-TVD schemes based on a fully multidimensional flux limiter. *Journal of Computational Physics*, 198:131–158, 2004.

[108] Sweby, P. High-resolution schemes using flux limiters for hyperbolic conservation-laws. *SIAM Journal on Numerical Analysis*, 21:995–1011, 1984.

[109] Berger, M., Aftosmis, M. J., and Murman, S. M. Analysis of slope limiters on irregular grids. *AIAA paper*, 490:1–22, 2005.

[110] Blunt, M. and Rubin, B. Implicit flux limiting schemes for petroleum reservoir simulation. *Journal of Computational Physics*, 102:194–210, 1992.

[111] Alhumaizi, K. Flux-limiting solution techniques for simulation of reaction-diffusion-convection system. *Communications in Nonlinear Science and Numerical Simulation*, 12:953–965, 2007.

[112] Liu, J., Delshad, M., Pope, G. A., and Sepehrnoori, K. Application of higher-order flux-limited methods in compositional simulation. *Transport in Porous Media*, 16:1–29, 1994.

[113] Smaoui, H., Zouhri, L., and Ouahsine, A. Flux-limiting techniques for simulation of pollutant transport in porous media: Application to groundwater management. *Mathematical and Computer Modelling*, 47:47–59, 2008.

[114] Hou, J., Simons, F., and Hinkelmann, R. Improved total variation diminishing schemes for advection simulation on arbitrary grids. *International Journal for Numerical Methods in Fluids*, 70:359–382, 2012.

[115] Harimi, I. and Pishevar, A. R. Evaluating the capability of the flux-limiter schemes in capturing strong shocks and discontinuities. *Shock and Vibration*, 20:287–296, 2013.

[116] Galiano, S. J. and Zapata, M. U. A new TVD flux-limiter method for solving nonlinear hyperbolic equations. *Journal of Computational and Applied Mathematics*, 234:1395–1403, 2010.

[117] Fazio, R. and Jannelli, A. Second order positive schemes by means of flux limiters for the advection equation. *IAENG International Journal of Applied Mathematics*, 39:1–11, 2009.

[118] Roe, P. Upwind differencing schemes for hyperbolic conservation-laws with source terms. *Lecture Notes in Mathematics*, 1270:41–51, 1987.

[119] Roe, P. Characteristic-based schemes for the Euler equations. *Annual Review of Fluid Mechanics*, 18:337–365, 1986.

[120] Leer, B. V. Towards ultimate conservative difference scheme .2. monotonicity and conservation combined in a second-order scheme. *Journal of Computational Physics*, 14:361–370, 1974.

[121] Albada, G. V., Leer, B. V., and Roberts, W. A comparative-study of computational methods in cosmic gasdynamics. *Astronomy & Astrophysics*, 108:76–84, 1982.

[122] Koren, B. A robust upwind discretization method for advection, diffusion and source terms. In *Advances in Data, Methods, Models and Their Applications in Oil/Gas Exploration*, pp. 117–138. Centrum voor Wiskunde en Informatica, Amsterdam, the Netherlands, 1993.

[123] Zhou, G. *Numerical simulations of physical discontinuities in single and multi-fluid flows for arbitrary Mach numbers*. PhD thesis, Chalmers University of Technology, Gothenburg, Sweden, 1995.

[124] Leer, B. V. Towards the ultimate conservative difference scheme: 5. second order sequel to Godunov's method. *Journal of Computational Physics*, 32:101–136, 1979.

4 Compositional simulation applied to EOR polymer flooding

4.1 Introduction

The demand for energy has been steadily increasing over the last 150 years, and along with it, the need to discover and/or develop new sources in order to fulfill the demand. Oil has been the main source of energy and also the main feedstock for the plastic industry [1]. The exploitation of oil reservoirs can be divided according to the mechanisms employed to recover it [2–4]. Primary recovery uses natural driven mechanisms present in the porous medium; then, water is injected in order to repressurize the medium as well as to sweep the remaining oil to the producer wells. It is usually considered that after these two stages around 45% to 55% of the original oil in place is still trapped underground [5]. This fact, along with increasing energy demand and the decreasing number of new fields being discovered, has led researchers and the oil industry to look for ways to increment the efficiency of existing fields. This marked the beginning of the tertiary oil recovery (EOR) techniques, which have become profitable over the last years due to the existing economic constraints [1]. There are many EOR methods, aimed at modifying the properties of the oil, water and/or rock formation. The employment of these techniques depends mainly on the physical characteristics of the crude oil trapped the porous medium. For medium and low viscosity oils, chemical EOR (CEOR) represents a good alternative to increase the lifespan of a reservoir [6]. The standard techniques in CEOR include polymers, surfactants or alkali injection in order to modify the rheological interfacial properties of the phases present in order to mobilize the remaining trapped oil. In this chapter a novel compositional numerical simulator is described and proposed for CEOR processes.

4.1.1 Polymer flooding

Water soluble polymers were one of the first methods developed to increase the efficiency of waterflooding processes, thanks to the pioneering work developed by Sandiford [7] and Pye [8]. The main objective with polymer molecules is to alter the rheological properties of the carrying phase (i.e., water/brine), and to reduce the mobility ratio, which is the relationship between the flowabilities of water and oil phases [6, 7, 9–11]. This can be accomplished in two different ways: modify the rock wettability and/or the phase viscosities. Even though a Newtonian behavior for medium and low viscosity oils can be adopted, this is not the case for the polymer carrying phase. The solution viscosity depends on the polymer concentration, architecture and molecular weight, the temperature, water salinity, total dissolved solids (TDS) and the concentration of divalent cations [10]. The possibility that the polymer can alter the microscopic effi-

https://doi.org/10.1515/9783110640250-004

ciency factor and with this, the residual oil saturation is also considered. The results, supported by the literature [12–14], indicate that the viscoelastic and interfacial effects of the polymer molecules affect the latter, and this is taken into account in the new simulator [10, 12–15].

A series of numerical simulations are then performed in order to understand how to increase the efficiency of both the process and the polymers used. The simulation of multiphase multicomponent flow in porous media involves solving a number of coupled, highly non-linear systems of equations dealing with temporal and spatial variations of pressure and mass concentrations. The compositional flow results in a suitable approach to study chemical EOR processes, which can be described as the mass transfer of a number of components (e.g., polymers, surfactants, salts, etc.) in a two- or three-phase system. This model can be also applied to underground flow in several disciplines, such as groundwater hydrology [16–18], storage [19], environmental [18, 20], and/or chemical engineering (involving chemical reactions between rock and the fluids) [21, 22].

However, the polymer flooding process is affected by a number of physical and chemical factors additional to those already known in mass transfer processes that must be taken into account during the simulation. The rheology of the injected solution is one of these key factors. The economic success of the whole process is susceptible to the injection rate, which is directly related to the viscosity of the polymeric solution [3]. The polymer solutions used in the industry have the characteristic of being non-Newtonian, different to the Newtonian rheology found during waterflooding. The viscosity depends on the polymer concentration and the shear-rate, which in the case of flow in porous media can reach high values. In laboratory tests a shear-thinning behavior has been reported. During flow in porous media, at high shear-rates, a shear-thickening rheology has been found, thus increasing the viscosity of the solution as the shear-rate increases. This was studied by many authors [10, 23] who developed coupled rheological models in order to fit the experimental tests. The viscosity depends, through the Flory and Mark–Houwink correlations, on the molecular weight of the polymer, which is a function of the length of the backbone chain. It is known from literature that polymers, during flow in porous media, are subjected to processes of thermal [24, 25], mechanical [26–28], chemical [3, 29, 30], and/or biological [10, 31, 32] degradation, which produce a chain-scission of the backbone, negatively affecting the viscosifying properties. In addition to the rheology, the presence of the polymer causes two phenomena that must be taken into account during the simulation: the permeability reduction of the aqueous phase and the inaccessible pore volume (IAPV). The first is caused by the polymer molecules adsorbed by the rock surface, which resist the flow of the aqueous phase. This can be modeled as an irreversible process reducing the relative permeability of latter. The second is caused because the polymer molecules, being much larger than water ones, can enter a limited region of the rock's poral volume. This is modeled as a "reduced pseudo-porosity" that only affects the mass conservation of the chemical component [10, 33].

4.1.2 Previous numerical work

The numerical modeling of the EOR processes in porous media can be categorized according to the approach used in *direct* (Navier–Stokes) or *continuum* (Darcy) modeling. The direct analysis involves solving the Navier–Stokes equations and thus its application in reservoir simulation is very limited. In this chapter the continuum approach is employed in order to simulate the recovery process in a reservoir. It is usually employed in CEOR techniques, the compositional method used to model the transport equations. This allows studying in a multiphase system, a number of chemical components present in the latter that affect the properties of fluids and rock. The mathematical simulation of such system has already been analyzed by several authors [34–38]. This has led to the development of commercial and academic simulators like UTCHEM and IPARS (University of Texas at Austin) [39], MRST (open code for MATLAB) [40–42], CMG (Stars) and Eclipse (Schlumberger), as well as several others that can be found in the literature [33, 43–51].

There are two main approaches concerning the treatment of the viscoelastic terms in polymer flooding. The first consists of solving the extra stress tensor in the Navier–Stokes equations using a known correlation (e.g., FENE, UCM, Oldroyd). The second, which is used in this simulator, consists of including the viscoelasticity in terms of shear thickening terms in the rheology behavior of the water phase. However, a correction to actual models is proposed in order to extend the viscoelasticity influence, considering it also in the calculation of the residual oil saturation.

Lee [52] employed UTCHEM as a tool to test polymer flooding without degradation mechanisms involved in heterogeneous layered formations and how the injection rate affects the recovery process. He reported that the shear-thinning behavior is beneficial from the point of view of the injection rate, due to the fact that viscosity near injection wells is lower due to a higher shear rate, which provides more favorable injectivity. This might be troublesome if the shear rates are high enough so pseudo-dilatant behavior in the polymeric solution appears. His objective was also to analyze how the degree of crossflow among the layers affects the flooding performance. The results showed that shear-thinning behavior did not affect the oil recovery in a homogeneous reservoir, but the latter decreases among layers as reservoir heterogeneity becomes larger. With an increase in the degree of crossflow, oil migrates to other layers more easily due to a higher sweep by the displacing fluid. The injectivity was also affected directly because different pressure gradients altered the viscosity of the displacing fluid. Yuan [53] also utilized well-known software to study polymer flooding in porous media, coupling the multiphase reactive transport module (TRCHEM) with IPARS. Polymer flow characteristics are: adsorption, non-Newtonian shear-thinning polymer viscosity (without degradation), polymer and electrolytes concentrations, permeability reduction, and inaccessible pore volume. His results showed that finer grids in the areal direction yielded higher recovery factors than those with a more vertical resolution whilst simulations with fewer vertical layers yielded higher oil recovery due to less permeability contrast. Wang [54] developed a mechanistic simulator, flooding with three phases and five components in order to take into account what the author called the viscoelastic behavior of the polymer solution. In order to numerically solve the system of non-linear equations, he used the IMPES method, considering the pseudo-dilatant characteristics of the solution at high shear rates, and moreover the degradation of the polymer molecules. However, as has already been presented in the literature, the phenomena associated with viscoelasticity are not only limited to this rheological feature, but other factors should be also taken into account [12–14, 23]. Nevertheless, to our best knowledge, it is the only one simulator present in the literature detailing the degradation phenomenon which adversely affects the efficiency of polymer flooding process. He also showed, as the literature has reported, that the earlier the polymer solution is injected, the shorter the whole process will take and therefore, the better the EOR flooding is in economical terms. The polymer solution should be injected as early as possible, according to the field conditions.

4.2 Aim of this chapter

The current numerical models used in polymer flooding simulator were presented in the previous section, describing their problems regarding the formulation of the physical properties. Most commercial simulators do not consider one or more of these aspects, so the purpose of this research is the development of a new simulator that includes all these phenomena. The model proposed and developed comprises a complete study of degradation due to breakage of the backbone chain, which to the extent of our knowledge, has not been reported yet in an appropriate way. The degradation process is also a parameter not considered so far in most of simulators. This plays a major role in polymer flooding since the rheological properties are based on the molecular weight. Even though in the literature there are models considering the degradation as mentioned before [54], these are based on models affecting the viscosity itself and not the molecular weight. This approach is not correct since it is not the rheological properties but the molecular structure the first parameter affected by the scission of the backbone, and subsequently the radius of gyration, relaxation times, and intrinsic, reduced and zero-shear viscosities. Recently, Lohne [55] developed an in-house simulator IORCoreSim® also considering the approach that the degradation takes place in the molecular weight and therefore affecting not only the viscosity. However, this model does not consider the possible influence of the polymer on the interfacial forces. Moreover, even though the effect of degradation on the shear-thickening was considered, its influence on the residual oil saturation due to the viscoelastic nature of the polymer was not modeled. If the viscoelasticity is considered in the model, it must be taken into account that the viscoelastic properties depend on the relaxation time, which is a function of the molecular structure. Considering the degradation affects only the viscosity and not the viscoelastic properties leaves a number of parameters related to the viscoelasticity and the residual oil saturation unaffected. That is the main reason a complete degradation mechanism is proposed in this new simulator, based on the variation of the molecular weight, hence affecting all the other related properties.

In addition to the previously mentioned physical phenomena, some problems of a numerical source have been studied when the mass conservation equation is discretized. It is well known that first order, linear schemes produce artificial diffusion of the different components which, in the case of the polymer, provokes a smearing of the chemical slug, decreasing the viscosity and thus the efficiency. Non-linear TVD schemes with flux limiting functions are proposed as a means of decreasing this undesired phenomena and allows a better tracking of the chemical front, without the appearance of spurious oscillations, commonly found in traditional finite difference method schemes [5]. The combination of the mentioned factors has resulted in a novel and complete simulator, which can be used for the design and screening of new polymers to be used in EOR.

4.3 Physical model

In order to model the process of polymer flooding a 2D geometrical model is used, with a geometric pattern usually found in the oil industry. The five-spot scheme is a good model which relatively satisfies the previous requirement. It consists of a square domain, with constant or variable properties, where an injection well is placed at the center, and four producing wells are located at the corners. During this analysis, a simplification of the model was performed in what is known as quarter five-spot (Fig. 4.1).

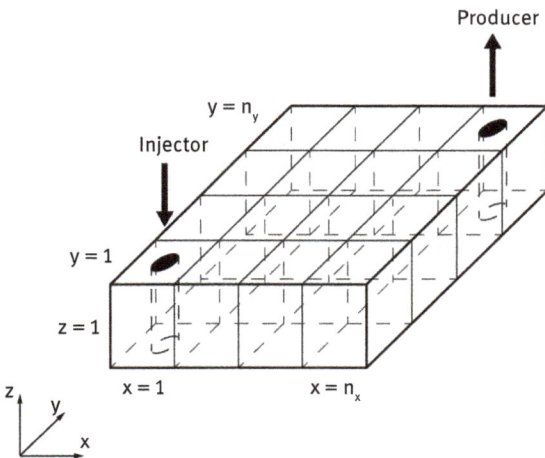

Fig. 4.1: Schematic representation of the quarter 5-spot used for the simulations [56].

A problem associated with these geometries is the high speed and pressure gradients occurring at the vicinity of the wells, which may cause problems during the numerical simulation.

The physical model is composed by a reservoir (Ω) of known geometric characteristics, with an absolute permeability (K) and porosity (ϕ), which can be considered constants or to have a statistical distribution. The flow is considered isothermal, incompressible and 2-dimensional (since it is assumed that the vertical permeability is negligible when compared to horizontal ones); it is also considered that the system is in local thermodynamic (phase) equilibrium. The domain is then discretized in a system of $n_x \times n_y$ blocks to perform the numerical simulation. Darcy's law is valid and gravitational forces are negligible compared to the viscous ones [57]. The fluids can be considered Newtonian or non-Newtonian, depending on the presence of the chemical in the corresponding phase. Although Darcy's Law is valid only for Newtonian fluids, literature shows that the approach of considering non-Newtonian in the whole domain but Newtonian in each cell yielded good results.

Polymer EOR flooding, as with other chemical techniques, involves the flow of fluids in two phases (aqueous and hydrocarbon), and various components (water, chemical, salt and petroleum). It is noteworthy that these can be mixtures of a number of pure components, since petroleum is a mixture of many hydrocarbons, water contains dissolved salts and the polymer is composed of a number of molecules of different lengths and architectures [2–4]. The recovery process involves injecting in as a first stage, an aqueous solution with the polymer and thereafter a water bank is injected in order to drive the chemical plug, sweeping the mobilized oil into the producing wells. This model is represented by a system of strongly non-linear partial differential equations (PDE) which are completed by a set of algebraic relationships representing physical properties of the fluid and the rock, namely: interfacial tension, residual phase saturations, relative permeabilities, rock wettability, disproportionate permeability reduction, inaccessible pore volume, phase viscosities, capillary pressure, adsorption on the formation, and dispersion. It is clear that most of these properties are strongly dependent on the polymer concentration, which alters the flowability of the aqueous phase. The interfacial tension depends on the former and it affects the residual saturations, which are dependent on the capillary number, a dimensionless parameter based on the IFT and the viscosity. The latter depends on the volumetric concentrations of each of the three components and the shear rate.

Another factor this new simulator will consider is the compressibility of the formation. Underground porous media are subject to internal and external stresses due to the forces acting on them. Internal stresses result mainly from the fluid pressure field whilst external stresses originate from the weight of the overburden and, if any, tectonic stresses. The combination of the aforementioned forces causes a corresponding stress-strain condition in the formation. External stresses tend to compact and therefore reduce the pore volume. Internal pressure forces exert an opposite stress condition, resisting pore volume reduction or increasing it. As time goes on, and the reservoir is exploited, the pressure decreases resulting in a porosity reduction. Similarly, in deeper formations the overburden pressure increases hence the porosity is reduced [5, 58, 59]. The numerical technique adopted for the resolution of these equations is the IMPEC method, which calculates pressures implicitly and the concentration for each of the component explicitly. In conclusion, it can be stated that the first step in the successful development of a numerical simulation of chemical flooding is to have a model that allows accurately predicting what the behavior of the phases present in the reservoir will be. All other properties are dependent on this, hence the recovered oil is a function of phase behavior [38, 60].

4.4 Mathematical model

4.4.1 Flow equations

The aim of this chapter is to develop a simulator based on the compositional system for the analysis of polymer flooding in porous media. Therefore, the equations used to describe the process are the Darcy equation for each phase and the mass conservation equation applied to each component [54, 61]. The compositional method was chosen because of its versatility in modelling the different physical properties of the phases according to the concentration of these components. This makes it an ideal approach for chemical EOR processes. These equations are then applied on a representative volume element (REV) of the porous medium. Considering Darcy's equation for each phase first,

$$\vec{u}^j = -\underline{\underline{K}} \cdot \frac{k_r^j}{\mu^j} \cdot \vec{\nabla} p^j ; \quad j = \text{o, a} \tag{4.1}$$

Considering Eq. (4.1) for each phase and adding for the number of phases is obtained,

$$\vec{u} = \vec{u}^o + \vec{u}^a = -\underline{\underline{\lambda}} \cdot \vec{\nabla} p^a - \underline{\underline{\lambda}}^o \cdot \vec{\nabla} p_c \tag{4.2}$$

where p_c is the capillary pressure for the water-oil system and the j-phase and total mobilities are also introduced and defined as

$$\underline{\underline{\lambda}} = \underline{\underline{\lambda}}^o + \underline{\underline{\lambda}}^a \quad \text{and} \quad \underline{\underline{\lambda}}^j = \underline{\underline{K}} \cdot \frac{k_r^j}{\mu^j} ; \quad j = \text{o, a} \tag{4.3}$$

Along with Darcy's equation, the mass balance is applied for each component within the REV, considering the prevailing mechanisms in the flow in porous media, that is, flow by advection and diffusion. The loss of the chemical component due to adsorption in the rock, and finally the source/sinks terms to represent the wells, are also considered in the formulation. Then, the equation is written as,

$$\frac{\partial (\phi z_i)}{\partial t} + \nabla \cdot \sum_j V_i^j \cdot \vec{u}^j - \nabla \cdot \sum_j \underline{\underline{D}}_i^j \cdot \nabla \cdot V_i^j = -\frac{\partial (\phi \text{Ad}_i)}{\partial t} + q_i ; \quad i = \text{p, c, w, s} \tag{4.4}$$

When the details of the phenomena causing the diffusive flux are studied, it can be inferred that it is caused by several sources. The simplest movement is molecular diffusion described by the random Brownian motion of molecules. Usually in reservoir simulations the latter is negligible when compared to other considered forces acting on the fluids. Mechanical dispersion is also present. Narrow channel flows experience parabolic diffusion along the fronts (Taylor dispersion) and the irregular pore networks disperse the mass at a microscale. Besides this, the phenomena of transverse and longitudinal (tortuosity effect) spreading are also present. The tensor of hydrodynamic dispersion taking into account the mentioned effects is expressed as follows [61–63]:

$$\underline{\underline{D}}_i^j = \mathrm{dm}_i^j \cdot \phi \cdot S^j \cdot \delta_i^j + \|\vec{u}^j\| \cdot \left[\frac{\mathrm{dl}^j}{\|\vec{u}^j\|^2} \cdot \begin{vmatrix} (u_x^j)^2 & u_x^j \cdot u_y^j \\ u_y^j \cdot u_x^j & (u_y^j)^2 \end{vmatrix} + \mathrm{dt}^j \cdot \begin{vmatrix} 1 - \frac{(u_x^j)^2}{\|\vec{u}^j\|^2} & -\frac{u_x^j \cdot u_y^j}{\|\vec{u}^j\|^2} \\ -\frac{u_y^j \cdot u_x^j}{\|\vec{u}^j\|^2} & 1 - \frac{(u_y^j)^2}{\|\vec{u}^j\|^2} \end{vmatrix} \right] \quad (4.5)$$

The above equations are complemented by a number of algebraic relations derived from the mass balance [63].

$$z_i = \sum_j V_i^j \cdot S^j \quad (4.6)$$

$$\sum_j S^j = 1 \quad (4.7)$$

$$\sum_i V_i^j = 1 \quad (4.8)$$

$$\sum_i z_i = 1 \quad (4.9)$$

$$\sum_i \underline{\underline{D}}_i^j \cdot \nabla \cdot V_i^j = 0 \quad (4.10)$$

One possible way to solve this system of equations is to use the pressure formulation and concentrations. This consists of obtaining the pressure of one phase (aqueous or hydrocarbon), by introducing the concept of capillary pressure. In this chapter, this is achieved by adding the mass conservation equation applied to each component and take into account the constraints established by Eqs. (4.2), (4.3), (4.7) to (4.10). Thus,

$$\phi c_r \frac{\partial p^a}{\partial t} + \vec{\nabla} \cdot (\lambda \cdot \nabla p^a) = \frac{\partial}{\partial t} \left(\phi \cdot \sum_i \mathrm{Ad}_i \right) - \vec{\nabla} \cdot (\lambda^o \cdot \nabla p_c) + q_t \quad (4.11)$$

$$\phi_{m,n}^{\langle n+1 \rangle} = \phi_{m,n}^{\mathrm{ref}} \left[1 + c_r \left(p_{m,n}^{a,\langle n+1 \rangle} - p^{\mathrm{ref}} \right) \right] \quad (4.12)$$

Equation (4.11) is the parabolic-type PDE usually found in reservoir simulation due to the formation and/or fluid compressibility. If the latter is assumed incompressible, the PDE becomes an elliptic-type (see Chapter 3). The resulting numerical model has for each REV 19 unknowns, they are: the Darcy velocities ($u^{o,a}$), pressures (p_c, $p^{o,a}$) and saturation ($S^{o,a}$) of each phase, volumetric ($V_{p,w,c}^{o,a}$) and overall concentrations

$(z_{p,w,c})$ of each component. Nevertheless, the model contains thus far only 15 equations (Eq. (4.1) for each phase, capillary pressure relationship, Eq. (4.4) for three components, Eq. (4.6) for each component, Eqs. (4.7), (4.8) for each phase, Eq. (4.9), and Eq. (4.11)). The remaining necessary equations, which are equal to $N_{comp} \cdot (N_{phase} - 1)$ for the system to be numerically determined, are obtained from algebraic relationships between the volumetric concentrations of the components on the phases [5]. These describe the phase behavior, of which its understanding and correct modeling is a vital concept in chemical EOR processes. These will be addressed in each chapter in particular since each CEOR process has its own characteristics in terms of how the chemical is distributed between the phases present in the reservoir.

4.4.2 Physical properties

Since the aim of this chapter is to define the numerical model to be used in CEOR processes, the goal is to discuss and present the general physical properties of a multicomponent CEOR simulator. These phenomena are generally present in every chemical recovery process and affect to a greater or lesser extent the efficiency of the EOR agent.

Chemical component partition

The most important part of a numerical simulation using a compositional model is to understand how the components distribute into the phase, what it is called the phase behavior of the system. The component partition in polymer flooding is relatively simple, since it is assumed that the hydrocarbon phase is purely composed by the petroleum component present in the system. Therefore, water and chemical components are only present in the aqueous phase. This is represented in Eq. (4.13) by the volumetric concentrations of petroleum, polymer and water in the hydrocarbon phase. With these relationships the system becomes numerically determined with a unique solution, and the parameters previously introduced can be calculated for each representative elementary volume (REV).

$$V_p^o = 1 \wedge V_c^o = V_w^o = V_s^o = 0 \tag{4.13}$$

Interfacial tension

The interfacial tension of the water-oil system depends on the presence and concentration of the polymer in the porous media. However, the effect of the polymer on the IFT is not the most important in the whole process and it is not as effective as in the case of surfactant flooding. Most importantly, the influence of the polymer also depends on its structure. When the latter is fully soluble in water, such as HPAM, the polymer does not influence the water-oil interfacial properties [64, 65]. However, when the polymer molecule has hydrophobic groups distributed along the backbone, these will affect the IFT and lower the interfacial energy [66–68]. In order to model the influence of these hydrophobically modified polymers, a simple correlation is introduced based on the work presented by Wever [69] and Pancharoen [70], in which the IFT varies linearly with the concentration of the polymer from a maximum value γ_{ow} to its minimum constant value for higher concentrations of polymer (Fig. 4.2). The reduction of IFT allows mobilizing the oil trapped in the reservoir and its influence is measured by the capillary number.

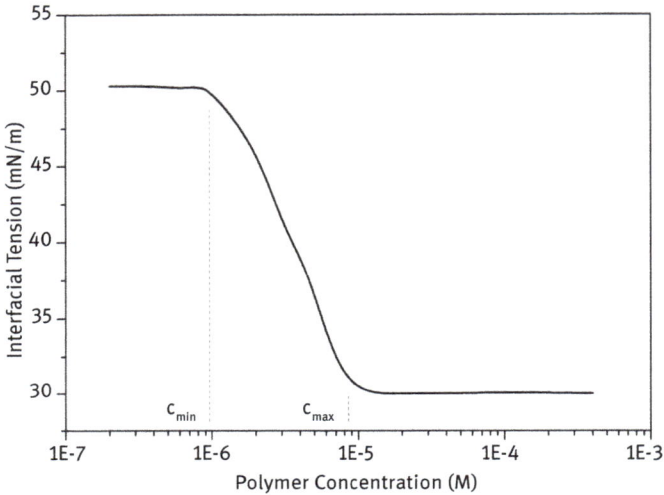

Fig. 4.2: Adopted model of the IFT as a function of polymer concentration [71].

Residual saturation

Residual saturations play an important role in oil recovery processes. They establish a certain limit to how much oil can be mobilized during the process. If such saturations can be reduced, this will increase the efficiency of the whole process. As explained in the previous section, they depend on the IFT in the water-oil two-phase system. The presence of the polymer can modify the residuals saturations in the porous medium. This relationship is ruled by a dimensionless group, the capillary number, defined by the following equation:

$$N_c = \frac{u \cdot K}{\lambda \cdot \gamma} \tag{4.14}$$

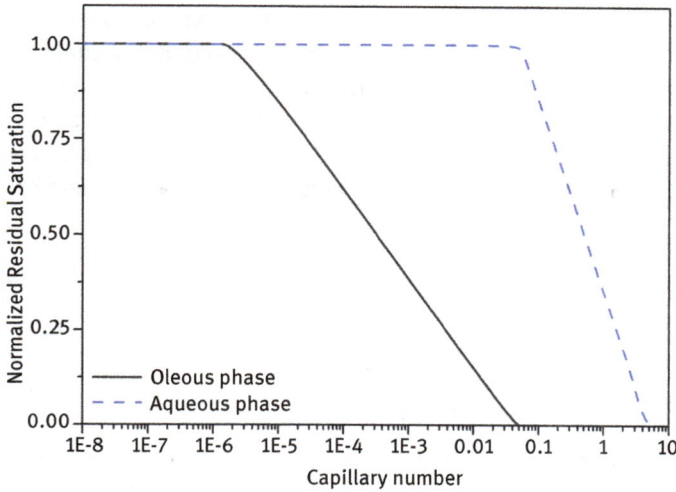

Fig. 4.3: Capillary desaturation curves (CDC) for non-wetting (hydrocarbon) and wetting (aqueous) phases used for this simulation [71].

The functionality between the capillary number and the residual saturation for both phases is described by the following model [57]:

$$\frac{S^{jr}}{S^{jrH}} = \begin{cases} 1 & \text{if} \quad N_c < 10^{(1/T_1^j)-T_2^j} \\ T_1^j \cdot \left[\log(N_c) + T_2^j\right] & \text{if} \quad 10^{(1/T_1^j)-T_2^j} \le N_c \le 10^{-T_2^j} \\ 0 & \text{if} \quad N_c > 10^{-T_2^j} \end{cases} \qquad (4.15)$$

The piecewise function is defined by constant parameters which depend on the fluids and the porous medium being simulated. The relationship between the residual saturation after chemical and waterflooding processes is known as normalized residual saturation of phase j. The form of Eq. (4.15) for both phases determines what is known as capillary desaturation curves (Fig. 4.3). At low capillary numbers, the behavior is similar to a process of waterflooding and the normalized residual saturation is not decreased. As the IFT decreases and/or the viscosity increases, the capillary number rises to higher values than those of the secondary recovery. It is for this reason that in areas of high speeds (i.e., nearby the wells) oil saturation values lower than those of waterflooding can be achieved. As can be seen, the aqueous phase requires much higher values of N_c to achieve a full desaturation [3].

However, several authors have discussed how the viscoelastic properties of polymer solutions influence the residual oil viscosity, and this is not only evidenced in the shear-thickening region polymer solutions develop at high shear-rates, but also in the microscopic recovery efficiency [11]. It is because of this reason that a modification to Eq. (4.15) is proposed in this chapter in order to take into account the effects of viscoelasticity in the polymer solution. With that purpose, when this equation is applied for the aqueous phase, the factor T_2^o is considered as a function of a parameter which takes into account the shear rate and the viscoelastic properties of the polymer (Eq. (4.16)). The Weissenberg number (Wi) relates both phenomena and it is therefore the factor modifying T_2^o (Eq. (4.17)).

$$T_{2,\text{mod}}^o = T_2^o \cdot \left(1 + T_2^{o,v} V_c^a \text{Wi}^n\right) \tag{4.16}$$

$$\text{Wi} = \lambda \frac{u^a}{L} \tag{4.17}$$

$$\lambda = \frac{6 M_w \left(\mu_{\text{0sr}} - \mu_{\text{water}}\right)}{\pi^2 R_g V_c^a T} \tag{4.18}$$

Where λ is the polymer relaxation time, R_g is the universal gas constant, T is the field temperature, and n and $T_2^{o,v}$ are constants to fit the experimental data. This is calculated according to the radius of gyration, which is a function of the molecular weight. With this parameter the effects of the viscoelasticity can be quantified in a typical oil/water desaturation curves (Fig. 4.4). The hydrocarbon line is then displaced to the

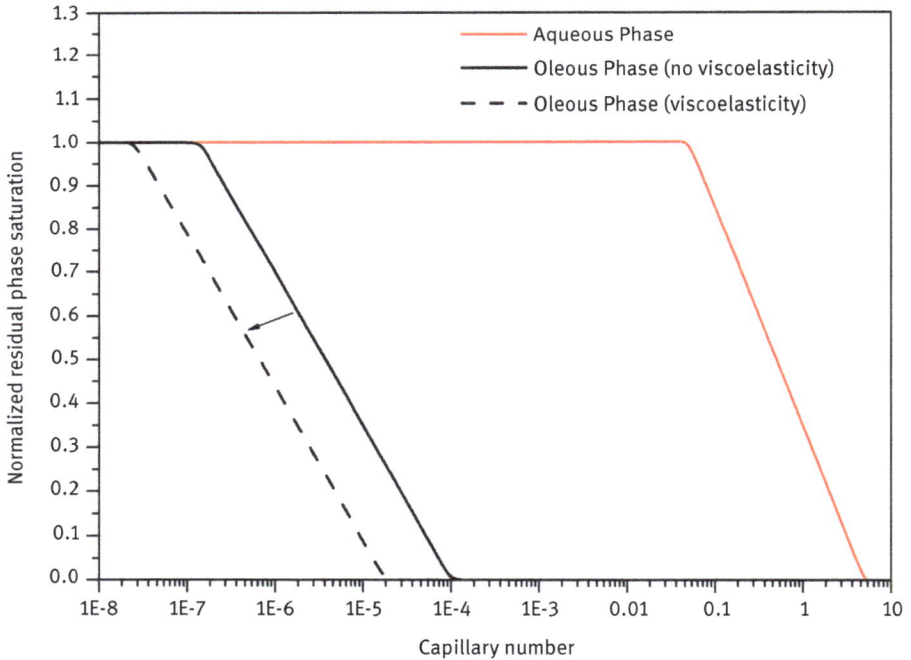

Fig. 4.4: Capillary desaturation curves for non-wetting (hydrocarbon) and wetting (aqueous) phases, showing the influence of the new model on the hydrocarbon residual saturation [71].

left, decreasing the normalized residual oil saturation for the same N_c. In waterflooding processes, the relaxation time of the water is negligible, rendering the waterflooding desaturation curve. When a polymer solution is injected, the viscoelasticity shifts the desaturation curve to the left, causing the hydrocarbon phase to be desaturated at lower capillary numbers [51]. Moreover, the relaxation time is a function of the molecular weight, so in order to fully couple the numerical model with the degradation process the relationship between both factors is taken into account (Eq. (4.18)) [12–15, 51, 72]. This model takes into account the degradation process to a full extent, with molecular weight, relaxation time, desaturation curves, and relative permeabilities varying as a function of time.

Relative permeabilities
Relative permeabilities influence Darcy's equation on the phase velocities, and therefore the efficiency of oil recovery. They depend on the residual saturations which were calculated in the previous section. The model used to calculate the relative permeabilities is taken from Camilleri [73, 74], which is used for most chemical flooding processes. By knowing beforehand the phase saturations, the relative permeabilities are calculated according to the following formula:

$$k_r^j = k_r^{j0} \cdot \left(\frac{S^j - S^{jr}}{1 - S^{jr} - S^{j'r}} \right)^{e^j} ; \quad j = o, a ; \; j \neq j' \tag{4.19}$$

Where k_r^{j0} and e^j represent the end point and the curvature of the function $k_r^j(S^j)$. These values are calculated by the following equations:

$$k_r^{j0} = k_r^{j0H} + (1 - k_r^{j0H}) \cdot \left(1 - \frac{S^{j'r}}{S^{j'rH}} \right) ; \quad j = o, a ; \; j \neq j' \tag{4.20}$$

$$e^j = e^{jH} + (1 - e^{jH}) \cdot \left(1 - \frac{S^{j'r}}{S^{j'rH}} \right) ; \quad j = o, a ; \; j \neq j' \tag{4.21}$$

where k_r^{j0H} and e^{jH} are the endpoint values of curvature and relative permeability function system for water-oil without the presence of chemical agents, respectively.

Inaccessible pore volume
This phenomenon was first reported by Dawson and Lantz [75], who noticed that polymer molecules traveled faster than other chemical species in the water phase. This is due to the size of the polymer molecules which cannot penetrate into the entire poral volume, whereas other small molecules (e.g., water, tracers) can access the whole domain. This is then influenced by the size and architecture of the polymer and the physical properties of the porous medium. The purpose of this parameter (IAPV) is to compensate the retention process causing the polymer to flow faster than the other

components. Since then, several authors in laboratory-scale experiments have confirmed this phenomenon [76–78]. This is quantified in the polymer mass conservation (Eq. (4.2)) where an extra-term is added affecting the porosity [33].

$$\phi_{IAPV} = 1 - \frac{IAPV}{\phi} \wedge \phi_{i=c} = \phi \cdot \phi_{IAPV} \tag{4.22}$$

Phase viscosities

Much has been written about the rheological characteristics of polymer solutions. They present a non-Newtonian behavior with a shear-thinning profile in rheometry experiments [52, 79]. The most used correlations to describe this are the Power-law and the Carreau–Yasuda law. Three different zones can be distinguished (Fig. 4.5), namely: the upper Newtonian where the viscosity remains somewhat constant and similar to the zero-shear viscosity; then, a shear-thinning region similar to the one described by the Power-law; and finally the lower Newtonian region at high shear rates where the viscosity is similar to the pure solvent viscosity. However, it has been reported that the rheological behavior of a polymer solution is different in a porous media, where shear rates can reach high values. At low- and medium-shear rates the solution has the same behavior as in the experiments. But after a critical value, a shear-thickening behavior is observed [80]. This is also called extensional flow, and the viscosity is described as the sum of the shear-thinning and the extensional (or elastic) viscosity. This phenomenon is caused by the elasticity of polymer molecules. It is noteworthy to mention that at even higher shear rates a maximum viscosity is achieved followed by a gradual decrease. This is due to the mechanical degradation phenomenon when the polymer chains are ruptured by the fluid acceleration field, lowering the molecular weight and the viscosity.

The first step to model the rheology of the solution is to adopt a model that allows quantifying and evaluating both phenomena. The unified viscosity model (UVM) developed by Delshad [23] is a correlation used and validated for the entire range of shear rates encountered in porous media. The overall viscosity consists of two terms, the shear-thinning viscosity, dominant at low and medium shear-rates, and the elastic viscosity which becomes important as the shear rate increases (Fig. 4.5). The shear thinning term is expressed according to the Carreau relationship,

$$\mu_{UVM} = \mu_{ST} + \mu_{ELAS} \tag{4.23}$$

$$\mu_{ST} = \mu_{0sr} + (\mu_w - \mu_{0sr}) \cdot \left[1 + \left(\frac{\dot{\gamma}}{\tau_r}\right)^2\right]^{\left(\frac{n-1}{2}\right)} \tag{4.24}$$

$$\mu_{ELAS} = \mu_{MAX} \cdot \left[1 - e^{-(\lambda_2 \tau_2 \dot{\gamma})^{n_2-1}}\right] \tag{4.25}$$

where μ_{0sr} is the viscosity of the solution at zero shear-rate, μ_w is the viscosity of water, $\dot{\gamma}$ is the shear rate, and β_1, β_2 and n are input parameters according to the rheological behavior, which defines the first critical shear rate, where the fluid passes from the

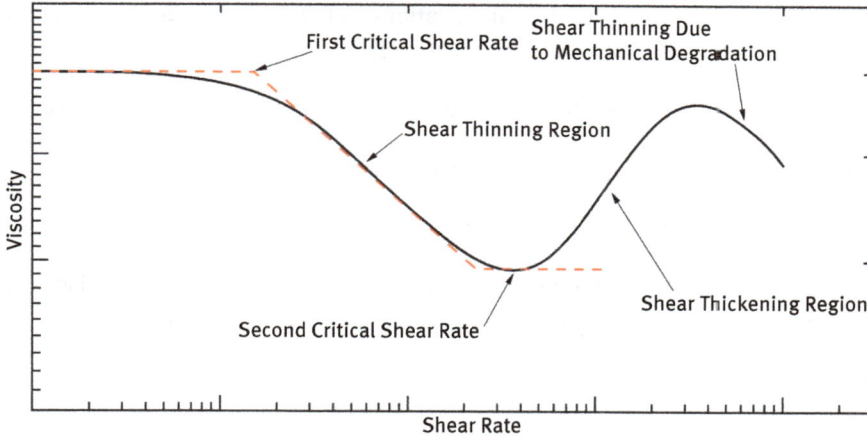

Fig. 4.5: Rheological model used in the simulation (UVM) compared to the Carreau–Yasuda regions [71].

upper Newtonian regime to the shear thinning region. The elastic viscosity is defined as,

$$\tau_r = \beta_1 \cdot e^{\beta_2 V_c^a} \tag{4.26}$$

$$\tau_2 = \tau_0 + \tau_1 \cdot V_c^a \tag{4.27}$$

$$\mu_{MAX} = \mu_w \left(AP_{11} + AP_{22} \cdot \ln V_c^a \right) \tag{4.28}$$

Where λ_2, n_2, τ_0, τ_1, AP_{11}, AP_{22} are the input parameters of the shear thickening region. The main concern of this formulation is then to find a way to calculate or measure the critical parameters, namely: μ_{MAX} and μ_{0sr}. For instance, Wang [72] suggested using a function to relate the elastic and the shear thinning viscosities. This approach will be followed in the model to express the critical parameters, accepting that the maximum shear thickening viscosity will be a function of the zero shear viscosity. The problem then has been reduced to finding a proper relationship for the zero shear viscosity [52, 81]. The modified Flory equation [82] is adopted to calculate this viscosity,

$$\mu_{0sr} = \mu_w \left[1 + \left(AP_1 V_c^a + AP_2 V_c^{a^2} + AP_3 V_c^{a^3} \right) C_{SEP}^{Sp} \right] \tag{4.29}$$

Where AP_1, AP_2, and AP_3 are input parameters which can be obtained from laboratory experiments. The term C_{SEP}^{Sp} takes into account the dependence of the polymer viscosity on the salinity and the percentage of divalent cations in the latter present in the porous medium. This can be written as the specific viscosity, and the terms expressed as functions of the intrinsic viscosity as follows,

$$C_{SEP} = \frac{V_s^a + (\beta_{pol} - 1) C_{DIV}^a}{V_w^a} \tag{4.30}$$

where C_{DIV}^a is the concentration of divalent cations in the water phase. The constant β_{pol} is obtained from laboratory measurements [33].

$$\mu_{sp} = \left(AP_1 V_C^a + AP_2 V_C^{a^2} + AP_3 V_C^{a^3} \right) C_{SEP}^{Sp} \tag{4.31}$$

where AP_i are expressed as a function of the intrinsic viscosity,

$$AP_1 = k_1 \cdot [\eta] \; ; \quad AP_2 = k_2 \cdot [\eta]^2 \; ; \quad AP_3 = k_3 \cdot [\eta]^3 \tag{4.32}$$

where k_i are constants affecting the specific viscosity. Finally, the intrinsic viscosity can be related to the average molecular weight using the Mark–Houwink formula,

$$[\eta] = K_{MH} \cdot M_w^{\alpha_{MH}} \tag{4.33}$$

A relationship between the rheological behavior of the polymer solution and the molecular weight has been reached which, to our best knowledge, has never been proposed in polymer simulators in the literature. It is well documented that polymer solutions undergo several degradation mechanisms in underground porous media, namely: mechanical (due to high shear rates), chemical (due to the presence of salts), thermal (related to high temperature reservoirs), and biological (bacteria affecting mostly biopolymers) [24–32]. These mechanisms cannot be excluded when longer flooding processes are being performed. The problem consists then in understanding how the polymer degrades as a function of the time it remains underground. This degradation will cause the scission of the backbone chain, modifying the molecular weight and therefore the rheological properties. In order to simulate the degradation, an exponential decay law will be assumed in the average molecular weight. This law was selected based on degradation experiments presented in the literature as well as numerical models developed to consider the degradation of polymer chains in porous media [83–88]. This yields,

$$\frac{dM_w}{dt} = -\lambda_{degmec} M_w \tag{4.34}$$

The degradation parameter (λ_{degmec}) will regulate how fast the (macro)molecules degrade and the molecular weight decreases with time (Fig. 4.6). For the purpose of this simulation, the decay parameter will be considered constant. However, this depends on many factors according to the mechanisms involved and the polymers used. For instance, the salinity or bacteria concentration are not constant throughout the whole domain, and these values affect seriously the degradation rate.

For the oil phase, since it was assumed that no water or chemicals are present in the phase, the rheology behavior can be considered as that of pure oil. According to the literature [2, 3, 6, 33], light and medium oil cuts exhibit Newtonian behavior while heavy oil might present a slight shear-thinning rheology [89]. For the purpose of this simulator, it will be considered that the hydrocarbon phase is a Newtonian fluid.

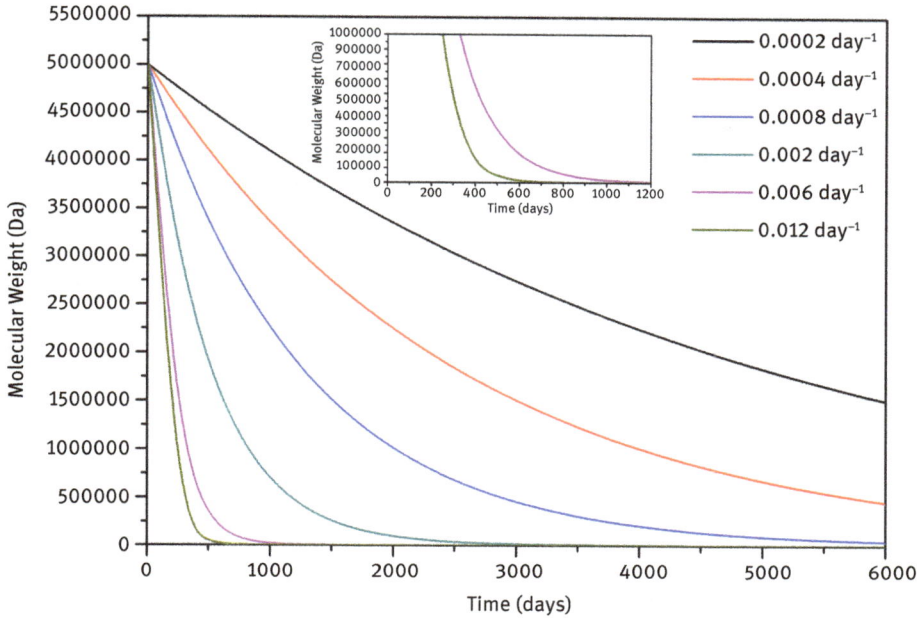

Fig. 4.6: Degradation rates (λ_{degmec}) and their influence on the polymer average molecular weight (degradation rates based on Wang [72] [71]).

Adsorption

The adsorption process occurs when polymer aggregates onto the surface of the formation rock. This irreversible phenomenon will cause a loss of polymer in the porous media, making the whole process economically unfeasible in case of high rates of adsorption. This is due to the fact that extra chemicals would be necessary and the viscosifying properties will be decreased. The adsorption isotherm is rather dependent on the type of polymer, the characteristics of the rock and the type of electrolytes present in the solution. The process starts with aggregates which are formed at the surface. The adsorption of the chemical component onto the rock used in this simulator is described by the Langmuir monolayer model: [33, 57]

$$\text{Ad}_c = \min \left[(Z_c + \text{Ad}_c) , \frac{a_{1,c} \cdot Z_c}{1 + a_{2,c} \cdot Z_c} \right] \tag{4.35}$$

$$a_{1,c} = (a_{11,c} + a_{12,c} \cdot C_{\text{SEP}}) \tag{4.36}$$

where a_1 and a_2 are the adsorption rate parameters, based on the salinity and the absolute permeability, and Ad_c is a dimensionless parameter representing the adsorbed volume of chemical component per unit of volume of the porous media. Since it was assumed the fluids are incompressible, adsorption is then formulated on a volume basis.

Disproportionate permeability reduction

As a result of the adsorption process, polymer molecules in the rock will resist the flow of the aqueous phase, which can be interpreted as a decrease in the relative permeability function. This phenomenon is denominated as the disproportionate permeability reduction (DPR) or relative permeability modification (RPM). It is noteworthy to point out that both DPR and RPM are terms which may refer to the same concept, but in the oil industry DPR is used more commonly for gels whilst RPM is used for water-soluble polymer solutions. However, in this book we refer to this concept using the term DPR. There is a direct relationship between the adsorbed molecules and the permeability reduction. This is an irreversible process, since the DPR does not decrease if the polymer concentration does, and it can be used as an indicator for the degree of channel blocking in the porous media. Then, the DPR factor can be modeled as [10, 33, 90, 91],

$$R_k = 1 + \frac{(R_{k,\max} - 1)\, b_{rk} V_C^a}{1 + b_{rk} V_C^a} \tag{4.37}$$

where $R_{k,\max}$ is the maximum permeability reduction factor and b_{rk} is an input parameter related to the adsorption process. This value can be obtained from core experiments or according to the following expression: [10, 33, 72]

$$R_{k,\max} = \min\left\{ \left[1 - \frac{c_{rk}\left(AP_1 C_{SEP}^{Sp}\right)^{1/3}}{\left(\frac{\sqrt{K_x K_y}}{\phi}\right)^{1/2}} \right]^{-4}, R_{k,\mathrm{cut}} \right\} \tag{4.38}$$

where c_{rk} is an input parameters related to the physical properties of the porous medium and the salinity present in the domain. The empirically-determined term $R_{k,\mathrm{cut}}$ is used as the upper limit of DPR and set in the simulator to a value of 10, although factors higher than this value were reported in low permeability fields (Fig. 4.7) [10, 92].

Capillary pressure

The capillary pressure is defined as the difference between the non-wetting (hydrocarbon) and the wetting (aqueous) phases. This parameter is usually defined as a function of the water saturation. In this chapter this relationship is described by the following power function [3]:

$$p_c = C \cdot \sqrt{\frac{\phi}{K}} \cdot \frac{\gamma}{\gamma^H} \cdot \left(\frac{1 - S^a - S^{or}}{1 - S^{ar} - S^{or}} \right)^n \tag{4.39}$$

where C is a constant parameter and n defines the curvature of the function. The capillary pressure parameter C relates the capillary forces in the three component system (petroleum, water and chemical) to the capillary forces in the oil-water system.

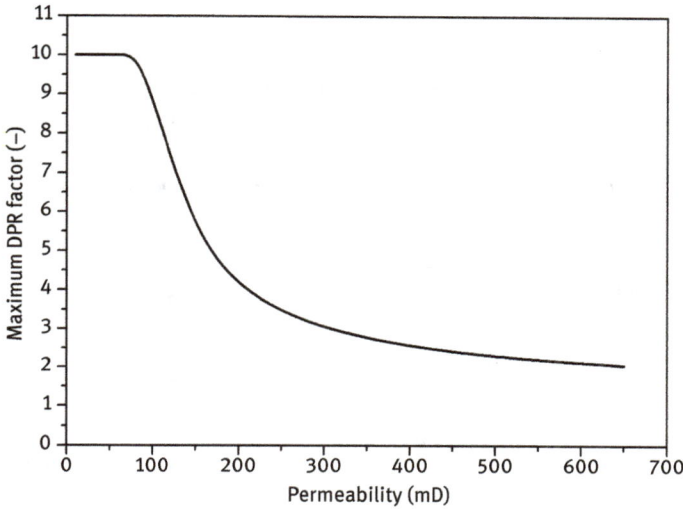

Fig. 4.7: Maximum DPR factor as a function of the absolute permeability [71].

4.4.3 Boundary conditions

At the start of the polymer flooding process, the residual saturation in the reservoir is either the result of a primary recovery process, or the saturation after a waterflooding which reached an economic limit of the fractional flow at the producing well. There is no chemical present and the initial pressure is constant throughout the reservoir. Thus:

$$t = 0 ; \quad \forall(x, y) \in \Omega : z_c = 0 ; \quad z_p = S^{orH} ; \quad p^a = p_i \qquad (4.40)$$

The flooding process begins injecting, for a certain period of time, a polymer or chemical solution with constant concentration. After this period, the chemical slug is followed by a bank of water in order to sweep the remaining oil. As the boundary conditions, 'no flow' is imposed on the contour (Γ), since it is assumed that the porous medium is surrounded by an impermeable rock layer. In the case of advective flux, this condition is satisfied if the transmissibilities (or mobilities) are zero on the boundary. In the case of diffusive flux, Fick's first law in the contour of the domain is applied, thus yielding:

$$\textbf{Injecting Well} \Rightarrow \begin{cases} 0 \leq t \leq t_{in} : z_c = z_{in} \\ t > t_{in} \quad : z_{in} = z_w, z_c = 0 \end{cases} \qquad (4.41)$$

$$\textbf{Boundaries} \Rightarrow \lambda_{m,n}^j = 0 \wedge \frac{\partial z_i}{\partial \check{n}_\Gamma} = 0 ; \quad i = p, c ; \quad \forall t \wedge \forall(m, n) \in \Gamma \qquad (4.42)$$

4.4.4 Nondimensionalization of the transport equations

It is important in every physical model to establish the influence and degree of dominance of the different phenomena involved. Thus, the dimensionless forms of Darcy and mass conservation equations were derived (Eqs. (4.43) and (4.44)) and expressed using the Capillary (Eq. (4.14)) and Peclet numbers (Eq. (4.45)). The dimensionless variables and derivatives are represented using a breve symbol ($\breve{\ }$).

$$\breve{\nabla} \cdot \left[\left(\frac{k_r^o}{N_c^o} + \frac{k_r^a}{N_c^a} \right) \cdot \breve{\nabla} \breve{p}^a \right] = \phi \cdot \frac{\partial}{\partial \breve{t}} \left(\sum_i \text{Ad}_i \right) - \breve{\nabla} \cdot \left(\frac{k_r^o}{N_c^o} \cdot \breve{\nabla} \breve{p}_c \right) + t_{ref} \cdot q_t \tag{4.43}$$

$$\phi \cdot \frac{\partial z_i}{\partial \breve{t}} + \breve{\nabla} \cdot \sum_j V_i^j \cdot \breve{u}^j - \breve{\nabla} \cdot \sum_j \frac{1}{Pe_i^j} \cdot \breve{\nabla} \cdot V_i^j = -\phi \cdot \frac{\partial \text{Ad}_i}{\partial \breve{t}} + t_{ref} \cdot q_i \tag{4.44}$$

$$Pe_i^j = \frac{l_{ref} \cdot u_{ref}^j}{D_i^j} \tag{4.45}$$

The capillary number describes the relationship between viscous and capillary forces and affects the pressure equation. The objective in a surfactant flooding is to make these forces of a similar order so the trapped oil can be displaced. The Peclet number defines the relative importance of the diffusion mechanism in the transport process. With negligible diffusion coefficients the Peclet number is high ($Pe_i^j \gg 1$) and then advection dominates the process. Increasing this coefficient renders lower Peclet numbers ($Pe_i^j \approx 1$ or $Pe_i^j < 1$) and diffusion mechanisms can no longer be neglected.

4.4.5 Discretization of the partial differential equations

The above formulation based on the Darcy's and mass conservation equations yielded a system of non-linear parabolic partial differential equations, which will be discretized and solved by the finite difference method. The first equation to analyze is the pressure of the aqueous phase (Eq. (4.11)), which is implicitly solved using a centered discretization scheme for the pressure terms and a second order Taylor approximation of the time derivatives. This scheme is often used in systems with a derived second order, with coefficients that are not constants in the domain in studio. Besides the Darcy equation, the discretization of the total and aqueous Darcy velocities is also presented, which are explicitly solved using a centered difference scheme.

Therefore, Eqs. (4.11), (4.1) and (4.2) are discretized as follows, respectively [93]:

$$
c_r \left(\phi + \frac{\Delta t}{2} \frac{\partial \phi}{\partial t} \right)_{m,n}^{\langle n+1 \rangle, [k]} \left(\frac{p_{m,n}^{a,\langle n+1 \rangle} - p_{m,n}^{a,\langle n \rangle}}{\Delta t} \right)^{[k+1]}
$$

$$
+ \frac{\lambda_{x,m+1/2,n}^{\langle n+1 \rangle, [k]}}{\Delta x^2} \cdot \left(p_{m+1,n}^{a} - p_{m,n}^{a} \right)^{\langle n+1 \rangle, [k+1]} - \frac{\lambda_{x,m-1/2,n}^{\langle n+1 \rangle, [k]}}{\Delta x^2} \cdot \left(p_{m,n}^{a} - p_{m-1,n}^{a} \right)^{\langle n+1 \rangle, [k+1]}
$$

$$
+ \frac{\lambda_{y,m+1/2,n}^{\langle n+1 \rangle, [k]}}{\Delta y^2} \cdot \left(p_{m,n+1}^{a} - p_{m,n}^{a} \right)^{\langle n+1 \rangle, [k+1]} - \frac{\lambda_{y,m-1/2,n}^{\langle n+1 \rangle, [k]}}{\Delta y^2} \cdot \left(p_{m,n}^{a} - p_{m,n-1}^{a} \right)^{\langle n+1 \rangle, [k+1]}
$$

$$
= \left(\phi + \frac{\Delta t}{2} \frac{\partial \phi}{\partial t} \right)_{m,n}^{\langle n+1 \rangle, [k]} \left(\frac{\mathrm{Ad}_{m,n}^{\langle n+1 \rangle} - \mathrm{Ad}_{m,n}^{\langle n \rangle}}{\Delta t} \right)^{[k+1]}
$$

$$
+ \left(\mathrm{Ad} + \frac{\Delta t}{2} \frac{\partial \mathrm{Ad}}{\partial t} \right)_{m,n}^{\langle n+1 \rangle, [k]} \left(\frac{\phi_{m,n}^{\langle n+1 \rangle} - \phi_{m,n}^{\langle n \rangle}}{\Delta t} \right)^{[k+1]}
$$

$$
+ \frac{\lambda_{x,m+1/2,n}^{o,\langle n+1 \rangle, [k]}}{\Delta x^2} \cdot \left(p_{m+1,n}^{c} - p_{m,n}^{c} \right)^{\langle n+1 \rangle, [k+1]} - \frac{\lambda_{x,m-1/2,n}^{o,\langle n+1 \rangle, [k]}}{\Delta x^2} \cdot \left(p_{m,n}^{c} - p_{m-1,n}^{c} \right)^{\langle n+1 \rangle, [k+1]}
$$

$$
+ \frac{\lambda_{y,m+1/2,n}^{o,\langle n+1 \rangle, [k]}}{\Delta y^2} \cdot \left(p_{m,n+1}^{c} - p_{m,n}^{c} \right)^{\langle n+1 \rangle, [k+1]} - \frac{\lambda_{y,m-1/2,n}^{o,\langle n+1 \rangle, [k]}}{\Delta y^2} \cdot \left(p_{m,n}^{c} - p_{m,n-1}^{c} \right)^{\langle n+1 \rangle, [k+1]}
$$

$$
+ q_{m,n}^{t,\langle n+1 \rangle, [k]} \tag{4.46}
$$

$$
\vec{u}_{m,n}^{\langle n+1 \rangle, [k+1]} = \left[-\frac{\lambda_{x,m,n}^{[k]}}{2 \cdot \Delta x} \cdot \left(p_{m+1,n}^{a} - p_{m-1,n}^{a} \right)^{[k+1]} \right.
$$

$$
\left. - \frac{\lambda_{x,m,n}^{o,[k]}}{2 \cdot \Delta x} \cdot \left(p_{c,m+1,n}^{a} - p_{c,m-1,n}^{a} \right)^{[k+1]} \right]^{\langle n+1 \rangle} \cdot \hat{i}
$$

$$
+ \left[-\frac{\lambda_{y,m,n}^{[k]}}{2 \cdot \Delta y} \cdot \left(p_{m,n+1}^{a} - p_{m,n-1}^{a} \right)^{[k+1]} \right.
$$

$$
\left. - \frac{\lambda_{y,m,n}^{o,[k]}}{2 \cdot \Delta y} \cdot \left(p_{c,m,n+1}^{a} - p_{c,m,n-1}^{a} \right)^{[k+1]} \right]^{\cdot n+1 \rangle} \cdot \hat{j} \tag{4.47}
$$

$$
\vec{u}_{m,n}^{a,\langle n+1 \rangle, [k+1]} = \left[-\frac{\lambda_{x,m,n}^{a,[k]}}{2 \cdot \Delta x} \cdot \left(p_{m+1,n}^{a} - p_{m-1,n}^{a} \right)^{[k+1]} \right]^{\langle n+1 \rangle} \cdot \hat{i}
$$

$$
+ \left[-\frac{\lambda_{y,m,n}^{a,[k]}}{2 \cdot \Delta y} \cdot \left(p_{m,n+1}^{a} - p_{m,n-1}^{a} \right)^{[k+1]} \right]^{\langle n+1 \rangle} \cdot \hat{j} \tag{4.48}
$$

In the discretized equations the notation is as follows: m, n represent the cells in the axes of the physical numerical domain $(x, y) = (m \cdot \Delta x, n \cdot \Delta y)$, respectively, $\langle n \rangle$ represents the temporal-step (time $= \langle n \rangle \cdot \Delta t$) in the simulation and $[k], \forall k \in \mathbb{N}^+$, is the iteration number within each time-step. One of the most sensitive points of reservoir models is to calculate the in- and outflows in the well blocks. Wells models are commonly used for reservoir simulation [94], and for the purpose of this book, it was adopted as the operating regime that producing wells operate at a constant total flow rate, and the injector will operate at a constant bottomhole pressure. For wells, both injectors and producers, in Cartesian coordinates, the following formula applies:

$$Q = \left[\text{PI}_{m,n}^{j,[k]} \cdot \left(p_{\text{wf}} - p_{m,n}^{j,[k+1]} \right) \right]^{\langle n+1 \rangle} \tag{4.49}$$

Where the productivity ratio is calculated for two-dimensional systems with the following formula:

$$\text{PI}_{m,n}^{j,[k]} = \frac{2 \cdot \pi \cdot \sqrt{k_x \cdot k_y} \cdot \Delta z}{0.15802 \cdot \left[\ln\left(\frac{r_o}{r_w} \right) + s \right]} \cdot \frac{k_{\text{r}}^j}{\mu_{m,n}^j} \tag{4.50}$$

The equivalent radius necessary in the previous equation is obtained using the Peaceman model for heterogeneous models: [94]

$$r_o = 0.28 \cdot \frac{\left[\left(\frac{k_x}{k_y} \right)^{1/2} \cdot \Delta y^2 + \left(\frac{k_y}{k_x} \right)^{1/2} \cdot \Delta x^2 \right]^{1/2}}{\left(\frac{k_x}{k_y} \right)^{1/4} + \left(\frac{k_y}{k_x} \right)^{1/4}} \tag{4.51}$$

Due to the scheme chosen (quarter five-spot), the wells are located in boundary blocks. This affects the value of the equivalent radius. The Peaceman model was extended to take into account this and other factors (e.g., non-square grids and non-Darcy effects). This correction has been already addressed in the literature [63] and it has been considered in the proposed model.

As the last step of the model, the analysis of the discretization of the mass conservation equation is analyzed. Equation (4.4) is the typical advection – diffusion PDE, which is employed in many phenomena in hydraulic and fluid studies in porous media. Advective terms are of a hyperbolic nature, and upwind discretization schemes cause a numerical diffusion/dispersion in the solution of the overall total compositions, as reported in the literature [95, 96]. In order to solve this, higher order schemes should be used [97]. In this chapter a fully second order explicit discretization scheme in time and space is derived, based on flux limiting techniques, as explained in Chapter 3. This allows increasing of the numerical accuracy of the simulator as well as decreasing the influence of numerical diffusion and dispersion on the recovery factor. The diffusive term is discretized using a centered second order scheme. In the present study, the longitudinal and transversal dispersive terms in the diffusion tensor will be neglected. The second order in time is achieved using a Taylor expansion of the second order [97]. Finally, a functional relationship is established between the gradient of the volumetric concentration and the limiting function ψ. Several second order methods have been proposed and studied. These depend on the ratio of the concentrations' consecutive gradients in the numerical mesh ($r_{x,i} = (V_{i,m,n}^{j,[k]} - V_{i,m-1,n}^{j,[k]})/(V_{i,m+1,n}^{j,[k]} - V_{i,m,n}^{j,[k]})$). All in all, the discretized mass conservation equation yields,

$$
\frac{\mathcal{C}_1}{\Delta t} z_i^{\langle n+1\rangle} = \mathcal{C}_2 z_i^{\langle n\rangle} + \frac{\mathcal{C}_3}{\Delta x} \cdot \sum_j F_{\text{LIM},x}^{j,\langle n+1\rangle,[k+1]} \left(u_{x,m,n}^{j,[k+1]} \cdot V_{i,m,n}^{j,[k]} - u_{x,m-1,n}^{j,[k+1]} \cdot V_{i,m-1,n}^{j,[k]} \right)^{\langle n+1\rangle}
$$

$$
+ \frac{\mathcal{C}_3}{\Delta y} \cdot \sum_j F_{\text{LIM},y}^{j,\langle n+1\rangle,[k+1]} \left(u_{y,m,n}^{j,[k+1]} \cdot V_{i,m,n}^{j,[k]} - u_{y,m,n-1}^{j,[k+1]} \cdot V_{i,m,n-1}^{j,[k]} \right)^{\langle n+1\rangle}
$$

$$
+ \frac{1}{\Delta x^2} \cdot \sum_j \left[\left(S^j \phi \mathrm{dm}_i^j \right)_{m+1/2,n} \cdot \left(V_{i,m+1,n}^j - V_{i,m,n}^j \right) \right.
$$

$$
\left. - \left(S^j \phi \mathrm{dm}_i^j \right)_{m-1/2,n} \cdot \left(V_{i,m,n}^j - V_{i,m-1,n}^j \right) \right]^{\langle n+1\rangle,[k]}
$$

$$
+ \frac{1}{\Delta y^2} \cdot \sum_j \left[\left(S^j \phi \mathrm{dm}_i^j \right)_{m,n+1/2} \cdot \left(V_{i,m,n+1}^j - V_{i,m,n}^j \right) \right.
$$

$$
\left. - \left(S^j \phi \mathrm{dm}_i^j \right)_{m,n-1/2} \cdot \left(V_{i,m,n}^j - V_{i,m,n-1}^j \right) \right]^{\langle n+1\rangle,[k]}
$$

$$
- \frac{1}{\Delta t} \left(\phi + \Delta t \frac{\partial \phi}{\partial t} \right)_{m,n}^{\langle n+1\rangle,[k+1]} \cdot \left(\mathrm{Ad}_i^{\langle n+1\rangle} - \mathrm{Ad}_i^{\langle n\rangle} \right)_{m,n}^{[k]} - \frac{\mathrm{Ad}_{i,m,n}^{\langle n+1\rangle,[k]}}{\Delta t} \cdot \left(\phi^{\langle n+1\rangle} - \phi^{\langle n\rangle} \right)_{m,n}^{[k]}
$$

$$
+ q_{i,m,n}^{\langle n+1\rangle,[k+1]} + \sum_j \frac{u_{x,m,n}^{j,[k+1]} \Delta t}{2\phi_{m,n}^{\langle n+1\rangle,[k+1]}} \cdot \left(\frac{\partial \mathrm{Ad}}{\partial x} \frac{\partial \phi}{\partial t} + \mathrm{Ad} \frac{\partial^2 \phi}{\partial t \partial x} + \frac{\partial \phi}{\partial x} \frac{\partial \mathrm{Ad}}{\partial t} \right)_{i,m,n}^{\langle n+1\rangle,[k]}
$$

$$
+ \sum_j \frac{u_{y,m,n}^{j,[k+1]} \Delta t}{2\phi_{m,n}^{\langle n+1\rangle,[k+1]}} \cdot \left(\frac{\partial \mathrm{Ad}}{\partial y} \frac{\partial \phi}{\partial t} + \mathrm{Ad} \frac{\partial^2 \phi}{\partial t \partial y} + \frac{\partial \phi}{\partial y} \frac{\partial \mathrm{Ad}}{\partial t} \right)_{i,m,n}^{\langle n+1\rangle,[k]}
$$

$$
- \sum_j \left(\frac{u_{x,m,n}^{j,[k+1]} \Delta t}{2} \frac{\partial^2 \mathrm{Ad}}{\partial t \partial x} + \frac{u_{y,m,n}^{j,[k+1]} \Delta t}{2} \frac{\partial^2 \mathrm{Ad}}{\partial t \partial y} \right)_{i,m,n}^{\langle n+1\rangle,[k]}
$$

$$
+ \sum_j \frac{\Delta t}{2\phi_{m,n}^{\langle n+1\rangle,[k+1]}} \left[\left(u_{x,m,n}^{j,[k+1]} \right)^2 \frac{\partial^2 V_{i,m,n}^{j,[k]}}{\partial x^2} + \left(u_{y,m,n}^{j,[k+1]} \right)^2 \frac{\partial^2 V_{i,m,n}^{j,[k]}}{\partial y^2} \right.
$$

$$
\left. + 2 u_{x,m,n}^{j,[k+1]} u_{y,m,n}^{j,[k+1]} \frac{\partial^2 V_{i,m,n}^{j,[k]}}{\partial x \partial y} \right]^{\langle n+1\rangle}
\tag{4.52}
$$

$$
\mathcal{C}_1 = \left(\phi_{m,n} + \Delta t \frac{\partial \phi}{\partial t} - \frac{u_{tx,m,n} \Delta t}{2\phi_{m,n}} \frac{\partial \phi}{\partial x} - \frac{u_{ty,m,n} \Delta t}{2\phi_{m,n}} \frac{\partial \phi}{\partial y} \right)^{\langle n+1\rangle,[k+1]}
$$

$$
\mathcal{C}_2 = \left(\frac{\phi_{m,n}}{\Delta t} - \frac{u_{tx,m,n}}{2\phi_{m,n}} \frac{\partial \phi}{\partial x} - \frac{u_{ty,m,n}}{2\phi_{m,n}} \frac{\partial \phi}{\partial y} + \frac{u_{tx,m,n} \Delta t}{2\phi_{m,n}} \frac{\partial^2 \phi}{\partial t \partial x} + \frac{u_{ty,m,n} \Delta t}{2\phi_{m,n}} \frac{\partial^2 \phi}{\partial t \partial y} \right)^{\langle n+1\rangle,[k+1]}
$$

$$
\mathcal{C}_3 = \left(1 - \frac{\Delta t}{2\phi_{m,n}} \frac{\partial \phi}{\partial t} \right)^{\langle n+1\rangle,[k+1]}
\tag{4.53}
$$

4.4.6 Solution algorithm

The discretization carried out in the previous section leads to a coupled system of strongly non-linear equations, which are solved by an implicit-explicit combined method known as IMPEC, calculating implicitly the pressures of the aqueous phase throughout the domain. Then, the hydrocarbon pressure, Darcy velocities and total concentrations of two components are obtained by an explicit method. The complexity of this problem lies in the fact that many of the auxiliary properties used in the formulas are functions of both the pressure and concentrations being calculated for the corresponding time step. This is solved using an iterative method in each time step. The difference between two consecutive iterations within the same time step in a parameter is required to be less than a preset error tolerance, which is set using a criterion based in the literature as well as in the numerical conditions adopted for the simulation [56, 57]. This difference is calculated using a specific vector norm for the overall concentrations, which can be also modified in the simulator. Then, with the data values of the parameters in the previous time step (n and $k = k_{n+1}^{max}$), are used as starting values for the next step ($n + \Delta t$ and $k = 1$) (Fig. 4.8).

Procedure

1. The aqueous pressure is calculated in the domain with the resulting system from Eq. (4.46).
2. The pressure of the oil phase is calculated with the capillary pressure relationship.
3. Darcy velocities are calculated with Eqs. (4.2), (4.47) and (4.48).
4. Flowrates in the wells are obtained for each component with Eqs. (4.49), (4.50) and (4.51).
5. The overall composition of the oil and chemical is calculated with Eqs. (4.52) and (4.53).
6. The water overall concentration, volume fractions of the components and phase saturations are calculated with Eqs. (4.6), (4.7), (4.8), (4.9) and the phase behavior equations.
7. Finally, having all parameters calculated at iteration $k + 1$, a set of norms are evaluated for the overall concentrations of $n - 1$ components with Eq. (4.54).

$$e_i = \max_{\forall m, n \in \Omega} \left| \left(z_{i,m,n}^{\langle n \rangle, [k+1]} - z_{i,m,n}^{\langle n \rangle, [k]} \right) \right| ; \quad i = \text{p, c, s} \tag{4.54}$$

Equation (4.54) represents the *max norm* of the error matrices. These norm can be adjusted in the model to change the convergence criteria. The $\ell_{1,2,\infty}$ norms may also be

Iterative IMPEC Simulator

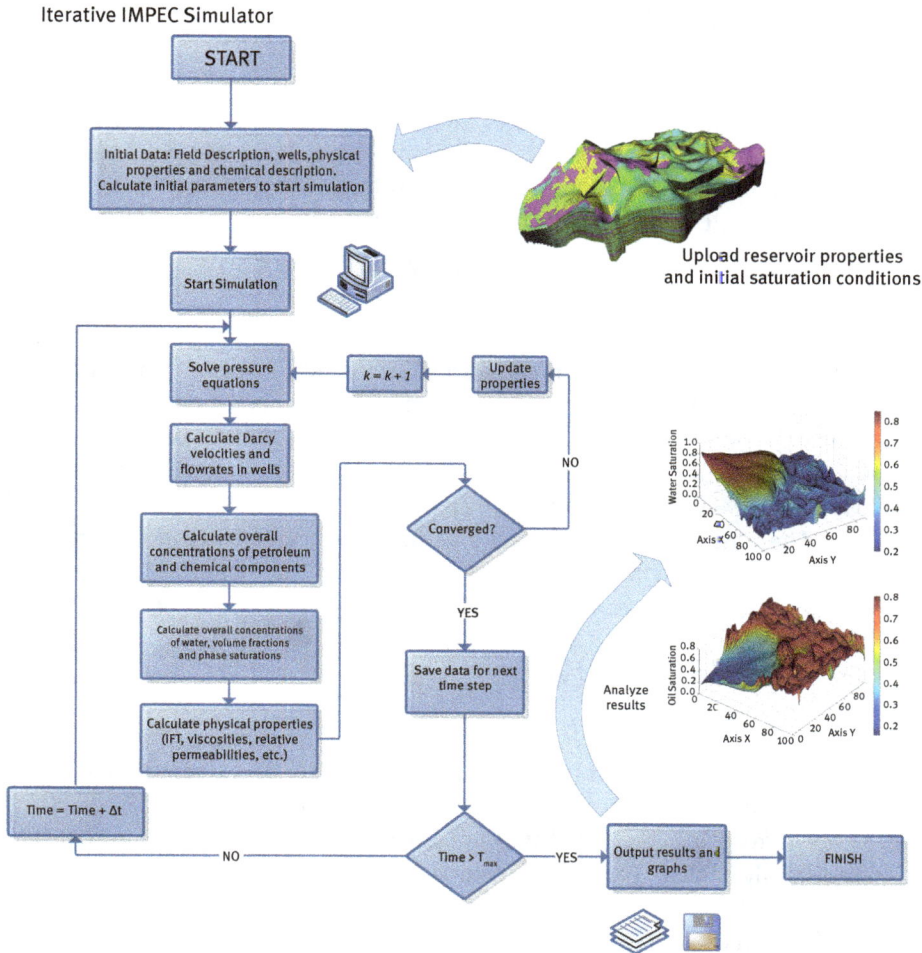

Fig. 4.8: Flowchart representing the steps of the iterative IMPEC model [93].

employed in the simulations to determine the convergence criterion. When both errors comply with a maximum allowable error given as the input, the calculation is concluded in a time step n and passed to $n + 1$, using as the next input, the data obtained in the last iteration of the previous time-step. If the errors do not meet the compliance, the values obtained are used as input data for a new iteration in the same time step.

4.5 Solution and validation

The simulator can be easily programmed to solve the non-linear system of equations that results from evaluating Eq. (4.11) in the domain. The assembly of the matrices (aqueous and capillary pressure terms) is solved using a direct or the sparse matrix tools, chosen due for its fast convergence to the solution. This is based on the size of the reservoir mesh. The system is then expressed as,

$$\underline{\underline{H}}_\Omega^{\langle n+1\rangle,[k]} \cdot \vec{P}_\Omega^{a,\langle n+1\rangle,[k+1]} = \underline{\underline{G}}_\Omega^{\langle n+1\rangle,[k]} \cdot \vec{P}_\Omega^{c,\langle n+1\rangle,[k]} + \vec{J}_T^{\langle n+1\rangle,[k]} \tag{4.55}$$

Due to the fact that the compositional approach employed in this simulator, secondary recovery processes (waterflooding) can also be simulated. The governing equations are easily modified to take into account this technique, reducing the system to the well-known implicit pressure and explicit saturation (IMPES) method. One point worth mentioning is that the traditional IMPES method is not iterative, unlike the model developed here, which is reflected in the processing time between the two methods (IMPES and iterative IMPEC), although the feature of being iterative confers better numerical stability than the IMPES method. The code has the option to perform the simulation of a secondary recovery based on an iterative or non-iterative method.

4.5.1 Validation of the model

In order to validate the new simulator, a two-dimensional field was considered, which is operated under a waterflooding scheme. The data and geometric characteristics were adopted from Najafabadi [98] and compared to the results obtained in UTCHEM and GPAS (a fully implicit, parallel EOS compositional reservoir simulator). The model dimensions are 201 m × 201 m × 30.5 m and it is discretized in a 10 × 10 gridblock scheme. The permeability is 100 mD and constant throughout the reservoir. At the beginning of the process, the water saturation is 0.3. Fig. 4.9 shows the result of the oil recovery factor and producing flowrates for the validation process and the comparison with the results reported by Najafabadi [98]. Although the new simulator presents larger numerical dispersion than UTCHEM, the final oil recovery factor is comparable to the latter. The major differences are observed when the waterfront reached the production well (≈ 0.5 PV) yielding an oil recovery factor slightly lower than the value in UTCHEM, mainly due to numerical errors. This numerical diffusion is not negligible but does not modify the final values in the new simulator and these are still accurate in terms of front location and cumulative oil production. This waterflooding test validates the behavior of the new simulator in two-dimensional fields.

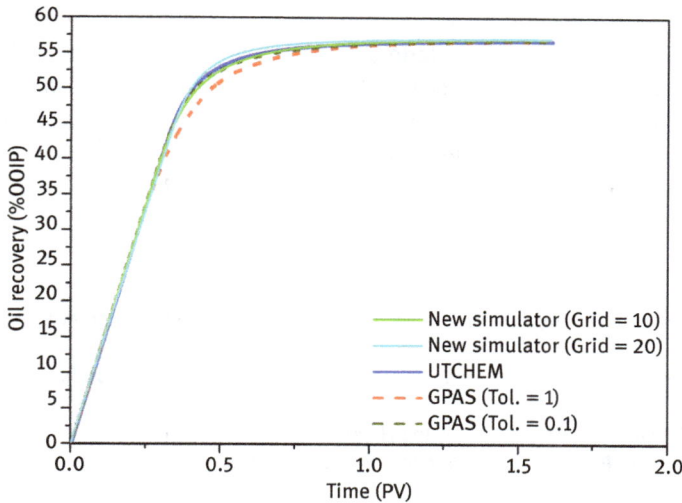

Fig. 4.9: Oil recovery factor during a waterflooding and comparison with the results obtained with UTCHEM and GPAS [93].

4.6 Conclusions

A new numerical simulator for a two-phase, four-component compositional flow has been presented and discussed, aimed at studying chemical EOR processes. The new simulator has been designed using the compositional approach in order to create a versatile source for the different chemical products used in industry. In order to diminish the numerical influence the differential equations were discretized using a fully second-order approach. This, coupled with a TVD flux-limiter function, allowed improving the front tracking of the chemicals being injected as well as reducing the numerical smearing of the latter, which causes a decrease in the recovery factor.

The validation process showed a good correspondence with commercial and academic simulators used for 2D waterflooding processes. The compositional method also allows simulations of secondary or tertiary recovery processes without further complications. This chapter was aimed at developing, validating and then testing the numerical code in waterflooding processes. The simulator can be numerically operated in two different ways (with constant or random permeability fields): iterative and non-iterative schemes. Due to the number of properties involved in EOR, only the iterative process will be used to simulate chemical tertiary recovery methods. However, waterflooding scenarios were also tested using the non-iterative method, faster than the iterative, but prone to be affected by numerical instabilities. There is a critical time step, a function of the simulation time and time-step, beyond which the system becomes unstable. This can be temporary and fade out as the simulation evolves, or it can cause the numerical crashing of the process. This simulator can be used as the base for developing further flooding process meant both to test and to set design standards for new different chemical products for EOR.

References

[1]   IEA. *Resources to Reserves 2013*. Organisation for Economic Co-operation and Development, 2013.
[2]   Dake, L. P. *Fundamentals of reservoir engineering*. Elsevier, Amsterdam, the Netherlands, 1978.
[3]   Lake, L. W. *Enhanced Oil Recovery*. Prentice-Hall Inc., Englewood Cliffs, USA, 1989.
[4]   Green, D. W. and Willhite, G. P. *Enhanced Oil Recovery*. Society of Petroleum Engineers, Richardson, USA, 1998.
[5]   Druetta, P., Tesi, P., Persis, C. D., and Picchioni, F. Methods in oil recovery processes and reservoir simulation. *Advances in Chemical Engineering and Science*, 6:39, 2016.
[6]   Donaldson, E. C., Chilingarian, G. V., and Yen, T. F. *Enhanced Oil Recovery, I: Fundamentals and Analyses*. Elsevier Science, Amsterdam, 1985.
[7]   Sandiford, B. B. Laboratory + field studies of water floods using polymer solutions to increase oil recoveries. *Journal of Petroleum Technology*, 16:917–&, 1964.
[8]   Pye, D. J. Improved secondary recovery by control of water mobility. *Journal of Petroleum Technology*, 16:911–&, 1964.
[9]   Seright, R. S., Seheult, M., and Talashek, T. Injectivity characteristics of eor polymers. *Spe Reservoir Evaluation & Engineering*, 12:783–792, 2009.
[10] Sheng, J. *Modern chemical enhanced oil recovery*. Elsevier, Amsterdam, the Netherlands, 2011.
[11] Wreath, D. G. A Study of Polymer Flooding and Residual Oil Saturation. Master's thesis, University of Texas at Austin, Austin, USA, 1989.
[12] Wang, D., Wang, G., Wu, W., Xia, H., and Yin, H. The influence of viscoelasticity on displacement efficiency – from micro to macro scale. In *SPE Annual Technical Conference and Exhibition*. Society of Petroleum Engineers, Anaheim, USA, 2007.
[13] Wang, D., Xia, H., Yang, S., and Wang, G. The influence of visco-elasticity on microforces and displacement efficiency in pores, cores and in the field. In *SPE EOR Conference at Oil & Gas West Asia*. Society of Petroleum Engineers, Muscat, Oman, 2010.
[14] Wei, B., Romero-Zerón, L., and Rodrigue, D. Mechanical properties and flow behavior of polymers for enhanced oil recovery. *Journal of Macromolecular Science, Part B*, 53:625–644, 2014.
[15] Lijuan, Z., Xiang'an, Y., and Fenqiao, G. Micro-mechanisms of residual oil mobilization by viscoelastic fluids. *Petroleum Science*, 5:56–61, 2008.
[16] Baehr, A. and Corapcioglu, M. A compositional multiphase model for groundwater contamination by petroleum-products. 2. numerical-solution. *Water Resources Research*, 23:201–213, 1987.
[17] Corapcioglu, M. and Baehr, A. A compositional multiphase model for groundwater contamination by petroleum-products. 1. theoretical considerations. *Water Resources Research*, 23:191–200, 1987.
[18] Smaoui, H., Zouhri, L., and Ouahsine, A. Flux-limiting techniques for simulation of pollutant transport in porous media: Application to groundwater management. *Mathematical and Computer Modelling*, 47:47–59, 2008.

[19] Wang, B. and Bauer, S. Pressure response of large-scale compressed air energy storage in porous formations. *Energy Procedia*, 125:588–595, 2017.

[20] Peter, B., Johannes, K., Robert, S., and Mary, W. *Simulation of Flow in Porous Media, Applications in Energy and Environment*. De Gruyter, Berlin, Germany, 2013.

[21] Ahusborde, E. and Ossmani, M. E. A sequential approach for numerical s mulation of two-phase multicomponent flow with reactive transport in porous media. *Mathematics and Computers in Simulation*, 137:71–89, 2017.

[22] Sleep, B. E. A method of characteristics model for equation of state compositional simulation of organic compounds in groundwater. *Journal of contaminant hydrology*, 17:189–212, 1995.

[23] Delshad, M. et al. Mechanistic interpretation and utilization of viscoelastic behavior of polymer solutions for improved polymer-flood efficiency. In *SPE Symposium on Improved Oil Recovery*. Society of Petroleum Engineers, Tulsa, USA, 2008.

[24] Maurer, J. J. and Harvey, G. D. Thermal degradation characteristics of poly(acrylamide-co-acrylic acid) and poly(acrylamide-co-sodium acrylate) copolymers. *Thermochimica Acta*, 121:295–306, 1987.

[25] Al-Sharji, H., Zaitoun, A., Dupuis, G., Al-Hashmi, A., and Al-maamari, R. Mechanical and thermal stability of polyacrylamide-based microgel products for EOR. In *SPE International Symposium on Oilfield Chemistry*. Society of Petroleum Engineers, The Woodlands, USA, 2013.

[26] Booth, C. The mechanical degradation of polymers. *Polymer*, 4:471–478, 1963.

[27] Argillier, J. F. et al. Impact of polymer mechanical degradation on shear and extensional viscosities: Toward better injectivity forecasts in polymer flooding operations. In *SPE International Symposium on Oilfield Chemistry*. Society of Petroleum Engineers, The Woodlands, USA, 2013.

[28] Farinato, R. and Yen, W. Polymer degradation in porous-media flow. *Journal of Applied Polymer Science*, 33:2353–2368, 1987.

[29] Shupe, R. D. Chemical-stability of polyacrylamide polymers. *Journal of Petroleum Technology*, 33:1513–1529, 1981.

[30] Kheradmand, H., François, J., and Plazanet, V. Hydrolysis of polyacrylam de and acrylic acid-acrylamide copolymers at neutral ph and high temperature. *Polymer*, 29:860–870, 1988.

[31] Bao, M., Chen, Q., Li, Y., and Jiang, G. Biodegradation of partially hydrolyzed polyacrylamide by bacteria isolated from production water after polymer flooding in an oil field. *Journal of hazardous materials*, 184:105–110, 2010.

[32] Fang, M. and Li, W. The molecular biology identification of a hydrolyzed polyacrylamide (hpam) degrading bacteria strain h5 and biodegradation product analysis. In *International Conference on Environmental Science and Technology*, pp. 135–141, 2007.

[33] Delshad, M., Pope, G., and Sepehrnoori, K. *UTCHEM version 9.0 Technical Documentation*. Center for Petroleum and Geosystems Engineering, The University of Texas at Austin, Austin, USA 78751, 2000.

[34] Lake, L., Pope, G., Carey, G., and Sepehrnoori, K. Isothermal, multiphase, multicomponent fluid-flow in permeable media .1. description and mathematical formulation. *In Situ*, 8:1–40, 1984.

[35] Pope, G. A., Carey, G. F., and Sepehrnoori, K. Isothermal, multiphase, multicomponent fluid flow in permeable media. part ii: Numerical techniques and solution. *In Situ*, 8:41–97, 1984.

[36] Helfferich, F. Theory of multicomponent, multiphase displacement in porous-media. *Society of Petroleum Engineers Journal*, 21:51–62, 1981.

[37] Fleming, P., Thomas, C., and Winter, W. Formulation of a general multiphase, multicomponent chemical flood model. *Society of Petroleum Engineers Journal*, 21:63–76, 1981.

[38] Hirasaki, G. J. Application of the theory of multicomponent, multiphase displacement to 3-component, 2-phase surfactant flooding. *Society of Petroleum Engineers Journal*, 21:191–204, 1981.

[39] Li, Z., Luo, H., Bhardwaj, P., Wang, B., and Delshad, M. Modeling dynamic fracture growth induced by non-newtonian polymer injection. *Journal of Petroleum Science and Engineering*, 147:395–407, 2016.

[40] Bao, K., Lie, K. A., Moyner, O., and Liu, M. Fully implicit simulation of polymer flooding with mrst. *Computational Geosciences*, 21:1219–1244, 2017.

[41] Krogstad, S., Lie, K. A., Nilsen, H. M., Berg, C. F., and Kippe, V. Efficient flow diagnostics proxies for polymer flooding. *Computational Geosciences*, 21:1203–1218, 2017.

[42] Mykkeltvedt, T. S., Raynaud, X., and Lie, K. A. Fully implicit higher-order schemes applied to polymer flooding. *Computational Geosciences*, 21:1245–1266, 2017.

[43] Bourgeat, A., Granet, S., and Smaï, F. Compositional two-phase flow in saturated-unsaturated porous media: benchmarks for phase appearance/disappearance. In *Simulation of Flow in Porous Media. Applications in Energy and Environment*, pp. 81–106. Walter de Gruyter, 2013.

[44] Braconnier, B., Preux, C., Flauraud, E., Tran, Q. H., and Berthon, C. An analysis of physical models and numerical schemes for polymer flooding simulations. *Computational Geosciences*, 21:1267–1279, 2017.

[45] Cao, R., Cheng, L., and Lian, P. Flow behavior of viscoelastic polymer solution in porous media. *Journal of Dispersion Science and Technology*, 36:41–50, 2015.

[46] Dang, C., Nghiem, L., Nguyen, N., Chen, Z., and Nguyen, Q. Evaluation of CO2 low salinity water-alternating-gas for enhanced oil recovery. *Journal of Natural Gas Science and Engineering*, 35:237–258, 2016.

[47] Ebaga-Ololo, J. and Chon, B. H. Prediction of polymer flooding performance with an artificial neural network: A two-polymer-slug case. *Energies*, 10:844, 2017.

[48] Hilden, S. T., Moyner, O., Lie, K. A., and Bao, K. Multiscale simulation of polymer flooding with shear effects. *Transport in Porous Media*, 113:111–135, 2016.

[49] Janiga, D., Czarnota, R., Stopa, J., Wojnarowski, P., and Kosowski, P. Performance of nature inspired optimization algorithms for polymer enhanced oil recovery process. *Journal of Petroleum Science and Engineering*, 154:354–366, 2017.

[50] Lotfollahi, M. et al. Mechanistic simulation of polymer injectivity in field tests. *Spe Journal*, 21:1178–1191, 2016.

[51] Lotfollahi, M., Koh, H., Li, Z., Delshad, M., and Pope, G. A. Mechanistic simulation of residual oil saturation in viscoelastic polymer floods. In *SPE EOR Conference at Oil and Gas West Asia*. Society of Petroleum Engineers, Muscat, Oman, 2016.

[52] Lee, K. S. Performance of a polymer flood with shear-thinning fluid in heterogeneous layered systems with crossflow. *Energies*, 4:1112–1128, 2011.

[53] Yuan, C., Delshad, M., and Wheeler, M. F. Modeling multiphase non-newtonian polymer flow in ipars parallel framework. *Networks and Heterogeneous Media*, 5:583–602, 2010.

[54] Wang, J., Liu, H. Q., and Xu, J. Mechanistic simulation studies on viscous-elastic polymer flooding in petroleum reservoirs. *Journal of Dispersion Science and Technology*, 34:417–426, 2013.

[55] Lohne, A., Nodland, O., Stavland, A., and Hiorth, A. A model for non-Newtonian flow in porous media at different flow regimes. *Computational Geosciences*, 21:1289–1312, 2017.

[56] Druetta, P., Yue, J., Tesi, P., Persis, C. D., and Picchioni, F. Numerical modeling of a compositional flow for chemical eor and its stability analysis. *Applied Mathematical Modelling*, 47:141–159, 2017.

[57] Bidner, M. S. and Savioli, G. B. On the numerical modeling for surfactant flooding of oil reservoirs. *Mecanica Computacional*, XXI:566–585, 2002.

[58] Jonsson, K. and Jonsson, B. Fluid-flow in compressible porous-media .1. steady-state conditions. *AIChE Journal*, 38:1340–1348, 1992.

[59] Jonsson, K. and Jonsson, B. Fluid-flow in compressible porous-media .2. dynamic behavior. *AIChE Journal*, 38:1349–1356, 1992.

[60] Bidner, M. and Porcelli, P. Influence of phase behavior on chemical flood transport phenomena. *Transport in Porous Media*, 24:247–273, 1996.

[61] Bear, J. *Dynamics of Fluids In Porous Media*. American Elsevier Publishing Company, New York, USA, 1972.

[62] Bear, J. and Bachmat, Y. Macroscopic modeling of transport phenomena in porous-media .2. applications to mass, momentum and energy-transport. *Transport in Porous Media*, 1:241–269, 1986.

[63] Chen, Z., Huan, G., and Ma, Y. *Computational Methods for Multiphase Flows in Porous Media*. Society for Industrial and Applied Mathematics, Philadelphia, USA, 2006.

[64] Li, M., Xu, M., Lin, M., and Wu, Z. The effect of hpam on crude oil/water interfacial properties and the stability of crude oil emulsions. *Journal of Dispersion Science and Technology*, 28:189–192, 2007.

[65] Bataweel, M. A. *Enhanced Oil Recovery in High Salinity High Temperature Reservoir by Chemical Flooding*. PhD thesis, Texas A&M University, College Station, USA, 2011.

[66] El-hoshoudy, A. N. et al. Evaluation of solution and rheological properties for hydrophobically associated polyacrylamide copolymer as a promised enhanced oil recovery candidate. *Egyptian Journal of Petroleum*, 26:779–785, 2017.

[67] Shen, M. et al. Relationship between the polymer structures and destabilization of polymer-containing water-in-oil emulsions. *Journal of Dispersion Science and Technology*, 28:1178–1182, 2007.

[68] Ye, Z., Guo, G., Chen, H., and Shu, Z. Interaction between aqueous solutions of hydrophobically associating polyacrylamide and dodecyl dimethyl betaine. *Journal of Chemistry*, 2014:932082, 2014.

[69] Wever, D. A. Z., Picchioni, F., and Broekhuis, A. A. Polymers for enhanced oil recovery: A paradigm for structure-property relationship in aqueous solution. *Progress in Polymer Science*, 36:1558–1628, 2011.

[70] Pancharoen, M. Physical Properties of Associative Polymer Solutions. Master's thesis, Stanford University, Stanford, USA, 2009.

[71] Druetta, P. *Numerical Simulation of Chemical EOR Processes*. PhD thesis, University of Groningen, Groningen, the Netherlands, 2018.

[72] Wang, J., Liu, H. Q., and Xu, J. Mechanistic simulation studies on viscous-elastic polymer flooding in petroleum reservoirs. *Journal of Dispersion Science and Technology*, 34:417–426, 2013.

[73] Camilleri, D. et al. Description of an improved compositional micellar/polymer simulator. *SPE Reservoir Engineering*, 2:427–432, 1987.

[74] Camilleri, D., Fil, A., Pope, G. A., Rouse, B. A., and Sepehrnoori, K. Comparison of an improved compositional micellar/polymer simulator with laboratory corefloods. *SPE Reservoir Engineering*, 2:441–451, 1987.

[75] Dawson, R. and Lantz, R. B. Inaccessible pore volume in polymer flooding. *Society of Petroleum Engineers Journal*, 12:448–&, 1972.

[76] Gilman, J. R. and MacMillan, D. J. Improved interpretation of the inaccessible pore-volume phenomenon. *SPE Formation Evaluation*, 2:442–448, 1987.

[77] Liauh, W. C., Duda, J. L., and Klaus, E. E. An investigation of the inaccessible pore volume phenomena. *Interfacial Phenomena in Enhanced Oil Recovery, AIChE Symposium Series*, 78:70–76, 1979.

[78] Pancharoen, M., Thiele, M. R., and Kovscek, A. R. Inaccessible pore volume of associative poly-mer floods. In *SPE Improved Oil Recovery Symposium*. Society of Petroleum Engineers, Tulsa, USA, 2010.

[79] Seright, R. Use of polymers to recover viscous oil from unconventional reservoirs. final report. Technical Report DE-NT0006555, US Department of Energy, 2011.

[80] Cao, R., Cheng, L., and Lian, P. Flow behavior of viscoelastic polymer solution in porous media. *Journal of Dispersion Science and Technology*, 36:41–50, 2015.

[81] Yacob, N. et al. Determination of viscosity-average molecular weight of chitosan using intrinsic viscosity measurement. *Journal of Nuclear and Related Technologies*, 10:39–44, 2013.

[82] Flory, P. J. *Principles of Polymer Chemistry*. Cornell University Press, Ithaca, USA, 1953.

[83] Gleadall, A. and Pan, J. Computer simulation of polymer chain scission in biodegradable poly-mers. *J Biotechnol Biomater*, 3:154–&, 2013.

[84] Gleadall, A., Pan, J., Kruft, M. A., and Kellomaki, M. Degradation mechanisms of bioresorbable polyesters. part 1. effects of random scission, end scission and autocatalysis. *Acta Biomateri-alia*, 10:2223–2232, 2014.

[85] Gleadall, A., Pan, J., Kruft, M. A., and Kellomaki, M. Degradation mechanisms of bioresorbable polyesters. part 2. effects of initial molecular weight and residual monomer. *Acta Biomateri-alia*, 10:2233–2240, 2014.

[86] McCoy, B. and Madras, G. Degradation kinetics of polymers in solution: Dynamics of molecular weight distributions. *AIChE Journal*, 43:802–810, 1997.

[87] Moreno, R., Muller, A., and Saez, A. Flow-induced degradation of hydrolyzed polyacrylamide in porous media. *Polymer Bulletin*, 37:663–670, 1996.

[88] Tayal, A. and Khan, S. Degradation of a water-soluble polymer: Molecular weight changes and chain scission characteristics. *Macromolecules*, 33:9488–9493, 2000.

[89] Ghannam, M. T., Hasan, S. W., Abu-Jdayil, B., and Esmail, N. Rheological properties of heavy & light crude oil mixtures for improving flowability. *Journal of Petroleum Science and Engineer-ing*, 81:122–128, 2012.

[90] Liang, B., Jiang, H., Li, J., Seright, R. S., and Lake, L. W. Further insights into the mechanism of disproportionate permeability reduction. In *SPE Annual Technical Conference and Exhibition*. Society of Petroleum Engineers, San Antonio, USA, 2017.

[91] Seright, R. S. Optimizing disproportionate permeability reduction. In *SPE/DOE Symposium on Improved Oil Recovery*. Society of Petroleum Engineers, Tulsa, USA, 2006.

[92] Nilsson, S., Stavland, A., and Jonsbraten, H. C. Mechanistic study of disproportionate perme-ability reduction. In *SPE/DOE Improved Oil Recovery Symposium*. Society of Petroleum Engi-neers, Tulsa, USA, 1998.

[93] Druetta, P. and Picchioni, F. Numerical Modeling and Validation of a Novel 2D Compositional Flooding Simulator Using a Second-Order TVD Scheme. *Energies*, 11:2280, 2018.

[94] Peaceman, D. W. Interpretation of well-block pressures in numerical reservoir simulation (in-cludes associated paper 6988). *Society of Petroleum Engineers Journal*, 18:183–194, 1978.

[95] Saad, N., Pope, G. A., and Sepehrnoorl, K. Application of higher-order methods in composi-tional simulation. *SPE Reservoir Engineering (Society of Petroleum Engineers)*, 5:623–630, 1990.

[96] Kamalyar, K., Kharrat, R., and Nikbakht, M. Numerical aspects of the convection-dispersion equation. *Petroleum Science and Technology*, 32:1729–1762, 2014.

[97] Liu, J., Delshad, M., Pope, G. A., and Sepehrnoori, K. Application of higher-order flux-limited methods in compositional simulation. *Transport in Porous Media*, 16:1–29, 1994.

[98] Najafabadi, N. F. *Modeling chemical EOR processes using IMPEC and fully IMPLICIT reservoir simulators*. PhD thesis, University of Texas at Austin, Austin, USA, 2009.

5 Nanotechnology in enhanced oil recovery

5.1 Introduction

The main objective in enhanced oil recovery processes is to alter the fluid and/or rock properties in order to diminish the oil saturation below the residual (S_{or}) after water-flooding [1]. Even though nanotechnology is not an EOR technique *per se*, the unique features found at the nanoscale allow boosting and improving of the performance of current methods, and modifying of parameters that result in an increase in the oil recovered. Therefore, the main objective of the nanotechnology assisted EOR processes is acting on one (or several) of the following factors: mobility control using viscosity-increasing water/polymer/nanoparticles solutions; altering the rock wettability; interfacial tension (IFT) reduction by adding surfactants; and lowering the oil viscosity by means of nanocatalysts which react at high temperatures, producing lighter fractions that are easier to recover. The use of nanotechnology in chemical EOR, such as polymer nano-composites (PNP's or polymer coated nanoparticles) [2–8] and silica nanoparticles have been reported [8–22].

5.1.1 Nanotechnology

Nanotechnology is defined as the science of manipulating matter on an atomic or molecular scale, comprising the design, characterization, production and application structures, devices and/or systems that have novel/superior properties and functions as a result of their physical size, with at least one dimension in the range from 1 to 100 nanometers (Fig. 5.1) [23, 24].

The first concept was laid down in 1959 by the physicist Richard Feynman when, in his renowned lecture [25], he discussed the capability of manipulating material at the scale of individual atoms and molecules, imagining the whole Encyclopedia Britannica written on the head of a pin and foreseeing the ability to examine and control matter at the nanoscale. Since the word "nanotechnology" was coined in 1974 when the term was utilized to refer to the capability of designing materials precisely at the nanometre scale [26], there has been a tremendous advance full of breakthroughs in several disciplines including medicine, materials, oil recovery and so forth [27, 28].

What makes nanotechnology attractive for research is not only the scale of things, but the properties objects show at these scales. Nanoparticles have, for instance, ultrahigh specific surface area ratio (per unit mass) [27, 29, 30], meaning more atoms are at or close to the surface, making them more weakly bonded and more reactive and giving them unique properties, such as high adsorption potential and heat conductivity. These particles are used in mixes with base fluids, leading to nanofluids, which show useful properties for their use in enhanced oil recovery, thus improving

https://doi.org/10.1515/9783110640250-005

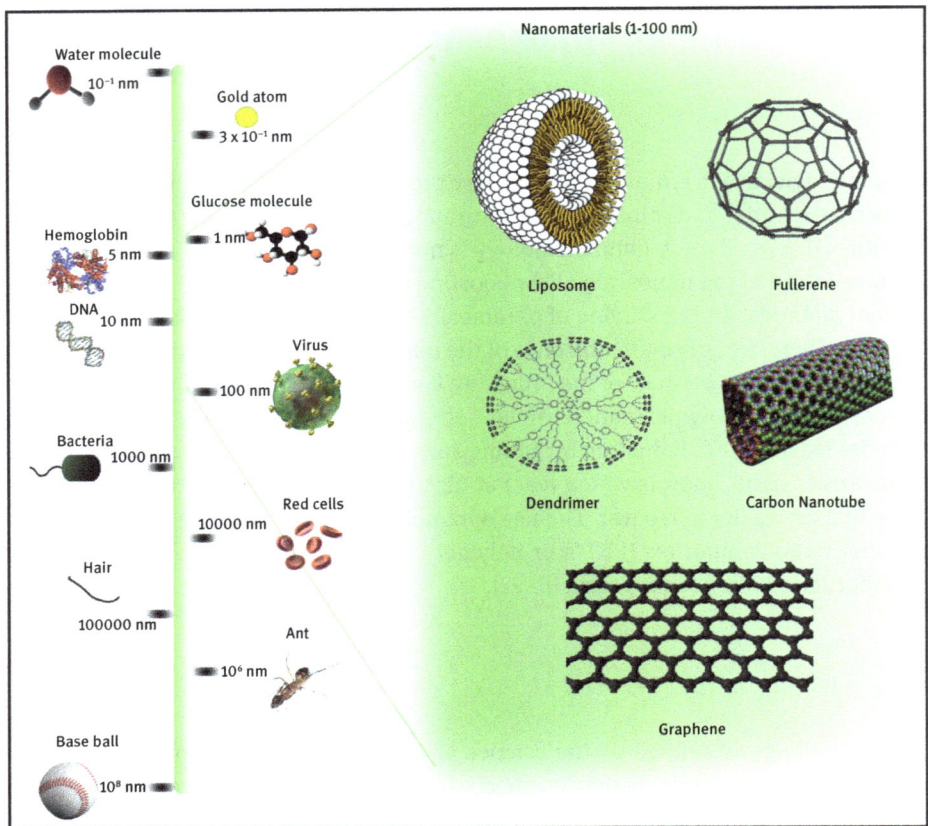

Fig. 5.1: Comparison of sizes between nanomaterials and common objects [27].

an oil field's performance. Also, since the size of structures is on same scale as single atoms or molecules, quantum mechanical effects take preponderance, namely: quantum confinement and fluorescence, wave-corpuscle duality, tunneling, surface plasmon resonance, size quantization, single-electron charging, metastable crystal phases, charge depletion, ballistic electron transport, enhanced catalytic activity and super paramagnetism [31–37], resulting in changes in electronic, mechanical, magnetic, chemical and optical properties [38].

5.1.2 Nanotechnology in EOR

The relationship between nanotechnology and the petroleum industry is not new but it has been developing for some time since now, for instance in downstream processes such as petroleum refining, where zeolites are now used to extract up to 40% more gasoline than their predecessors in fuel catalytic cracking (FCC) or in hydrocracking units [39–46]. As far as upstream operations are concerned, the first application of nanotechnology was the development of nanoenhanced agents [47] to provide the oil industry with strong and stable materials [48]. By means of nanotechnology and the enhanced and unique properties developed, lighter equipment could be produced, capable of withstanding harsh conditions. Nanotechnology could also help develop new metering techniques with nanosensors for improved temperature and pressure ratings in deep wells and hostile environments, or for reservoir characterization, fluid flow monitoring, and fluid type recognition or new imaging and computational techniques to allow better discovery, sizing and characterization of reservoirs [48–56].

This chapter is focused on the use of nanotechnology for enhanced oil recovery applications. Then, the mechanisms previously mentioned such as changing the properties of the displacing agent; altering the wettability of the porous media; lowering the interfacial tension (IFT) or the oil viscosity in-situ by means of catalysts; emulsion improvement; and increasing the mobility of the capillary-trapped oil are all properties that can be enhanced using nanotechnology [19, 47, 57–59]. In tight oil fields, the surface interactions play a major role where these mechanisms are especially relevant. Moreover, the scale of the nanomaterials makes for suitable injection into porous media, even in low permeabilities fields, since it has been established statistically that the pore throat openings commonly range between 100 and 10,000 nanometers in width. That is large enough for nanofluids to flow through relatively freely. Customized nanoparticles have the ability to enhance oil recovery, improve exploration, and be useful in formation scale control. Nanoparticles can be tailored to alter reservoir properties such as wettability, improve mobility ratio, or control formation fines migration. Nanofluids have been successfully developed in laboratories, and the upcoming challenge is to develop techniques for cost-efficient industrial-scale production of nanofluids [60]. Nanotechnology has the possibility to improve these methods beyond current applications. With ultra-small size and high surface area to volume ratio, nanoparticles have the ability to penetrate pores where conventional recovery methods are unable to. Recently, studies have explored the potential of Al_2O_3, MgO, Fe_2O_3 in addition to SiO_2 nanoparticles, observing that some combinations have yielded better recoveries than SiO_2 nanoparticles [16, 61, 62].

Another field of reservoir engineering where nanotechnology is being utilized is in the developing of new types of nanofluids or "smart fluids" for IOR/EOR, such as nanofluids of surfactants/polymers, nanoemulsions and colloidal dispersion gels (CDG). These fluids are composed of small volumetric fractions of nanoparticles in a liquid in order to enhance or improve some of the fluid's properties. Nanofluids can be designed to be compatible with reservoir fluids/rocks and be environmentally friendly [50]. Some newly developed nanofluids have shown highly improved properties in such applications as drag reduction, binders for sand consolidation, gels, wettability alteration and anticorrosive coatings [47, 63–67]

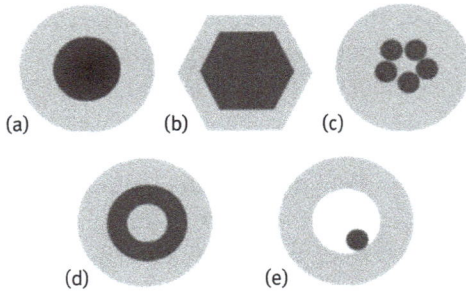

Fig. 5.2: Different core/shell nanoparticles: (a) Spherical core/shell, (b) Hexagonal core/shell, (c) Multiple small core materials coated by single shell material, (d) Nanomatryushka material, and (e) Movable core within hollow shell material [72].

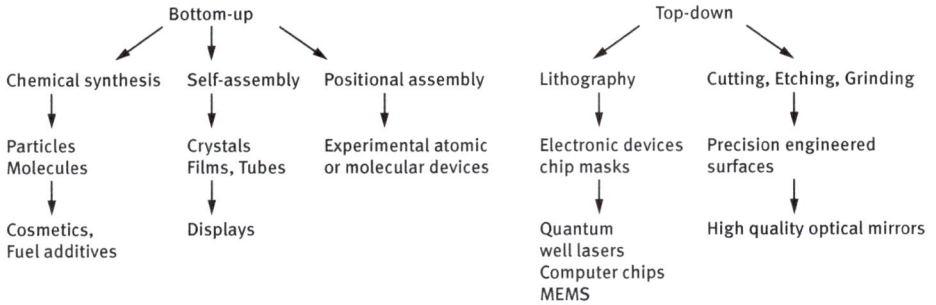

Fig. 5.3: Bottom-Up and Top-Down techniques used in nanotechnology [23].

5.2 Nanofluids

5.2.1 Introduction

Nanofluids are dilute liquid colloidal suspensions of nanoparticles (around 0.0001–10%) with at least one of their principal dimensions smaller than 100 nm [68]. They are characterized by the fact that Brownian agitation overcomes gravitational settlement. Nanofluids are mixtures of disperse nanoparticles in a base fluid with the purpose of enhancing one or several of its properties [69]. Since Choi [70] coined the term, there has been numerous publications and research about the many applications of nanofluids in different fields [30, 69]. Nanoparticles have, broadly speaking, a core/shell structure [71], with different geometric configurations and material combinations which allow them to fulfill a broad spectrum of requirements. Chaudhuri [72] reviewed the different possible combinations as well as their synthesis procedure and characteristics (Fig. 5.2).

There are several techniques or processes capable of creating these nanostructures with different degrees of quality, speed and cost. These approaches can be classified into two categories: Bottom-up and Top-down (Fig. 5.3) [23, 73–75]. Several preparation techniques for nanoparticles based on these two approaches can be found in the literature [76–81, 81, 82].

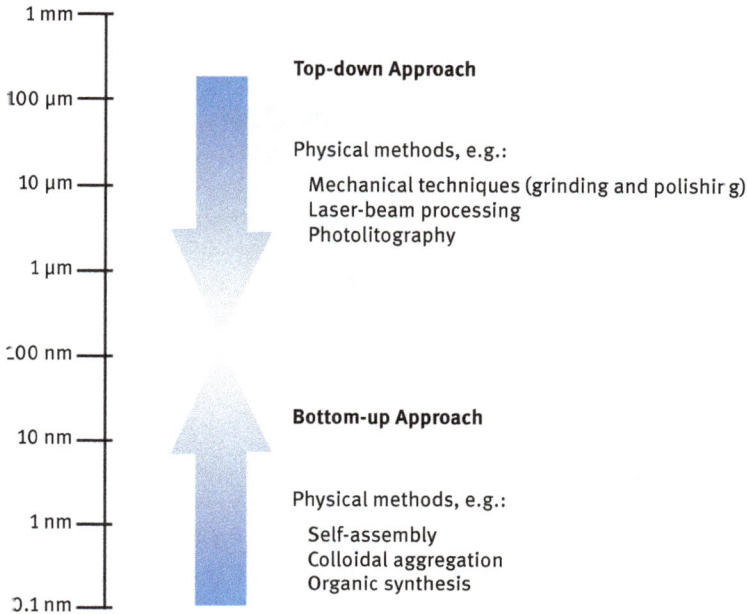

Top-down Approach

Physical methods, e.g.:

    Mechanical techniques (grinding and polishing)
    Laser-beam processing
    Photolitography

Bottom-up Approach

Physical methods, e.g.:

    Self-assembly
    Colloidal aggregation
    Organic synthesis

Fig. 5.4: Methods to prepare nanoparticles [83].

Top-down methods are those in which nanoparticles are directly prepared from bulk materials through the generation of isolated atoms, using distribution techniques involving physical methods such as milling or grinding, laser beam processing, repeated quenching and photolithography [82]. Generally speaking, Top-down techniques start with a larger piece of material and create by means of etching, milling or machining a nanostructure by removing material. This can be done by using methods such as precision engineering and lithography, which has been developed and improved by the semiconductor industry over the past 40 years. As for advantages, Top-Down methods offer reliability and device complexity, but they are generally higher in energy usage, and produce more waste than Bottom-Up procedures (Fig. 5.4) [23].

Bottom-Up methods involve molecular components as the starting materials linked to each other by means of chemical reactions, nucleation and growth processes so as to promote the formation of nanoparticles [76, 84–94]. Bottom-up processes involve the building of structures in a way that could be described as atom-by-atom or molecule-by-molecule. These techniques can be split into three categories: chemical synthesis, self-assembly, and positional assembly. So far according to the literature, positional assembly (with its many practical drawbacks as a manufacturing tool) is the only technique in which single atoms or molecules can be placed deliberately one-by-one. More typically, large numbers of atoms, molecules or particles are used or created by chemical synthesis, and then arranged through naturally occurring processes into a desired structure (Fig. 5.4) [23].

5.2.2 Properties of nanoparticles

Structural properties

One of the most important features of nanoparticles is their high surface area to volume ratio. Decreasing particle size, which increases the total surface area, leads to changes in interatomic spacing. This effect, according to Engeset [60], can be related to the compressive strains induced by internal pressure as a consequence o the small radius of curvature in the nanoparticles. There is also an apparent stability of metastable structures in small nanoparticles and clusters. What it is noticeable is that these and nano-dimensional layers, may adopt different crystal structures than normal bulk material [74].

Chemical properties

As pointed out previously, the reduction of the particle size increases the total surface/volume ratio. The chemical reactivity increases because of this increase in the surface area to volume ratio. Nanocatalysts, using finely divided nanoscale systems, can increase the rate, selectivity and efficiency of chemical reactions such as combustion or synthesis, whilst simultaneously significantly reducing waste and pollution. Furthermore, nanoscale catalytic supports with controlled pore sizes can select the products and reactants of chemical reactions based on their physical size and thus ease of transport to and from internal reaction sites within the nanoporous structure [74]. It was also reported [60] that nanoparticles present a different chemical behavior from their respective bulk material. A substance may for example not be soluble in water at a micro scale, but will dissolve easily when at the nanostructure scale [74].

Mechanical properties

Mechanical properties are strongly dependent on both the ease of the formation, or the presence of defects within a material. When the particle size decreases, the ability to support such defects becomes more difficult, and mechanical properties are significantly altered [74]. Nanostructures, which are very different from bulk structures in terms of the atomic structural arrangement, will obviously show very different mechanical properties. For instance, single-walled carbon nanotubes (SWCNT's) have proved to be stronger than steel due to its high mechanical strengths [95]. In addition, many nanostructured metals and ceramics are observed to be super elastic. They have the ability to undergo extensive deformation without necking or fracture [74]. These are properties that extend the current strength-ductility of conventional materials, and give nanomaterials a great advantage when it comes to mechanical properties.

5.2.3 Polysilicon nanoparticles in porous media

When the solution of PSNP and solvent is injected in a porous media, four phenomena will occur with the nanoparticles: adsorption, desorption, blocking and transportation [96]. Since PSNPs can be considered as Brownian particles, then different forces are responsible for the interactions between PSNP and porous walls: the attractive potential energy of London-Van der Waals, gravity, inertia, electrostatic forces between the particles, repulsion energy of electric double layers, Born repulsion, buoyancy, acid-base interaction, and hydrodynamics [96–101]. The effects of these forces on the phenomena previously mentioned are presented below.

Since the mass of nanosized particles is practically negligible, the gravity force is much smaller than the others existing in the system. Nevertheless, due to the attractive forces, the particles can undergo agglomeration and form bigger particles or clusters with significant mass, which may result in deposition and destabilization of the suspension. An opposing force to gravity is the buoyancy force, which depends on the volume of the particle and the solvent density. This force is also insignificant for nanoparticles due to their very small volume.

If particles and porous media have opposite charges, then, depending on the position of particles, there might be a significant other gravity-opposing force. This is the attractive electrostatic force between the particle and the matrix grain, or the attractive Van der Waals force with the particles above it. Also, the same attractive forces between the particle and the matrix/particle below it may result in a faster deposition. Thus, there are both repulsive and attractive electrostatic forces influencing the particles. If only the bulk of the suspension is considered, far from the matrix rock, the two significant forces that affect the stability of particles are the attractive Van der Waals and repulsive electrostatic forces, which are described by the classical DLVO (Derjaguin, Landau, Verwey and Overbeek) theory. The magnitude of the attractive force depends on the size, shape, and type of the particles. However, the magnitude of the electrostatic repulsive force can be changed by modifying the surface charges of the medium [101].

When the resultant of these forces is negative, the attraction is larger than repulsion between PSNP and the porous wall, and this will cause the PSNPs to be adsorbed onto the porous wall. On the contrary, when the repulsion is larger than attraction, then desorption from the porous wall will occur. These phenomena are then the result of a dynamic balance controlled by the total energy between the particle and the porous media. Blocking is due to physical or geometric reasons. It will take place when the diameter of the PNSP is larger than the size of pore throat, or when several PNSP bridge at the pore throat. Finally, transportation of PNSP in the porous media is governed by diffusion and convection.

The phenomena examined above have two important impacts on the properties of flow through the reservoir rock. The wettability of the rock due to the adsorption will be altered, hence these changes will affect the relative permeabilities. Also, the adsorption of the PSNPs and blocking at the small pore throats will result in a decrease in the porosity and absolute permeability of the porous media. The first impact is favorable for improving water flooding, but the second, however, has an unfavorable effect on the enhancement of water flooding.

Lipophobic and hydrophilic polysilicon (LHP)

LHP nanoparticles can alter the wettability of a rock turning an oil-wet rock to a water-wet, or make a water-wet rock strongly water-wet. As mentioned previously, a fluid is needed for injecting the nanoparticles into the rock matrix. LHP particles are one kind of hydrophilic nanopowder. Because of this feature, water is a common choice for LHP injection. When the nanoparticles are injected into porous media, the hydrophobic behavior of pore walls will be changed to hydrophilic due to the adsorption of LHP's. This alteration of wettability will cause a change in the flowing conditions in the porous media, since the relative permeability of the oil phase (k_{ro}) increases, decreasing the resistance to oil flow, while at the same time, the relative permeability of the water phase (k_{rw}) declines considerably. Furthermore, oil trapped in the small pores will be displaced due to LHP adsorption and wettability change, and the effective pore diameters for oil flow in the porous medium may, in turn, be enlarged [96, 102–104].

Hydrophobic and lipophilic polysilicon (HLP)

HLP nanoparticles have the opposite behavior than LHPs. They alter the wettability of the rock making a water-wet rock into oil-wet, or enhancing the wettability features of an already oil-wet rock. Because HLP nanoparticles have hydrophobic characteristics, water cannot be used as a dispersing agent, so in this case organic solvents are preferred. By injection of HLP suspension into the reservoir rock, the wettability of the pore surface goes from hydrophilic to hydrophobic. Then, the relative permeability of the water phase (k_{rw}) increases, hence the resistance of water to flow decreases to a certain extent. Secondly, water film on the surface of pores will be displaced by HLP adsorption and wettability change. Moreover, the effective pore diameters for water flow in porous media may be increased [9, 96, 102–105].

Neutral wet polysilicon (NWP)

NWP nanoparticles present intermediate characteristics. They can change either oil- or water-wet formations to a mixed state because NWPs is composed of both hydrophilic and hydrophobic nanoparticles. Due to its features, NWPs are only partially dispersed in water, thus water will not be a suitable fluid for dissolving its particles. Actually, some of these are absorbed and suspended in water, while the rest will keep floating. So, a bipolar carrier fluid like ethanol must be used for dispersing NWP nanoparticles. The mechanisms of EOR using NWPN are a reduction of interfacial tension by the improved quality of ethanol and wettability alteration. Onyekonwu [104] showed that the displacement efficiency of NWP is higher than other types of polysilicon nanoparticles when used with light oils. However, its displacement efficiency is not as efficient as HLP in intermediate to heavy oils [9, 96, 102, 103, 105].

5.2.4 Effect of nanoparticles on reservoir and fluid properties

Formation damage is an undesirable operational and economical problem that can occur during the various phases of oil and gas recovery from subsurface reservoirs, including production, drilling, hydraulic fracturing, and workover operations. The immediate effect of this problem is a detriment to the productivity of the oil field/wells. Formation damage can be caused by different unfavorable processes including chemical, physical, biological, and thermal interactions of rock and fluids. The indicators of formation damage include permeability impairment, skin damage, and deformation of formation under stress and fluid shear [106, 107]. Nanoparticles can also be a source of formation damage, since once they are injected, retention in porous media can damage formation properties, and is one of the major issues regarding nanoparticle transport that should be analyzed before the beginning of EOR operations.

Retention in porous media

Porous media is composed of a complex and random structure of pore bodies and throats covering a wide variety of sizes. Particle retention in porous media has been a serious issue for many industries, since the transport of particles is limited to the degree to which these are retained, by means of different mechanisms. Nanoparticles are transported through a porous media through diffusion, convection and hydrodynamics. Reservoir rocks then can be severely affected by particle invasion [60, 108, 109]. Li [110, 111] performed core flooding experiments and reported that nanofluids have a tendency to reduce the porosity and permeability of a porous rock. During the early stage of flooding, adsorption and desorption of nanoparticles will occur at the pore wall [112]. This is a dynamic balancing process which will eventually reach an equilibrium state, where a nanofluid can travel through the pore system without significant adsorption and diffusion [113].

Particle movement in a porous media is a very complex process due to the complexity of forces controlling the solid movement in porous media. Flow and retention of solid particles in porous media is an intricate process ruled by several factors such as particle size and shape, flowrate, chemistry of the carrying fluid, properties of the rock and concentration suspended particles [56, 114–116]. Four different mechanisms can lead to formation damage: [106–108, 113, 117–119] (1) adsorption of nanoparticles

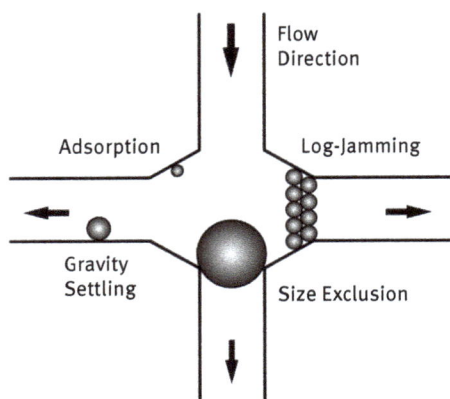

Fig. 5.5: Mechanisms of nanoparticles entrapment in porous media [83].

onto the rock surface due to particles Brownian motion and their electrostatic interactions with the surface of the porous rock, (2) mechanical entrapment, or deep-bed filtration, where the size of the nanoparticle is larger than the pore throat, (3) sedimentation or gravity settling when the densities of moving particles and the carrying fluid are very dissimilar, and finally (4) log-jamming where particles (smaller than the pore throat) move at lower velocities compared to the carrying fluid and accumulate at the pore throats, eventually leading to the blockage of the channel (Fig. 5.5).

The adsorption of nanoparticles can take place both on the surface of the reservoir rock and also at the interfaces between oil and water phases. Mechanical entrapment, also known as straining, leads to blocking of narrow pore throats by larger particles. The evidence for mechanical entrapment is taken to be either that the particle concentration in the effluent does not reach the injected concentration, or that it would do so only after injecting a large volume of particles [120, 121]. Even though mechanical entrapment was listed as a mechanism of retention, it is noteworthy that pore throats are usually significantly larger than nanoparticle sizes, which means that few particles are retained from mechanical entrapment [8, 122]. Log-jamming is similar to straining, but in this case the size of particles might not be larger than the particle size. Due to density differences between moving particles and the carrying fluid, sedimentation or gravity settling will also take place. When pore throats narrow, flow velocity will increase. Water molecules will then accelerate faster than heavier particles, and accumulation will occur. Due to gravity settling the pore throat will gradually be reduced and eventually blocked. The main factors governing the log-jamming effect are particle concentration and effective hydrodynamic size, pore size distribution and flow rate [60].

Dahle [113] reported that temperature has a negligible effect on the particle retention, with two percent points greater at 80 °C when compared to 21 °C [123]. However, the existence of salt ions in the carrying fluid has been observed to significantly delay nanoparticle breakthrough time and to increase retention [56].

Effect on permeability and porosity due to adsorption of nanoparticles

As a direct consequence of the previous point, the retention of nanoparticles in the porous media will alter some reservoir properties [123, 124]. Ju [66, 67] studied the wettability and permeability changes caused by adsorption of nanometer particles onto rock surfaces, and evaluated the changes of porosity and absolute permeability caused by particle injection and subsequent retention. The reduction in the porosity can be expressed as follows,

$$\phi = \phi_0 - \sum \Delta\phi \tag{5.1}$$

where ϕ_0 represents the initial porosity and $\sum \Delta\phi$ the variation caused by retention. In addition to this formula, a modification of Xianghui and Civan's model for permeability is presented as an expression for instantaneous permeability [60].

$$K = K_0 \cdot \left[(1-f) \cdot k_f + \frac{f\phi}{\phi_0} \right]^n \tag{5.2}$$

Here K_0 is the initial absolute permeability, k_f is given as a constant for fluid seepage allowed by the plugged cores, the value of the exponent n ranges from 2.5 to 3.5 [125], and f is a flow efficiency factor of the cross-section area open to flow [67].

Ju [67] also analyzed how the relative permeabilities vary as a function of the wettability alteration, a product of the nanoparticles retention in the porous media. He considered that these would be gradually adsorbed by the rock, altering the wettability and the permeabilities. This retention would come to a limit when all of the surface of the rock gets covered by the nanoparticles, reaching the maximum wettability alteration in the porous media. From that point onwards, further retention of nanoparticles will affect the porosity and absolute permeability. Both ratios (K/K_0 and ϕ/ϕ_0) decline as the volume of nanofluid injected increases. Also, numerical solutions show that porosity and permeability ratios are smaller close to the inlet than to the outlet. Both ratios are functions of dimensionless distance (in 1D models), and increase gradually toward the initial value when the dimensionless distance approaches one. These results imply that nanoparticle adsorption at pore walls, and pore throat blocking, occur at a higher frequency closer to the inlet [60, 66].

Effect on rheology and viscoelasticity

Experiments have shown that adding nanoparticles to water increases the shear viscosity. Water molecules layered at the nanoparticles' surface decrease the fraction of adjacent mobile fluid molecules, increasing the shear viscosity and elastic properties [126, 127]. This effect becomes more noticeable in polymer/nanoparticle solutions in which the storage and loss moduli are strongly affected by the nanoparticle concentration. This enhances the viscoelastic properties of the sweeping agent and therefore increases the oil recovery factor [128, 129]. The former can be increased by either increasing the nanoparticle concentration, or by modifying the size of the particles [130]. Einstein was the first to study the influence of particles on the viscosity of a fluid at very low volume fractions ($c < 0.02$), predicting a linear increase with the particle volume concentration. Since Einstein's formula, new correlations were published taking into account the particle size and shape, higher volume concentration, temperature, pH, and size distribution [131, 132].

$$\mu_{nf} = \mu_{cf} \cdot (1 + 2.5c) \tag{5.3}$$

Here μ_{nf} is the viscosity of the nanofluid, μ_{cf} is the viscosity of the carrier fluid and c is the volume fraction of the particle in suspension. This formula shows a linear increase in viscosity with particle volume concentration, which shows a good agreement with experimental results at low concentrations. Nevertheless, this formula has some limitations, as it does not consider structure and particle-particle interaction within the solution and higher particle concentrations. The new numerical correlations for the viscosity of nanofluids take into account the effect of particle size and shape, the volume concentration (for higher values than Einstein's formula), the temperature, the particle aggregation, the effect of pH, and the size distribution [132]. It is important to note that there are contradictory results in the literature showing different trends with respect to the viscosity dependency on the particle size: both an increase and a decrease were observed with smaller particle sizes as well as an independence of the rheological properties on the particle size [131, 133]. However, it is considered, in accordance with Meyer [131], that the increase in viscosity in nanofluids comes from two major sources: fluid-particle and particle-particle interactions. These are dependent on the overall surface area of the nanoparticles. At a same concentration, smaller particles show a larger surface area, and thus the viscosity should be higher. Aggregation of nanoparticles is then one of the major detrimental effects in the rheological properties of nanofluids.

Effect on IFT

Oil and water are immiscible fluids, this means that the IFT between them is high. Several authors reported that introducing silica hydrophilic nanoparticles to the system has been observed to decrease the IFT, which may potentially lead to the production of more oil. The nanoparticles will structure themselves at the oil/brine interface, reducing the contact between the two phases. The layer of particles generates a lower IFT between the two phases, in a similar way to how surfactants work, but using a different principle. The IFT is reported to be sensitive to nanofluid concentration: as the latter increases, the IFT decreases [110, 111, 113, 134–136]. Moreover, Frijters [137] described the mechanisms behind the adsorption of neutral particles by using the Lattice–Boltzmann method, and compared it with surfactants. What the latter (amphiphiles) do is to adsorb at the interface due to their hydrophilic head and hydrophobic tail, whilst neutral wetting nanoparticles adsorb by maintaining a particle-fluid interface that requires less energy. Neutral wetting nanoparticles were reported to change the interfacial free energy by taking away energetically expensive fluid-fluid

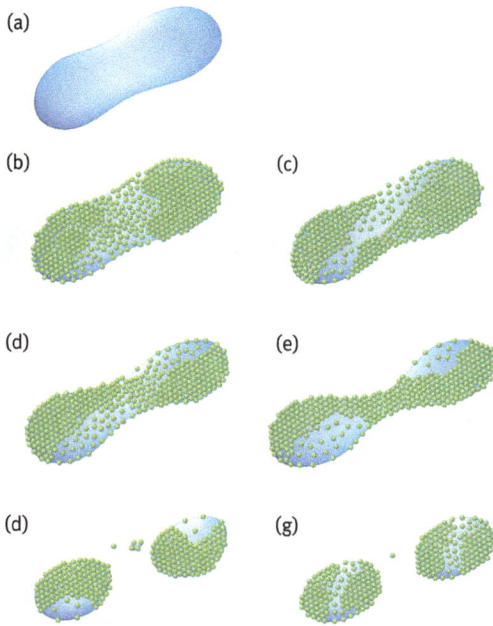

Fig. 5.6: Oil droplet breakup in presence of nanoparticles [83].

interfaces and replacing them with a cheaper particle-fluid interface [113]. This behavior can be explained using the free energy term, expressed as a function of the IFT and the interface area.

$$F_\gamma = \oint_{\partial D} \gamma dA \qquad (5.4)$$

The reduction of the interfacial free energy requires either a decrease of the IFT, which is done by adding surfactants, or the reduction of the area of contact, which is the effect of adsorbed particles. This formulation shows that neutral wetting nanoparticles can reduce the overall interfacial free energy not by reducing the IFT itself, but by removing parts of the energetically unfavorable fluid-fluid interface area. For emulsions, the assembly of particles on the oil droplet's surface is favorable because it blocks destabilization by Ostwald ripening (larger droplets grow at the expense of smaller ones). It can also break up oil droplets (Fig. 5.6), making it easier for the emulsion to migrate through the porous media [113, 137, 138].

Effect on the rock wettability and surface wetting

As mentioned previously, different types of nanoparticles can alter the wettability depending on their surface coating. Most of the particles that have been utilized in EOR applications are polysilicon nanoparticles. The untreated LHPN turns an already

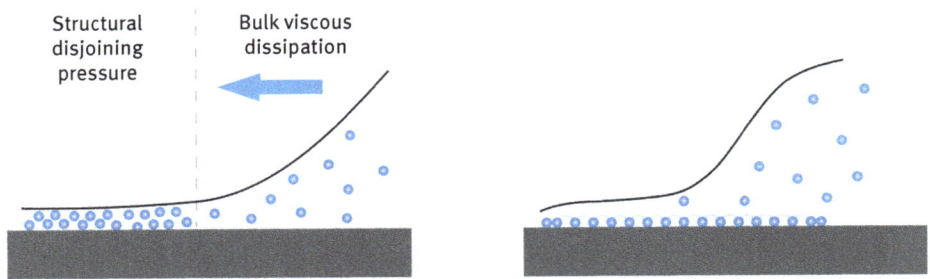

Fig. 5.7: Illustration of nanoparticle ordering in the wedge of a spreading meniscus (left) and the lubrication effect caused by nanoparticle adsorption on the solid (right) [83].

water-wet rock strongly water-wet or makes an oil-wet rock water-wet. HLPN is treated with single layer organic compound, and can alter a water-wet rock to be oil-wet, or make an already oil-wet rock strongly oil-wet. While NWNP is treated with silane, and can achieve mixed wettability conditions by making a rock either strongly oil-wet and strongly water-wet at the same time, or make the rock neither oil or water-wet [104, 138–140]. For these nanoparticles to change the wettability of a rock surface, they need to be adsorbed onto the rock formation. Furthermore, in order to achieve an optimal distribution, parameters such as concentration and particle size take relevance. The degree of dispersion also plays an important role in the change of contact angle and wettability, and has been widely addressed by various authors [60, 141].

Vafaei [142] studied the effect of nanoparticles on the sessile droplet contact angle. His results indicated that the concentration and size of nanoparticles in solution have an important role in the variation of the droplet contact angle. With increasing concentration, the latter increases linearly for the same droplet volume until it reaches a peak, before decreasing with increasing concentration. Observations from the study also show that smaller nanoparticles were more effective in raising the contact angle. Sefiane [143] suggested that the improvement in contact line motion affected by the presence of a nanoparticle solution may have two potential underlying mechanisms. These could be either the pressure gradient within the nanofluid which is created due to the nanoparticles forming a solid-like ordering in the fluid "wedge" in the vicinity of the three-phase contact line (Fig. 5.7 – left), or the nanoparticle adsorption on the solid surface and the resulting reduction in friction could be also contributing to the observed enhancement (Fig. 5.7 – right).

Wasan [144, 146] stated that the wetting and spreading behavior of liquids over solid surfaces changes if the former contain nanoparticles or surfactant micelles, globular proteins and macromolecules. He discussed the progress made in the wetting and spreading of nanofluids over solid surfaces with an emphasis on the interactions between the particles and with the solid substrate, as well as the spreading of thin nanofluid films containing nanoparticles on hydrophilic surfaces driven by the struc-

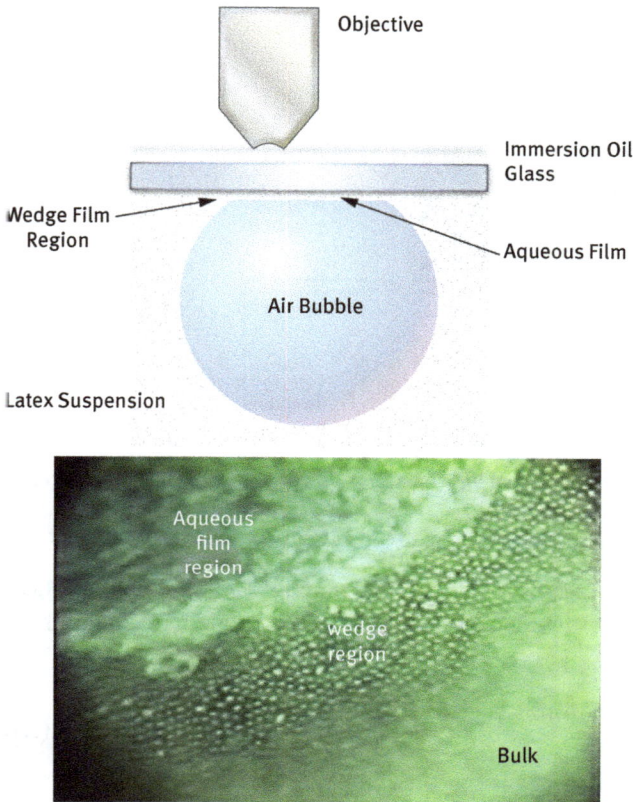

Fig. 5.8: Nanoparticles in a wedge film, the scheme of the experimental set-up (top) [83], and the nanoparticle distribution in a wedge film (bottom). Latex particles had diameter of 1 µm [144].

tural disjoining pressure gradient. The latter can be defined as the pressure produced when two surface layers reciprocally overlap, and it is caused by the total effect of forces of a different nature: electrostatic, the forces of "elastic" resistance of solvated, or adsorbed solvated, films, and the forces of molecular interaction can act as components of the disjoining pressure [60]. He also reported in this study that the driving force for the spreading of a nanofluid is the structural disjoining pressure or film tension gradient ($\Delta\gamma$) direct towards the wedge from the bulk solution (Figs. 5.8 and 5.9). The film tension is high near the vertex, because the particles are structuring into a wedge confinement. As the tension on the film increases towards the top of the wedge, it will cause the nanofluid to spread at the wedge tip. This will enhance the dynamic spreading behavior of the nanofluid. The result of this process is that the nanoparticles will exert a large pressure through the wedge film relative to the bulk solution. This effect, also called disjoining pressure, will eventually separate the two phases from each other [60]. An analytical expression for this pressure, based on the Ornstein–Zernike

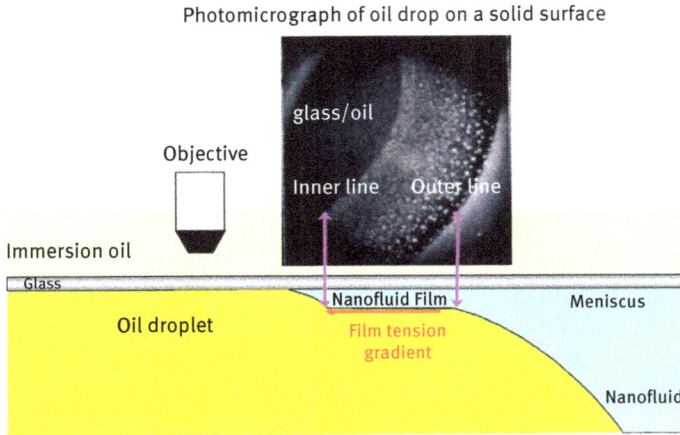

Fig. 5.9: Photomicrograph taken using reflected-light interferometry depicting the inner and outer contact lines and the nanofluid film region [145].

equation, was introduced by Wasan [146], from Trokhymchuk [147], which is applicable for hard sphere particles in vacuum, confined between two rigid hard walls which form symmetric films.

$$\Pi_{st}(h) = \begin{cases} \Pi_1 \cos(\omega h + \phi_2) e^{-\kappa h} + \Pi_2 e^{-\delta(h-d)} & \text{for } h \geq d \\ -P & \text{for } 0 \leq h < d \end{cases} \tag{5.5}$$

where d is the diameter of the nanoparticle and the other parameters ($\Pi_1, \phi_2, \omega, \kappa$) are fitted as cubic polynomials in terms of the nanoparticle's volume fraction. The term P refers to the bulk osmotic pressure of the nanofluid. The film-meniscus microscopic contact angle, θ_e, is related to the disjoining pressure, given by the Frumkin–Derjaguin equation [144, 146].

$$\Pi_0(h_e) h_e + \int_{h_e}^{\infty} \Pi(h)dh = \gamma_{o/nf}(\cos\theta_e - 1) = S \tag{5.6}$$

Here S is the spreading coefficient, $\gamma_{o/nf}$ is the IFT between the oil and nanofluid, h_e is the equilibrium thickness of a thin film, Π_0 is represented by the sum of the capillary pressure and the hydrostatic pressure of the droplet, and Π is the disjoining pressure represented by the terms, $\Pi = \Pi_{vw} + \Pi_d + \Pi_{st}$. The first represents the short-range Van der Waals force, the second the forces which are electrostatic or steric in nature, and the last one represents the long range structural forces arising from the ordering of the nanofluid's particles in the wedge film.

Wasan [144], Kondiparty [145, 148], and Nikolov [149] were able to use different reflected-light digital video microscopy techniques to study the mechanism of spreading dynamics in a liquid containing different latex and silicon nanoparticles (Figs. 5.8

and 5.9). They were able to demonstrate the two dimensional crystal-like formation of the nanoparticles in water and how this phenomenon increases the spreading dynamics of a micellar fluid at the three-phase region. When an oil drop is surrounded by the nanofluid, the nanoparticles will concentrate and reorder around the drop creating a wedge-like region between the surface and the oil drop (Fig. 5.9). The nanoparticles then diffuse into the wedge film and cause an increase in concentration and hence an increment in the disjoining pressure around the film region. Due to this, the oil-solution interface moves forward allowing the nanoparticles to spread along the surface. It is this mechanism that causes the oil drop to eventually detach completely from the surface.

The wetting and spreading of nanofluids composed by liquid suspensions of nanoparticles might have significant applications in EOR. Recent studies have revealed that the spreading of liquids without nanoparticles is not as effective as the spreading of nanofluids on solid surfaces, and this is due to the action of the structural disjoining pressure.

Kondiparty [148] presented experimental observations and results of the statics analysis based on the augmented Laplace equation, which includes a term for the contribution of the structural disjoining pressure, taking into account the effects of several parameters such as the nanoparticle concentration and size, contact angle, and drop size. Furthermore, he examined the effects on the displacement of the drop-meniscus profile and spontaneous spreading of a nanofluid as a film on a solid surface. Their analysis showed that a suitable combination of the previous parameters can result not only in the displacement of the three-phase contact line, but also in the spontaneous spreading of the nanofluid film on the surface. Moreover, he also showed that the complete wetting and spontaneous spreading of the nanofluid film, driven by the structural disjoining pressure gradient, is possible by decreasing the nanoparticle size and the interfacial tension. This ordering of the nanoparticles inside the wedge is a consequence of the fact that this increases the entropy of the overall dispersion by permitting greater freedom for the nanoparticles in the bulk liquid. The electrostatic repulsion between the particles will rise as the size of the nanoparticles decreases, therefore increasing the structural disjoining pressure. Also, the amount of particles is directly related to the force working on the wedge film. On the basis of their results, they concluded that a nanofluid with an effective particle size (including the electrical double layer) of about 40 nm, a low equilibrium contact angle (< 3°), and a high effective volume concentration (> 30 vol.%) is desirable for the dynamic spreading of a nanofluid system with an interfacial tension of 0.5 $^{mN}/_m$.

Wasan [144] also showed that the spreading behavior increased with decreasing film thickness, this is the number of particle layers in the film (Fig. 5.10). The force will be at a maximum at the tip of the wedge (Figs. 5.11 and 5.12). The magnitude of this pressure depends, on parameters such as the particle size and volume fraction, polydispersity, temperature, salinity and rock properties. The presence of more electrolytes will lower the disjoining pressure. Increasing salt concentration will lower the

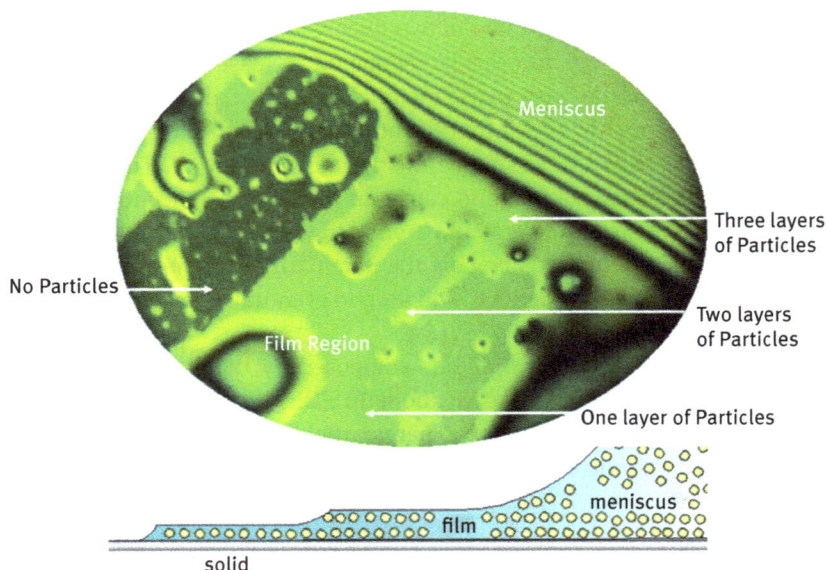

Fig. 5.10: Photomicrograph depicting particle layering of a 10 vol.% aqueous silica nanofluid with a particle diameter d = 19 nm on a solid surface (film size = 868 μm) [149].

repulsive forces between nanoparticles and hence reduce the pressure that drives the wedge film. Because of this, an increase in salinity have a negative effect on oil removal in the case of nanofluids [113, 150].

Feng-Chao Wang [151] studied the oil droplet detachment from solid surfaces immersed in charged nanoparticle suspensions via molecular dynamics simulations. The results obtained indicated that the surface wettability of the nanoparticles plays an essential role in the oil removal processes. An increase in the interactions between nanoparticles and water molecules would obstruct the oil droplet detachment. According to the results, suspensions of charged hydrophobic nanoparticles can be considered to be high-performance agents in removing oil droplets from solid surfaces in EOR applications. The process of detachment can be divided into three stages [151–153]: (1) the contact line shrinks due to the decrease of IFT induced by the adsorption of surfactant at the oil-water interface; (2) the diffusion of water disjoins the oil from the solid substrate in the vicinity of the contact line; and (3) the contact radius becomes sufficiently small and the droplet detaches.

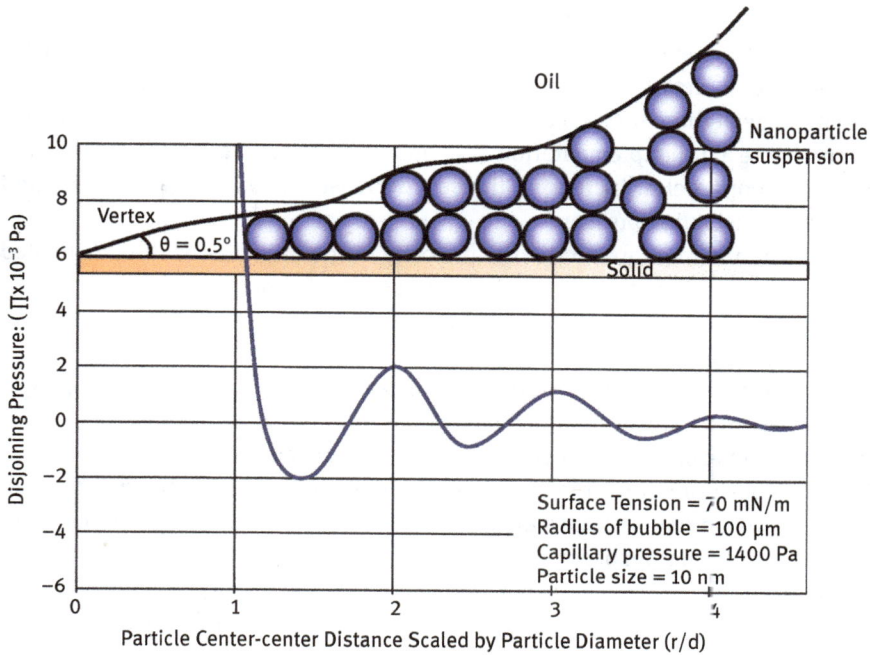

Fig. 5.11: Pressure on the walls of wedge for 0.5° contact angle at the vertex as a function of radial distance. Particle volume fraction $V_m = 0.36$ and particle diameter $d = 10$ nm [148].

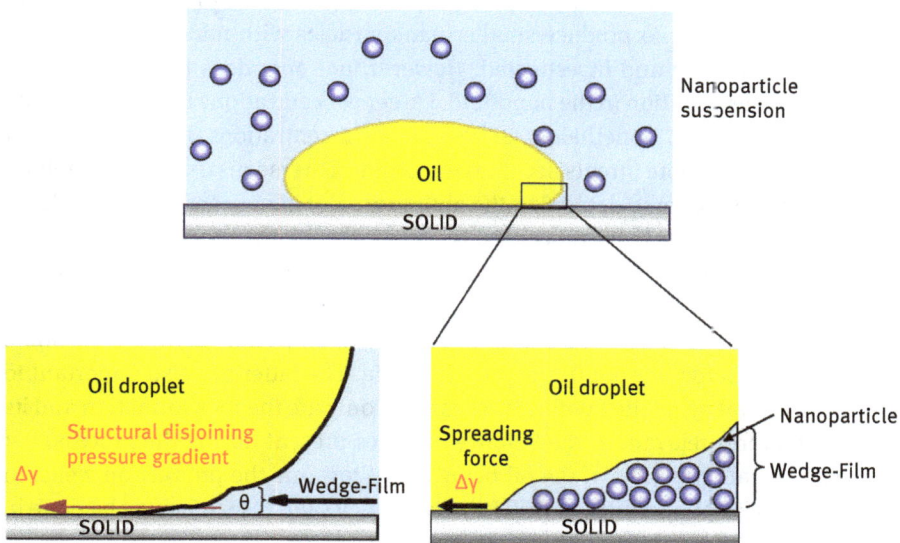

Fig. 5.12: Scheme of the disjoining pressure acting onto an oil droplet [148].

Summary of factors affecting nanofluid EOR applications

As reviewed in the previous points, EOR flooding techniques involving nanoparticles depend on several factors which must be taken into account before carrying out any operation. Some of these topics have not been fully addressed and further research is necessary before a full implementation of nanoparticles flooding in porous media.

- Size of nanoparticles: increasing the size decreases the surface/area ratio, and lowers the disjoining pressure.
- Concentration of nanoparticles: the viscosity of the nanofluid is directly related to the concentration of nanoparticles, among other factors.
- Type of nanoparticles: as discussed, the particles alter the wettability of the rock, which is relevant to the recovery efficiency.
- Brine concentration: increasing the TDS decreases the disjoining pressure, negatively affecting the recovery process.
- Brine pH: increasing the pH decreases the disjoining pressure.
- Rock composition: since the rock has a charged surface and the gravitational forces acting on nanoparticles are neglected, the interactions between the latter and the rock formation become significant.
- Oil composition: the concentration of nanoparticles necessary to achieve an efficient sweeping process depends on the crude oil composition in the reservoir.

The particle size is, by far, the most important parameter in the previous list. If the particle size becomes too large, the special characteristics of the nanofluid decrease and become negligible. This has special relevance during the preparation techniques, presented in this section. The improvement of existing methods or the development of newer tools in order to produce smaller nanoparticles with minimum variability is still to be addressed and investigated. However, not only does their size matter, but also their concentration in the nanofluid. Larger concentrations increase viscosity and recovery efficiency. Nonetheless, at very large concentrations, the probability of blocking/logging the pore throats becomes important. This means that each nanofluid should be tested in order to determine the optimum valor of concentration in order to reach a balance between these processes. In chemical EOR, the concentration of salts (and divalent cations) and the acidity negatively affects the fluid properties. Further research is necessary on this topic, in order to develop resistant nanofluids, for long period of time, to the action of a saline environment, as well as be less susceptible to the pH of the fluids present in the porous media. The latter are also important to determine the strategy of the flooding process. Not only are the rock formation and its geological features relevant in the balance of forces present during the process, but also the chemical properties of the oil being swept towards the production wells. In summary, the use of nanofluids in EOR is in its infancy sand research must be done in order to improve the performance of the whole process.

5.3 Nanoemulsions

5.3.1 Introduction

According to the literature [154, 155], an emulsion consists of at least two immiscible liquids, where one of these liquids is dispersed as small droplets in the other [156–160]. Standard emulsions usually have droplets with mean radii of between 100 nm and 100 μm. The purpose of this section is to analyze the role of emulsions at a nanometric scale. A nanoemulsion it is defined generally as an emulsion that contains very small droplets, with a mean radii between 10 to 100 nm [161–165] or, according to different sources and authors, covering the size range of 50 to 500 nm [166–174]. Nanoemulsions exhibit the typical properties of all emulsions, but with some specific ones which distinguish them from standard macroemulsions [161, 169, 175, 176]. The particles can exist as an oil-in-water (o/w) or water-in-oil (w/o) form, where the core of the particle is either oil or water, respectively [177]. However, such as happens with nanoparticles, the scale size in nanoemulsions provokes different and new properties to appear.

The droplet size in nanoemulsions is smaller than the wavelength of light ($r \ll \lambda$) that make them transparent, or to present a slight turbidity. Furthermore, the droplet size leads to a better stability in gravitational separation and aggregation than conventional emulsions [161, 162, 175]. Since this interaction depends strongly on to the drop size, changes in visual aspects will occur as soon as there is some evolution in the drop size. Because of their drop size, nanoemulsions, as nanoparticles, present a higher surface area and consequently higher quantities of surfactant are required to stabilize them. Furthermore, gravitational forces have a negligible effect on nanoemulsions, whilst Brownian motion is likely to put the droplets into motion and provide them with a driving force for destabilization processes, such as flocculation and coalescence. If the Laplace pressure excess inside the drops is quite large due to their size, then the drops are difficult to deform (shear, elongation or break), provided the IFT is very low. This pressure also results in a strong osmotic driving force from smaller to larger droplets, that it is called Ostwald ripening [175].

Unlike microemulsions (size range < 50 nm), which are also transparent or translucent and present thermodynamic stability, nanoemulsions are only kinetically stable. Nonetheless, their long-term physical stability (with no apparent flocculation or coalescence) make some authors describe them as "approaching thermodynamic stability". However, if they are not adequately prepared and stabilized against Ostwald ripening, nanoemulsions will lose their transparency and properties with time, as a result of an increasing droplet size. As mentioned, nanoemulsions are thermodynamically unstable since the separated oil and water phases have a lower free energy than the emulsified ones. Thus, they have a tendency to lose their properties over time due to several factors such as gravitational separation, flocculation, coalescence or Ostwald ripening [154, 155]. Therefore, researchers are focused on developing na-

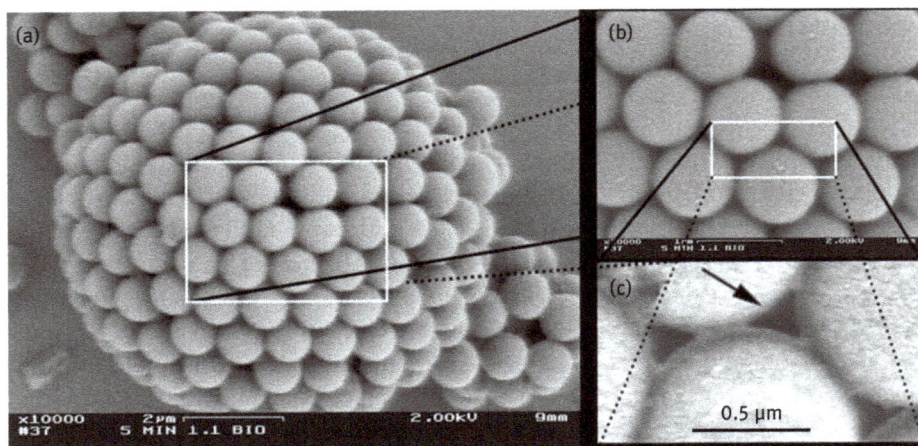

Fig. 5.13: Scanning electron microscope image of a dried, 10 μm-diameter colloidosome composed of 0.9 μm-diameter polystyrene spheres, sintered at 105 °C for 5 min. The colloidosome was formed with an oil droplet, containing 50 vol% vegetable oil and 50 vol% toluene. The water phase contained 50 vol% glycerol to increase its boiling temperature to allow sintering (a). Close-ups of the first image: the arrow points to one of the 0.15 μm holes that define the permeability. To view these colloidosomes with the electron microscope, they were washed with ethanol and dried in vacuum (b and c) [207].

noemulsions with a long kinetic stability for commercial applications, among them, EOR processes. This stability can be improved by controlling their composition, microstructure or by incorporating substances known as stabilizers.

Nanoemulsions are generally stabilized by the use of surfactants, but solid particles dissolved in one phase can also stabilize emulsions. Emulsions stabilized by solid particles have been known for more than a century and were named after Pickering, who discovered that coalescence of droplets is suppressed when solid particles are adsorbed at the oil-water interface [178], which are known as Pickering emulsions (Figs. 5.13 and 5.14). The solid particles adsorbs at the interface of the two liquids, thereby creating a physical barrier that hinders coalescence [179, 180]. It is widely accepted that this suppression in the coalescence is a kinetic effect caused by a combination of the formation of a rigid interfacial film and the increase in viscosity of the continuous phase [181–197]. One of the attractive features of Pickering emulsions is the great ease in breaking the emulsion and recovering the two phases when needed [187, 198]. The stability of Pickering emulsions is greatly affected by several parameters, among them, the composition of the organic and aqueous phases, contact angle among the phases, particle size, concentration and particle/particle interaction at the interface [181–189, 199–205]. Clay minerals and other nanoparticles are reported in the literature to produce very stable Pickering emulsions [206].

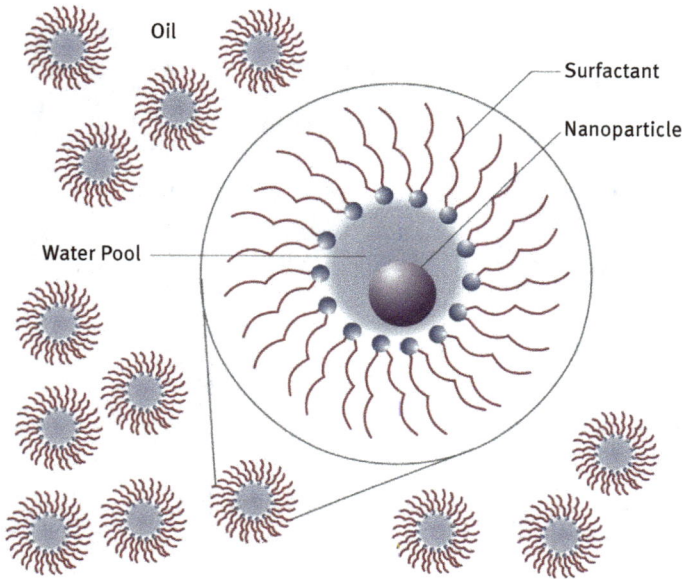

Fig. 5.14: Scheme of a water-in-oil emulsion stabilized with surfactant and nanoparticles [81].

The use of nanoparticles in nanoemulsions is not only applied to the cases previously analyzed. It is well-known that, because of the low density and more importantly very low viscosity of the supercritical and liquid CO_2, the macroscopic recovery efficiency of CO_2-related EOR processes is poor. The surfactant stabilized CO_2 foams is one way to approach and solve the mobility problem, but a different one arises: the long-term stability. The surfactant-stabilized foams require continuous regeneration; moreover, a fraction of the surfactant is adsorbed on the rock, increasing material related costs and besides, the surfactant is sensitive to harsh reservoir conditions. Thus, the use of nanotechnology proved to be a way to tackle this issue by using nanoparticles in stabilizing CO_2 foams (Fig. 5.15). Several results [123, 208–218] have shown improved viscosity properties, lowering of the mobility ratio and thereby improving the volumetric sweep efficiency and, more importantly, achieving stability even at high temperatures or with salts present in the aqueous phase. These studies have shown that nanoparticle foams are significantly more stable than surfactant foam because of the high adsorption energy of the nanoparticles at the gas-liquid interface. Furthermore, the attraction between nanoparticles and the gas-liquid interface is believed to help in minimizing nanoparticle loss to the rock surface, and silica nanoparticles are expected to withstand reservoir conditions better than surfactants [210]. Furthermore, commercial silica nanoparticles can be obtained at lower costs than commercial surfactants. Another advantage in using nanoparticles is their strong and selective ad-

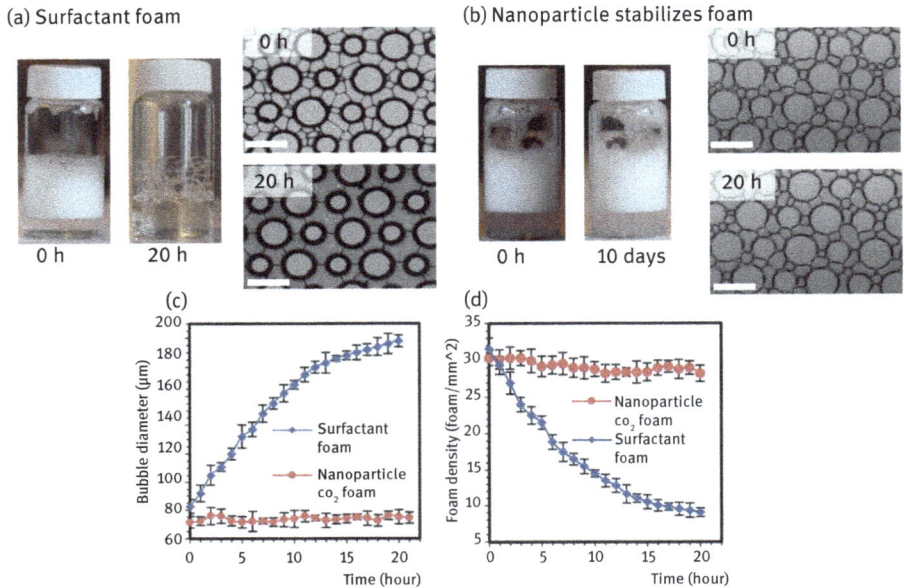

Fig. 5.15: Foam stability results in (a) SDS surfactant foam, (b) nanoparticle-stabilized CO_2 foam, (c) foam coalescence measurement as a function of changes in the bubble diameter and (d) foam quality measurement as a function of the bubble density (bubbles/mm^2). Scale bars = 400 μm [210].

sorption at fluid-to-fluid interfaces. As mentioned previously (see nanofluids), their shell can be tailored to enhance the creation of CO_2/water foams without creating oil/water emulsions [209].

According to Engeset [60], the use of nanoparticles in EOR processes may be valuable, since they are solid and around two orders of magnitude smaller than colloidal particles. Nanoparticles can stabilize emulsions droplets (Pickering emulsions), as they are small enough to pass typical pores and flow throughout the reservoir rock without much retention [123, 208, 219]. Spherical silica nanoparticles with a diameter in the range of several to tens of nanometers are the most commonly used (see previous section). With hydrophilic nanoparticles, a stable oil-in-water emulsion will be formed. On the other hand, if the silica particles are hydrophobic, they will form a water-in-oil emulsion [123, 208, 220]. Nanoemulsions of nanoparticles are very stable over time, and resistant to coalescence and the exchange of the dispersed phase between droplets [221]. The nanoparticles are also able to stabilize supercritical CO_2-in-water [195] and water-in-supercritical CO_2 emulsions [222].

As mentioned earlier, emulsions and microemulsions are well known in the oil and gas industry [172, 173]. Nanoemulsions containing oilfield chemicals may be applicable to a wide number of applications such as well treatments (scale inhibition, fracture acidizing, etc.), flow assurance (multiple additive packages), deposit removal/

clean-up and also for EOR operations. Their long-term stability and ease of preparation are compatible with the demands from the oil industry. The stability and characteristics of nanoemulsions depend upon the preparation methodology, the order of
addition of the components and the nature of the phases generated during the emulsification process, since nanoemulsions do not form spontaneously [161, 172, 173, 223–
225].

Nanoemulsions can be tailored to withstand the harsh conditions present in
reservoirs, for long periods of time. Something to be taken into account is the difference between surfactants and nanoemulsions is that the nanoparticles attach to the
fluid/fluid interface, and the required energy to attach nanoparticles to the interface
is higher than with surfactants, thus the process can be considered irreversible.

5.3.2 Nanoemulsion stability

Nanoemulsions will eventually evolve according to several physicochemical mechanisms susceptible to affect any kind of emulsion (Fig. 5.16), but in their particular case,
some particularities will arise [175]. Emulsions will tend to present an unstable behavior over time due to these mechanisms, such as creaming, sedimentation, flocculation,
coalescence and Ostwald ripening [154–157, 159, 160, 226].

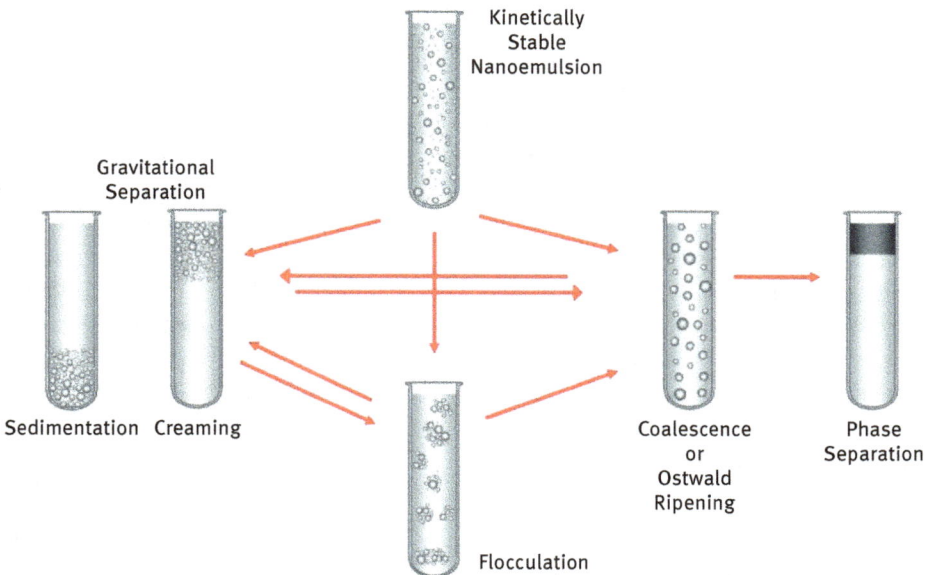

Fig. 5.16: Schematic diagram of the most common instability mechanisms that occur in emulsions:
creaming, sedimentation, flocculation, coalescence, Ostwald ripening and phase inversion [155].

Generally speaking, the small droplet size of nanoemulsions confers a better physicochemical stability than for conventional emulsions because of Brownian motion, and consequently the diffusion rate is higher than the phenomena caused by the gravity field. Ostwald ripening or molecular diffusion, which arises from emulsion polydispersity and the difference in solubility between small and large droplets, is the main mechanism for nanoemulsion destabilization [169]. Compared to conventional emulsions, nanoemulsions present better stability to gravitational separation (sedimentation or creaming) and droplet aggregation (flocculation and coalescence), but they are more susceptible to Ostwald ripening, because of the influence of their small particle size on the colloidal interactions [154, 161]. Creaming is the upward movement of the droplets due to a lower density than the surrounding liquid, whereas sedimentation represents the opposite phenomenon.

Furthermore, nanoemulsions may be more susceptible to chemical degradation since they have very large specific surface areas, and any chemical degradation reaction taking place at the oil-water interface may be promoted [154, 155]. In addition, since nanoemulsions are transparent or slightly turbid, UV and visible light can easily penetrate into them, which may trigger light-sensitive chemical degradation reactions [227]. Thus, it will be necessary to take additional steps to improve the chemical stability of the components within nanoemulsions [154, 155, 228].

Ostwald ripening

Ostwald ripening is an important physical process that occurs because larger particles are, energetically speaking, more stable than smaller ones. The mean size of the droplets in an emulsion increases over time due to diffusion of oil molecules from small to large droplets as the system tries to lower its overall energy [154, 155, 229–231]. The mechanism involved in this process is the Kelvin effect, whereby small droplets have higher local oil solubility than larger ones due to the difference in Laplace pressures [162]. The molecules on the surface of small particles will tend to continuously detach, diffuse through the solution and then finally attach to the surface of the larger particles. As time goes on, the number of smaller ones diminishes, whilst larger ones will steadily grow. This is the major destabilization mechanism of nanoemulsions [229, 232]. This effect is one of the main problems for their stability, which results from the difference in solubility between small and large droplets. The difference in chemical potential of the dispersed phase between different sized droplets was given by Lord Kelvin [161, 233].

$$c(r) = c(\infty)e^{\frac{2\gamma V_m}{rRT}} \tag{5.7}$$

Here $c(r)$ is the solubility surrounding a particle of radius r, $c(\infty)$ is the bulk phase solubility and V_m is the molar volume of the dispersed phase. The quantity $(2\gamma V_m/(rRT))$ is referred to as the *characteristic length*. Theoretically speaking, the process should

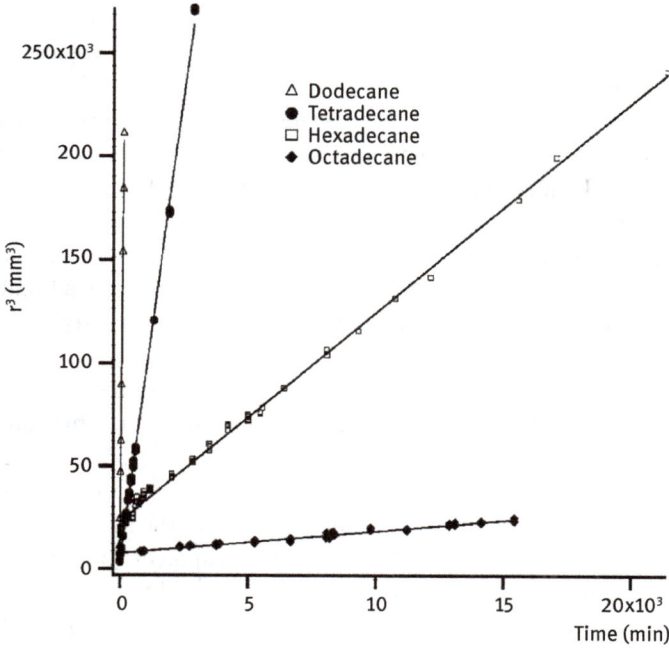

Fig. 5.17: Ostwald ripening plots for alkane oil nanoemulsions stabilized by the SDS-PEG surfactant system [162].

lead to the formation of one single drop. This does not take place in practice since the rate of this phenomenon decreases as the droplet size increases [161]. Considering two droplets of radii r_1 and r_2 (where $r_1 < r_2$),

$$\left(\frac{RT}{V_m}\right) \ln\left[\frac{c(r_1)}{c(r_2)}\right] = 2\gamma\left(\frac{1}{r_1} - \frac{1}{r_2}\right) \tag{5.8}$$

Then, when the radii of the particles tend to be of the same magnitude, the process will slow down its conversion rate. Ostwald ripening can be quantitatively assessed from plots of the cube of the radius as a function of time t (Lifshitz–Slyozov–Wagner Theory) [234] [235] (Fig. 5.17),

$$\omega = \frac{dr^3}{dt} = \frac{8}{9}\left[\frac{c(\infty)\gamma V_m D}{\rho RT}\right] \tag{5.9}$$

where D is the diffusion coefficient of the disperse phase in the continuous phase and ρ is the density of the disperse phase. In order to prevent Ostwald ripening, two methods were proposed [230, 231, 236].

– The addition of a second disperse phase component that is insoluble or nearly insoluble in the continuous phase. Then, a partitioning between different droplets occurs; the component with low solubility in the continuous phase will concentrate in the smaller droplets. If the Ostwald ripening phenomenon takes place in a two-component disperse phase system, an equilibrium is created when the chemical potential from different droplets is balanced by the difference in the potential resulting from partitioning of the two components. If the secondary component has zero solubility in the continuous phase, the size distribution will not deviate from the initial one ($\omega = 0$). For a nearly insoluble secondary component, there will be a growth rate, but lower than the one registered for the more soluble component [161]. The applicability of this method is limited since it requires a highly insoluble oil as the secondary phase, which is miscible with the primary phase (Fig. 5.17).

- Modifying the interfacial film at the O/W interface. According to the previous equations, a reduction in y will lead to a decrease of this phenomenon. This alone is not enough since this reduction should be significant. Walstra [237] theorized that using polymeric surfactants adsorbed at the O/W interface, which do not desorb during ripening, could lower the rate, and hence reduce the latter (Gibbs–Marangoni Effect). Tadros [161] suggested that A-B-A block copolymers soluble in the oil phase and insoluble in the continuous phase are useful in achieving the above effect. The polymeric surfactant should enhance the lowering of y by the emulsifier.

5.3.3 Preparation of nanoemulsions

According to the literature, stable nanoemulsions can be prepared either by high or low energy techniques [154, 155, 161, 238–243]. Since nanoemulsions are non-equilibrated systems [165, 177, 239, 244], their elaboration demands energy, surfactants and in some cases both. The presence of the latter will help with lowering the IFT between the phases. Small molecules such as non-ionic surfactants lower surface tension more than polymeric surfactants [177]. Surfactant mixtures are used in the practice yielding good results in lowering IFT. One of the goals of the emulsifier is to prevent shear induced coalescence during emulsification.

High energy processes use mechanical devices capable of generating disruptive forces that mix and disrupt oil and water phases leading to the formation of tiny oil droplets [154, 155, 162, 164, 240, 245–247]. On the other hand, low energy approaches depend on the spontaneous formation of tiny oil droplets when the solution or environmental conditions are altered [161, 163, 239, 248, 249]. The final size of the droplets will depend then on the technique, the operating conditions and the composition of the system.

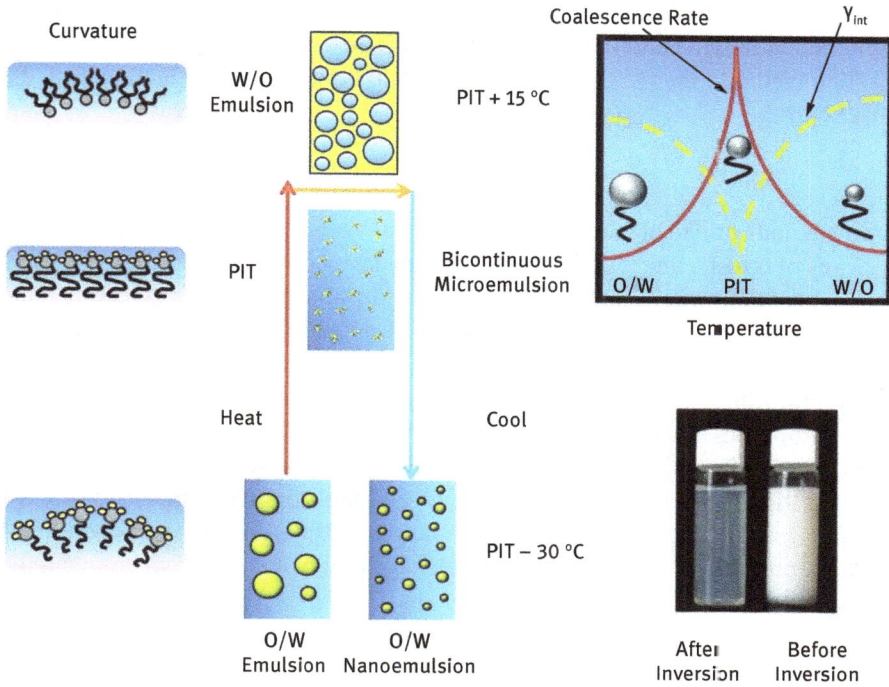

Fig. 5.18: Schematic diagram of the formation of nanoemulsions by the PIT method [155].

Low energy methods

These approaches are mainly dependent on the variation of interfacial phenomenon/ phase transitions and physicochemical properties of the surfactants and oil to produce the nanoemulsion [177, 250–252]. Low energy processes were developed taking into account the phase behavior and properties of the involved components, leading to the formation of nanodroplets [169, 253–256]. For instance, the condensation method is based on the phase transitions occurring during the emulsification process [176, 257]. These can be achieved, according to Chime [177], by either by changing the spontaneous curvature of non-ionic surfactants with changes in the temperature (keeping a constant composition), that turns out to be phase inversion temperature (PIT) [258, 259] (Fig. 5.18), or by changing the composition of the system (at a constant temperature), which is the emulsion inversion point (EIP) method [224, 260, 261]. An advantage of this is that it uses the available energy of the whole system to form the nanoemulsion. On the other hand, the disadvantages are, complexity, a precise approach is required, and the use of synthetic surfactants [177]. The most commonly used low-energy emulsification methods include: spontaneous emulsification, PIT, the solvent displacement method and the phase inversion composition method (self-nanoemulsification method), membrane emulsification and liquid-liquid nucleation [239, 252, 260, 262–272].

High energy methods

These methods utilize high mechanical energy devices which create nanoemulsions by means of transferring it to the system as kinetic energy. These generate disruptive forces which eventually will break up the oil and water phases, thus creating nanosized droplets. The most common equipment used are ultrasonicators, high pressure homogenizers and microfluidizers (Fig. 5.19) [165, 274–277]. The final size in the nanoemulsion will depend on the instruments utilized, their operating conditions and sample properties and composition [278]. From all the high energy techniques, high-pressure homogenization is the most commonly process used. Several effects, such as hydraulic shear, turbulence and cavitation, create emulsions with nanosized droplets. Microfluidization employs a high-pressure positive displacement pump operating at very high pressures [177], forcing the emulsion through a series of micro-channels.

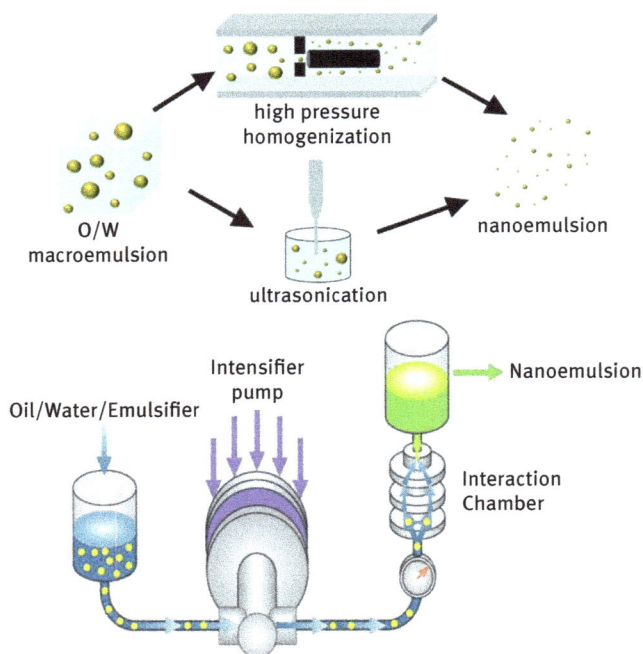

Fig. 5.19: High-pressure homogenizer (top) [243] and the basic concept of a microfluidizer (bottom) [273].

The collisions resulting from this flow create very fine nanoemulsions. Moreover, ultrasonic emulsification uses waves to disintegrate the emulsion droplets by means of cavitation [274, 276, 278]. According to the literature, high-pressure homogenization and microfluidization are used for fabrication of nanoemulsions on laboratory and industrial scales, whereas ultrasonic is used only on a laboratory scale. One disadvantage of these processes is that they usually require sophisticated instruments and a high energy input, thus increasing the cost of the final product. An advantage is that they allow a greater control of particle size and composition, which in the end controls the stability, rheology and turbidity of the emulsion. Examples of these techniques include high-pressure microfluidic homogenization, ultrasonic emulsification, flow focusing and satellite droplets [162, 240, 266, 279–290].

5.4 Conclusions

This chapter was aimed at presenting how nanotechnology can be employed in EOR applications, focusing on the application of nanoparticles to create nanofluids and nanoemulsion in order to increase the productivity and efficiency of existing recovery techniques. Many authors agree that nanotechnology allowed the addressing of EOR issues from a novel point of view. This is due to the unique properties presented at the nanoscale, such as the particle size and surface area to volume ratio. Nanoparticles present a distinctive advantage over other techniques since the scale of pore throats in porous media causes some problems regarding the flowability of micro- and macroparticles. Nanoparticles can flow freely since its size is one order of magnitude smaller than normal pore throats and channels, allowing them to reach regions of the oil field previously unreachable without the risk of blockage. This chapter presented a brief historical introduction to the origins of nanotechnology, followed by its potential uses and how it can be applied to oil recovery methods (upstream and downstream).

Clearly, nanotechnology is a new branch of science, and although much progress has been made lately, there are still many questions to be answered. This can be appreciated in that most of the literature found on direct applications at the nanoscale are recent, and numerous investigations are still being carried out. For instance, there are several environmental concerns on the application of nanoparticles not only in EOR but in all branches of science. How these particles affect our health and the environment is still being discussed. Production methods are also being researched nowadays in order to produce smaller sizes with less variability in their size, since the particle diameter is one of the most important factors in the success or failure of the technique being employed. Out of the possible uses in EOR, the employment of nanoparticles to create "smart fluids" or nanofluids is by far the most developed. This chapter discussed their particularities, for instance, how particles can alter the rheology of the injected fluid or the wettability of the rock formation. Furthermore, the disjoining mechanism of the oil drops from the porous medium, thanks to the pioneering work carried out by Wasan and Nikolov, was also analyzed. Several research developments have also been presented, both in laboratory-scale projects as well as in field tests. The observations show that nanoparticles can actually increase the oil recovery factor by means of enhancing traditional techniques, such as waterflooding, polymer flooding, or thermal/chemical EOR processes.

The procedure carried out in this chapter for nanofluids was adopted to analyze nanoemulsions, which in laboratory tests have demonstrated improved stability characteristics, either by using surfactants or nanoparticles (Pickering emulsions). The main problem reported about the nanoemulsions, that should be subjected to further investigation, is their stability under the harsh conditions usually found in reservoirs (high pressure, temperature and the presence of dissolved salts). Similarly to the analysis done with the nanoparticles, the production methods were also discussed, along with their advantages, disadvantages and problems which still required further research.

As presented in the literature, it is considered that EOR is not an exclusive field belonging to a single discipline. The use of nanotechnology in EOR as a tool to enhance traditional techniques has demonstrated a promising future, although there are still many unanswered questions. Nanotechnology cannot be addressed as an independent recovery process in EOR, but it was developed as a way to enhance standard recovery techniques. This cannot be considered as a new branch in petroleum engineering but, such as every process in EOR, a combination of different disciplines working together for the greater good. It is the intersection of the already well-known disciplines in EOR together with nanotechnology, which collaborates with its unique features allowing the development of a whole new branch of research in the oil industry. This chapter showed clearly how using new technologies to deal with old problems can be an interesting approach to increase or enhance the productivity or performance of current processes. This does not apply exclusively to the oil industry, but it is an example of a novel course of action applicable to all branches of science. It is to our best understanding that future research, not only on the laboratory scale but also in real field tests, will allow nanotechnology to give a boost to facilities which were considered to be at the limit of their operational life.

References

[1]   Morrow, N. R. A review of the effects of initial saturation, pore structure and wettability on oil recovery by waterflooding. In *Proceedings of the North Sea Oil and Gas Reservoirs Seminar*, pp. 179–191, Graham and Trotman, London, UK, 1987.
[2]   Pourjavadi, A. and Doulabi, M. Improvement in oil absorbency by using modified carbon nanotubes in preparation of oil sorbents. *Advances in Polymer Technology*, 32:n/a–n/a, 2013.
[3]   Pourjavadi, A., Doulabi, M., and Soleyman, R. Novel carbon-nanotube-based organogels as candidates for oil recovery. *Polymer International*, 62:179–183, 2013.

[4]   Wang, L. et al. Preparation of microgel nanospheres and their application in EOR. In *International Oil and Gas Conference and Exhibition in China*. Society of Petroleum Engineers, Beijing, China, 2010.

[5]   Kanj, M. Y., Rashid, M. H., and Giannelis, E. Industry first field trial of reservoir nanoagents. In *SPE Middle East Oil and Gas Show and Conference*. Society of Petroleum Engineers, Manama, Bahrain, 2011.

[6]   Qiu, F. and Mamora, D. D. Experimental study of solvent-based emulsion injection to enhance heavy oil recovery in Alaska North Slope area. In *Canadian Unconventional Resources and International Petroleum Conference*. Society of Petroleum Engineers, Calgary, Canada, 2010.

[7]   ShamsiJazeyi, H., Miller, C. A., Wong, M. S., Tour, J. M., and Verduzco, R. Polymer-coated nanoparticles for enhanced oil recovery. *Journal of Applied Polymer Science*, 131:40576, 2014.

[8]   Zhang, H., Nikolov, A., and Wasan, D. Enhanced oil recovery (EOR) using nanoparticle dispersions: Underlying mechanism and imbibition experiments. *Energy Fuels*, 28:3002–3009, 2014.

[9]   Roustaei, A., Saffarzadeh, S., and Mohammadi, M. An evaluation of modified silica nanoparticles' efficiency in enhancing oil recovery of light and intermediate oil reservoirs. *Egyptian Journal of Petroleum*, 22:427–433, 2013.

[10] Hendraningrat, L., Li, S., and Torster, O. A coreflood investigation of nanofluid enhanced oil recovery. *Journal of Petroleum Science and Engineering*, 111:128–138, 2013.

[11] Suleimanov, B. A., Ismailov, F. S., and Veliyev, E. F. Nanofluid for enhanced oil recovery. *Journal of Petroleum Science and Engineering*, 78:431–437, 2011.

[12] Maghzi, A., Kharrat, R., Mohebbi, A., and Ghazanfari, M. H. The impact of silica nanoparticles on the performance of polymer solution in presence of salts in polymer flooding for heavy oil recovery. *Fuel*, 123:123–132, 2014.

[13] Maghzi, A., Mohebbi, A., Kharrat, R., and Ghazanfari, M. H. Pore-scale monitoring of wettability alteration by silica nanoparticles during polymer flooding to heavy oil in a five-spot glass micromodel. *Transport in Porous Media*, 87:653–664, 2011.

[14] Zhu, D., Han, Y., Zhang, J., Li, X., and Feng, Y. Enhancing rheological properties of hydrophobically associative polyacrylamide aqueous solutions by hybriding with silica nanoparticles. *Journal of Applied Polymer Science*, 131:40876, 2014.

[15] Zhu, D., Wei, L., Wang, B., and Feng, Y. Aqueous hybrids of silica nanoparticles and hydrophobically associating hydrolyzed polyacrylamide used for eor in high-temperature and high-salinity reservoirs. *Energies*, 7:3858–3871, 2014.

[16] Ogolo, N. A., Olafuyi, O. A., and Onyekonwu, M. O. Enhanced oil recovery using nanoparticles. In *SPE Saudi Arabia Section Technical Symposium and Exhibition*. Society of Petroleum Engineers, Al-Khobar, Saudi Arabia, 2012.

[17] Zargartalebi, M., Barati, N., and Kharrat, R. Influences of hydrophilic and hydrophobic silica nanoparticles on anionic surfactant properties: Interfacial and adsorption behaviors. *Journal of Petroleum Science and Engineering*, 119:36–43, 2014.

[18] Zeyghami, M., Kharrat, R., and Ghazanfari, M. H. Investigation of the applicability of nano silica particles as a thickening additive for polymer solutions applied in eor processes. *Energy Sources Part A-Recovery Utilization and Environmental Effects*, 36:1315–1324, 2014.

[19] JPT. *Technology tomorrow – jpt article series*. 2006. J2: SPE-160929-MS.

[20] Ayatollahi, S. and Zerafat, M. M. Nanotechnology-assisted EOR techniques: New solutions to old challenges. In *SPE International Oilfield Nanotechnology Conference and Exhibition*, Society of Petroleum Engineers, Noordwijk, the Netherlands, 2012.

[21] Rahimi, K. and Adibifard, M. Experimental study of the nanoparticles effect on surfactant absorption and oil recovery in one of the iranian oil reservoirs. *Petroleum Science and Technology*, 33:79–85, 2015.

[22]   Khezrnejad, A., James, L. A., and Johansen, T. E. Water enhancement using nanoparticles in water alternating gas (WAG) micromodel experiments. In *SPE Annual Technical Conference and Exhibition*, Society of Petroleum Engineers, Amsterdam, the Netherlands, 2014.

[23]   The Royal Society and The Royal Academy of Engineering. *Nanoscience and Nanotechnologies: Opportunities and Uncertainties*. The Royal Society, London, UK, 2004.

[24]   Druetta, P., Raffa, P., and Picchioni, F. Plenty of Room at the Bottom: Nanotechnology as a Solution to an Old Issue in Enhanced Oil Recovery. *Applied Sciences*, 8:2596, 2018.

[25]   Feynman, R. P. There's plenty of room at the bottom. *Engineering and Science*, 23:22–36, 1960.

[26]   Taniguchi, N. On the basic concept of 'nano-technology'. In *International Conference on Production Engineering*, Japan Society for Precision Engineering, Tokyo, Japan, 1974.

[27]   Panneerselvam, S. and Choi, S. Nanoinformatics: Emerging databases and available tools. *International Journal of Molecular Sciences*, 15:7158–7182, 2014.

[28]   Foster, L. E. *Nanotechnology: Science, Innovation, and Opportunity*. Prentice Hall PTR, Upper Saddle River, USA, 2005.

[29]   Bell, T. E. *Understanding risk assessment of nanotechnology*. National Nanotechnology Initiative, Alexandria, VA, USA, 2007.

[30]   Taylor, R. et al. Small particles, big impacts: A review of the diverse applications of nanofluids. *Journal of Applied Physics*, 113:011301, 2013.

[31]   Perez, J. M. Iron oxide nanoparticles – hidden talent. *Nature Nanotechnology*, 2:535–536, 2007.

[32]   Loss, D. Quantum phenomena in nanotechnology. *Nanotechnology*, 20:430205, 2009.

[33]   Hodes, G. When small is different: Some recent advances in concepts and applications of nanoscale phenomena. *Advanced Materials*, 19:639–655, 2007.

[34]   Kim, B. Y. S., Rutka, J. T., and Chan, W. C. W. Current concepts: Nanomedicine. *New England Journal of Medicine*, 363:2434–2443, 2010.

[35]   Koehler, E. O. Zeolite catalysts reduce the aromatics surplus in refinery and petrochemical streams. In *17th World Petroleum Congress*, World Petroleum Congress, Rio de Janeiro, Brazil, 2002.

[36]   Ju-Nam, Y. and Lead, J. R. Manufactured nanoparticles: An overview of their chemistry, interactions and potential environmental implications. *Science of The Total Environment*, 400:396–414, 2008.

[37]   Bhushan, B. *Springer Handbook of Nanotechnology*. Springer, Würzburg, Germany, 2007.

[38]   Murty, B. S., Shankar, P., Raj, B., Rath, B. B., and Murday, J. *Textbook of Nanoscience and Nanotechnology*. Springer Heidelberg, Berlin, Germany, 2013.

[39]   Ratner, M. A. and Ratner, D. *Nanotechnology: A Gentle Introduction to the Next Big Idea*. Prentice Hall, Upper Saddle River, USA, 2003.

[40]   Wilson, M., Kannangara, K., Smith, G., Simmons, M., and Raguse, B. *Nanotechnology: Basic Science and Emerging Technologies*. CRC Press, Singapore, 2002.

[41]   Eastwood, S. C., Plank, C. J., and Weisz, P. B. New developments in catalytic cracking. In *8th World Petroleum Congress*, World Petroleum Congress, Moscow, USSR, 1971.

[42]   Letszch, W. S., Michaelis, D. G., and Pollock, J. D. New lz-210 zeolites produce superior FCC performance. *Journal of Canadian Petroleum Technology*, 28:102–105, 1989.

[43]   Chekriy, P. S., Chernykh, S. P., Bitman, G. L., and Loktev, A. S. [11]P3 novel petrochemical alkylation processes based on zeolites. In *14th World Petroleum Congress*, World Petroleum Congress, Stavanger, Norway, 1994.

[44]   Loktev, A. and Chekriy, P. Alkylation of Binuclear Aromatics with Zeolite Catalysts. *Elsevier Science*, 84, 1994.

[85]   Bock, C., Paquet, C., Couillard, M., Botton, G., and MacDougall, B. Size-selected synthesis of PtRu nano-catalysts: Reaction and size control mechanism. *Journal of the American Chemical Society*, 126:8028–8037, 2004.

[86]   Murray, C., Norris, D., and Bawendi, M. Synthesis and characterization of nearly monodisperse CdE (E = S, Se, Te) semiconductor nanocrystallites. *Journal of the American Chemical Society*, 115:8706–8715, 1993.

[87]   Alivisatos, A. Semiconductor clusters, nanocrystals, and quantum dots. *Science*, 271:933–937, 1996.

[88]   Shen, S. C., Hidajat, K., Yu, L. E., and Kawi, S. Simple hydrothermal synthesis of nanostructured and nanorod Zn-Al complex oxides as novel nanocatalysts. *Advanced Materials*, 16:541–545, 2004.

[89]   Li Yao, Y., Ding, Y., Ye, L. S., and Xia, X. H. Two-step pyrolysis process to synthesize highly dispersed Pt-Ru/carbon nanotube catalysts for methanol electrooxidation. *Carbon*, 44:61–66, 2006.

[90]   Pluym, T. et al. Palladium metal and palladium oxide particle-production by spray pyrolysis. *Materials Research Bulletin*, 28:369–376, 1993.

[91]   Pluym, T. et al. Solid silver particle-production by spray pyrolysis. *Journal of Aerosol Science*, 24:383–392, 1993.

[92]   Hellweg, T. Phase structures of microemulsions. *Current Opinion in Colloid & Interface Science*, 7:50–56, 2002.

[93]   Gurav, A., Kodas, T., Pluym, T., and Xiong, Y. Aerosol processing of materials. *Aerosol Science and Technology*, 19:411–452, 1993.

[94]   Altavilla, C. and Ciliberto, E. *Inorganic Nanoparticles: Synthesis, Applications, and Perspectives*. CRC Press, Boca Raton, USA, 2010.

[95]   Dalton, A. B. et al. Super-tough carbon-nanotube fibres. *Nature*, 423:703–703, 2003.

[96]   Bagherzadeh, H., Roostaie, A., and Shahrabadi, A. A comprehensive study of polysilicon nanoparticles applications in enhanced oil recovery. In *3rd National Iranian Chemical Engineering Congress*, pp. 1–12, 1998.

[97]   Ahmadi, M., Habibi, A., Pourafshari, P., and Ayatollahi, S. Zeta potential investigation and mathematical modeling of nanoparticles deposited on the rock surface to reduce fine migration. In *SPE Middle East Oil and Gas Show and Conference*. Society of Petroleum Engineers, Manama, Bahrain, 2011.

[98]   Ahmadi, M. A. and Shadizadeh, S. R. Nanofluid in hydrophilic state for eor implication through carbonate reservoir. *Journal of Dispersion Science and Technology*, 35:1537–1542, 2014.

[99]   Ruths, M. and Israelachvili, J. N. Surface forces and nanorheology of molecularly thin films. In *Nanotribology and Nanomechanics II*, pp. 107–202. Springer, 2011.

[100]   Shokrlu, Y. H. and Babadagli, T. In-situ upgrading of heavy oil/bitumen during steam injection by use of metal nanoparticles: A study on in-situ catalysis and catalyst transportation. *SPE Reservoir Evaluation & Engineering*, 16:333–344, 2013.

[101]   Mcelfresh, P. M., Holcomb, D. L., and Ector, D. Application of nanofluid technology to improve recovery in oil and gas wells. In *SPE International Oilfield Nanotechnology Conference and Exhibition*, Society of Petroleum Engineers, Noordwijk, the Netherlands, 2012.

[102]   Torsater, O., Engeset, B., Hendraningrat, L., and Suwarno, S. Improved oil recovery by nanofluids flooding: An experimental study. In *SPE Kuwait International Petroleum Conference and Exhibition*, Society of Petroleum Engineers, Kuwait City, Kuwait, 2012.

[103]   Torsater, O., Li, S., and Hendraningrat, L. Effect of some parameters influencing enhanced oil recovery process using silica nanoparticles: An experimental investigation. In *SPE Reservoir*

Characterization and Simulation Conference and Exhibition, Society of Petroleum Engineers, Abu Dhabi, UAE, 2013.

[104] Onyekonwu, M. O. and Ogolo, N. A. Investigating the use of nanoparticles in enhancing oil recovery. In *Nigeria Annual International Conference and Exhibition*, Society of Petroleum Engineers, Calabar, Nigeria, 2010.

[105] Roustaei, A., Moghadasi, J., Bagherzadeh, H., and Shahrabadi, A. An experimental investigation of polysilicon nanoparticles' recovery efficiencies through changes in interfacial tension and wettability alteration. In *SPE International Oilfield Nanotechnology Conference and Exhibition*, Society of Petroleum Engineers, Noordwijk, the Netherlands, 2012.

[106] Civan, F. *Reservoir Formation Damage*. Elsevier Science, Boston, USA, 2016.

[107] Civan, F. Formation damage mechanisms and their phenomenological modeling- an overview. In *European Formation Damage Conference*, Society of Petroleum Engineers, Scheveningen, the Netherlands, 2007.

[108] Gao, C. Factors affecting particle retention in porous media. *Emirates Journal for Engineering Research*, 12:1–7, 2007.

[109] Khilar, K. C. and Fogler, H. S. *Migrations of Fines in Porous Media*. Springer, Dordrecht, the Netherlands, 1998.

[110] Li, S., Hendraningrat, L., and Torsaeter, O. Improved oil recovery by hydrophilic silica nanoparticles suspension: 2-phase flow experimental studies. In *International Petroleum Technology Conference*, Society of Petroleum Engineers, Beijing, China, 2013.

[111] Li, S., Kaasa, A. T., Hendraningrat, L., and Torsæter, O. Effect of silica nanoparticles adsorption on the wettability index of Berea sandstone. In *International Symposium of the Society of Core Analysts*, Society of Petroleum Engineers, Napa Valley, USA, 2013.

[112] Li, K. and Xie, R. Effect of heterogeneity on production performance in low permeability reservoirs. In *SPE EUROPEC/EAGE Annual Conference and Exhibition*, Society of Petroleum Engineers, Vienna, Austria, 2011.

[113] Dahle, G. Investigation of how Hydrophilic Silica Nanoparticles Affect Oil Recovery in Berea Sandstone. Master's thesis, Norwegian University of Science and Technology, Trondheim, Norway, 2014.

[114] Todd, A. C., Somerville, J. E., and Scott, G. The application of depth of formation damage measurements in predicting water injectivity decline. In *SPE Formation Damage Control Symposium*, Society of Petroleum Engineers, Bakersfield, USA, 1984.

[115] Vetter, O. J., Kandarpa, V., Stratton, M., and Veith, E. Particle invasion into porous medium and related injectivity problems. In *SPE International Symposium on Oilfield Chemistry*, Society of Petroleum Engineers, San Antonio, USA, 1987.

[116] Moghadasi, J., Müller-Steinhagen, H., Jamialahmadi, M., and Sharif, A. Theoretical and experimental study of particle movement and deposition in porous media during water injection. *Journal of Petroleum Science and Engineering*, 43:163–181, 2004.

[117] Huh, C., Lange, E. A., and Cannella, W. J. Polymer retention in porous media. In *SPE/DOE Enhanced Oil Recovery Symposium*, Society of Petroleum Engineers, Tulsa, USA, 1990.

[118] Bolandtaba, S. F. et al. Pore scale modelling of linked polymer solution (LPS). In *15th European Symposium on Improved Oil Recovery 2009*, Paris, France, 2009.

[119] Markus, A. A., Parsons, J. R., Roex, E. W. M., de Voogt, P., and Laane, R. W. P. M. Modeling aggregation and sedimentation of nanoparticles in the aquatic environment. *Science of the Total Environment*, 506:323–329, 2015.

[120] Skauge, T., Spildo, K., and Skauge, A. Nano-sized particles for EOR. In *SPE Improved Oil Recovery Symposium*, Society of Petroleum Engineers, Tulsa, USA, 2010.

[162] Wooster, T. J., Golding, M., and Sanguansri, P. Impact of oil type on nanoemulsion formation and ostwald ripening stability. *Langmuir*, 24:12758–12765, 2008.

[163] Bouchemal, K., Briancon, S., Perrier, E., and Fessi, H. Nano-emulsion formulation using spontaneous emulsification: solvent, oil and surfactant optimisation. *International journal of pharmaceutics*, 280:241–251, 2004.

[164] Velikov, K. P. and Pelan, E. Colloidal delivery systems for micronutrients and nutraceuticals. *Soft Matter*, 4:1964–1980, 2008.

[165] Mason, T. G., Wilking, J. N., Meleson, K., Chang, C. B., and Graves, S. M. Nanoemulsions: formation, structure, and physical properties. *Journal of Physics-Condensed Matter*, 18:R635–R666, 2006.

[166] Tadros, T. F. *Applied surfactants: principles and applications*. Wiley-VCH Verlag GmbH & Co. KGaA, Weinheim, Germany, 2006.

[167] Nakajima, H., Tomomasa, S., and Okabe, M. Preparation of nanoemulsions. In *First Emulsion Conference*, vol. 1, 1993.

[168] Solans, C. and Kunieda, H. *Industrial applications of microemulsions*, vol. 66. CRC Press, New York, USA, 1996.

[169] Solans, C., Izquierdo, P., Nolla, J., Azemar, N., and Garcia-Celma, M. Nano-emulsions. *Current Opinion in Colloid & Interface Science*, 10:102–110, 2005.

[170] Ugelstad, J., Elaasser, M., and Vanderho, J. Emulsion polymerization – initiation of polymerization in monomer droplets. *Journal of Polymer Science Part C-Polymer Letters*, 11:503–513, 1973.

[171] Lovell, P. A. *Emulsion polymerization and emulsion polymers*. Wiley, Chichester, UK, 2011.

[172] Mandal, A. and Bera, A. Surfactant stabilized nanoemulsion: Characterization and application in enhanced oil recovery. In *Proceedings of World Academy of Science, Engineering and Technology. International Journal of Chemical, Molecular, Nuclear, Materials and Metallurgical Engineering*, vol. 6, pp. 537–542, 2012.

[173] Mandal, A., Bera, A., Ojha, K., and Kumar, T. Characterization of surfactant stabilized nanoemulsion and its use in enhanced oil recovery. In *SPE International Oilfield Nanotechnology Conference and Exhibition*, Society of Petroleum Engineers, Noordwijk, the Netherlands, 2012.

[174] Starov, V. M. *Nanoscience: Colloidal and Interfacial Aspects*. CRC Press, Boca Raton, USA, 2011.

[175] Salager, J., Forgiarini, A., and Marquez, L. *Nanoemulsions*. FIRP Booklet nE237A Laboratorio FIRP, Escuela De Ingenieria Quimica, Universidad de Los Andes, Merida, Venezuela, 2006.

[176] Mittal, K. L. and Shah, D. O. *Adsorption and aggregation of surfactants in solution*. CRC Press, Boca Raton, USA, 2002.

[177] Chime, S., Kenechukwu, F., and Attama, A. Nanoemulsions – Advances in Formulation, Characterization and Applications in Drug Delivery. In *Application of Nanotechnology in Drug Delivery*, chapter 3, pp. 77–126. InTech, Rijeka, Croatia, 2014.

[178] Shen, M. and Resasco, D. E. Emulsions stabilized by carbon nanotube-silica nanohybrids. *Langmuir*, 25:10843–10851, 2009.

[179] Ramsden, W. Separation of solids in the surface-layers of solutions and 'suspensions' (observations on surface-membranes, bubbles, emulsions, and mechanical coagulation).– preliminary account. *Proceedings of the Royal Society of London*, pp. 156–164, 1903.

[180] Pickering, S. U. CXCVI – emulsions. *Journal of the Chemical Society, Transactions*, 91:2001–2021, 1907.

[181] Binks, B. P. and Kirkland, M. Interfacial structure of solid-stabilised emulsions studied by scanning electron microscopy. *Physical Chemistry Chemical Physics*, 4:3727–3733.

[182] Binks, B. P. and Lumsdon, S. O. Pickering emulsions stabilized by monodisperse latex particles: effects of particle size. *Langmuir*, 17:4540–4547, 2001.

[183] Binks, B. P. and Clint, J. H. Solid wettability from surface energy components: Relevance to pickering emulsions. *Langmuir*, 18:1270–1273, 2002.

[184] Binks, B., Murakami, R., Armes, S., and Fujii, S. Effects of pH and salt concentration on oil-in-water emulsions stabilized solely by nanocomposite microgel particles. *Langmuir*, 22:2050–2057, 2006.

[185] Binks, B., Philip, J., and Rodrigues, J. Inversion of silica-stabilized emulsions induced by particle concentration. *Langmuir*, 21:3296–3302, 2005.

[186] Binks, B. and Whitby, C. Nanoparticle silica-stabilised oil-in-water emulsions: improving emulsion stability. *Colloids and Surfaces A-Physicochemical and Engineering Aspects*, 253:105–115, 2005.

[187] Binks, B. and Lumsdon, S. Stability of oil-in-water emulsions stabilised by silica particles. *Physical Chemistry Chemical Physics*, 1:3007–3016, 1999.

[188] Binks, B. P., Liu, W., and Rodrigues, J. A. Novel stabilization of emulsions via the heteroaggregation of nanoparticles. *Langmuir*, 24:4443–4446, 2008.

[189] Binks, B. P., Kirkland, M., and Rodrigues, J. A. Origin of stabilisation of aqueous foams in nanoparticle-surfactant mixtures. *Soft Matter*, 4:2373–2382, 2008.

[190] Hunter, T. N., Pugh, R. J., Franks, G. V., and Jameson, G. J. The role of particles in stabilising foams and emulsions. *Advances in Colloid and Interface Science*, 137:57–81, 2008.

[191] Arditty, S., Whitby, C., Binks, B., Schmitt, V., and Leal-Calderon, F. Some general features of limited coalescence in solid-stabilized emulsions. *European Physical Journal E*, 11:273–281, 2003.

[192] Horozov, T., Aveyard, R., Binks, B., and Clint, J. Structure and stability of silica particle monolayers at horizontal and vertical octane-water interfaces. *Langmuir*, 21:7405–7412, 2005.

[193] Kralchevsky, P., Ivanov, I., Ananthapadmanabhan, K., and Lips, A. On the thermodynamics of particle-stabilized emulsions: Curvature effects and catastrophic phase inversion. *Langmuir*, 21:50–63, 2005.

[194] Stocco, A., Rio, E., Binks, B. P., and Langevin, D. Aqueous foams stabilized solely by particles. *Soft Matter*, 7:1260–1267, 2011.

[195] Dickson, J., Binks, B., and Johnston, K. Stabilization of carbon dioxide-in-water emulsions with silica nanoparticles. *Langmuir*, 20:7976–7983, 2004.

[196] Son, H., Kim, H., Lee, G., Kim, J., and Sung, W. Enhanced oil recovery using nanoparticle-stabilized oil/water emulsions. *Korean Journal of Chemical Engineering*, 31:338–342, 2014.

[197] Du, Z. et al. Outstanding stability of particle-stabilized bubbles. *Langmuir*, 19:3106–3108, 2003.

[198] Tambe, D. and Sharma, M. The effect of colloidal particles on fluid-fluid interfacial properties and emulsion stability. *Advances in Colloid and Interface Science*, 52:1–63, 1994.

[199] Drexler, S., Faria, J., Ruiz, M. P., Harwell, J. H., and Resasco, D. E. Amphiphilic nanohybrid catalysts for reactions at the water/oil interface in subsurface reservoirs. *Energy & Fuels*, 26:2231–2241, 2012.

[200] Ashby, N. P. and Binks, B. P. Pickering emulsions stabilised by laponite clay particles. *Physical Chemistry Chemical Physics*, 2:5640–5646, 2000.

[201] Giermanska-Kahn, J., Schmitt, V., Binks, B. P., and Leal-Calderon, F. A new method to prepare monodisperse pickering emulsions. *Langmuir*, 18:2515–2518, 2002.

[202] Paunov, V. N., Binks, B. P., and Ashby, N. P. Adsorption of charged colloid particles to charged liquid surfaces. *Langmuir*, 18:6946–6955, 2002.

[203] Binks, B. P. and Lumsdon, S. O. Effects of oil type and aqueous phase composition on oil-water mixtures containing particles of intermediate hydrophobicity. *Physical Chemistry Chemical Physics*, 2:2959–2967, 2000.

[204] Binks, B. P. and Lumsdon, S. O. Catastrophic phase inversion of water-in-oil emulsions stabilized by hydrophobic silica. *Langmuir*, 16:2539–2547, 2000.

[205] Binks, B. and Rodrigues, J. Inversion of emulsions stabilized solely by ionizable nanoparticles. *Angewandte Chemie-International Edition*, 44:441–444, 2005.

[206] Khosravani, S., Alaei, M., Rashidi, A. M., Ramazani, A., and Ershadi, M. O/w emulsions stabilized with gamma-alumina nanostructures for chemical enhanced oil recovery. *Materials Research Bulletin*, 48:2186–2190, 2013.

[207] Dinsmore, A. et al. Colloidosomes: Selectively permeable capsules composed of colloidal particles. *Science*, 298:1006–1009, 2002.

[208] Zhang, T., Davidson, D., Bryant, S. L., and Huh, C. Nanoparticle-stabilized emulsions for applications in enhanced oil recovery. In *SPE Improved Oil Recovery Symposium*, Society of Petroleum Engineers, Tulsa, USA, 2010.

[209] Jikich, S. J. CO_2 EOR: Nanotechnology for mobility control studied. *JPT Technology Update*, 64:28–31, 2012.

[210] Nguyen, P., Fadaei, H., and Sinton, D. Pore-scale assessment of nanoparticle-stabilized CO_2 foam for enhanced oil recovery. *Energy & Fuels*, 28:6221–6227, 2014.

[211] Espinoza, D. A., Caldelas, F. M., Johnston, K. P., Bryant, S. L., and Huh, C. Nanoparticle-stabilized supercritical CO_2 foams for potential mobility control applications. In *SPE Improved Oil Recovery Symposium*, Society of Petroleum Engineers, Tulsa, USA, 2010.

[212] Aroonsri, A. et al. Conditions for generating nanoparticle-stabilized CO_2 foams in fracture and matrix flow. In *SPE Annual Technical Conference and Exhibition*, Society of Petroleum Engineers, New Orleans, USA, 2013.

[213] Carpenter, C. Gelled emulsions of CO_2 water, and nanoparticles. *Journal of Petroleum Technology*, 66:135–137, 2014.

[214] Zargartalebi, M., Kharrat, R., and Barati, N. Enhancement of surfactant flooding performance by the use of silica nanoparticles. *Fuel*, 143:21–27, 2015.

[215] Worthen, A. J., Bryant, S. L., Huh, C., and Johnston, K. P. Carbon dioxide-in-water foams stabilized with nanoparticles and surfactant acting in synergy. *AIChE Journal*, 59:3490–3501, 2013.

[216] Worthen, A. J. et al. Nanoparticle-stabilized carbon dioxide-in-water foams with fine texture. *Journal of colloid and interface science*, 391:142–151, 2013.

[217] Yu, J., An, C., Mo, D., Liu, N., and Lee, R. L. Foam mobility control for nanoparticle-stabilized supercritical CO_2 foam. In *SPE Improved Oil Recovery Symposium*, Society of Petroleum Engineers, Tulsa, USA, 2012.

[218] Yu, J., Liu, N., Li, L., and Lee, R. L. Generation of nanoparticle-stabilized supercritical CO_2 foams. In *Carbon Management Technology Conference*, Carbon Management Technology Conference, Orlando, USA, 2012.

[219] Avila, J. N. L. D., Grecco Cavalcanti De Araujo, L. L., Drexler, S., de Almeida Rodrigues, J., and Nascimento, R. S. V. Polystyrene nanoparticles as surfactant carriers for enhanced oil recovery. *Journal of Applied Polymer Science*, 133:43789, 2016.

[220] Zhang, K., Wu, W., Meng, H., Guo, K., and Chen, J. F. Pickering emulsion polymerization: Preparation of polystyrene/nano-SiO2 composite microspheres with core-shell structure. *Powder Technology*, 190:393–400, 2009.

[221] Kong, X. and Ohadi, M. Applications of micro and nano technologies in the oil and gas industry – overview of the recent progress. In *Abu Dhabi International Petroleum Exhibition and Conference*, Society of Petroleum Engineers, Abu Dhabi, UAE, 2010.

[222] Adkins, S. S., Gohil, D., Dickson, J. L., Webber, S. E., and Johnston, K. P. Water-in-carbon dioxide emulsions stabilized with hydrophobic silica particles. *Physical Chemistry Chemical Physics*, 9:6333–6343, 2007.

[223] Morales, D., Gutierrez, J., Garcia-Celma, M., and Solans, Y. A study of the relation between bicontinuous microemulsions and oil/water nano-emulsion formation. *Langmuir*, 19:7196–7200, 2003.

[224] Forgiarini, A., Esquena, J., Gonzalez, C., and Solans, C. Formation of nano-emulsions by low-energy emulsification methods at constant temperature. *Langmuir*, 17:2076–2083, 2001.

[225] Uson, N., Garcia, M., and Solans, C. Formation of water-in-oil (w/o) nano-emulsions in a water/mixed non-ionic surfactant/oil systems prepared by a low-energy emulsification method. *Colloids and Surfaces A-Physicochemical and Engineering Aspects*, 250:415–421, 2004.

[226] Setya, S., Talegaonkar, S., and Razdan, B. Nanoemulsions: Formulation methods and stability aspects. *World J. Pharm. Pharm. Sci*, 3:2214–2228, 2013.

[227] Huang, Q. *Nanotechnology in the Food, Beverage and Nutraceutical Industries*. Woodhead Publishing, Cambridge, UK, 2012.

[228] Mao, L. et al. Effects of small and large molecule emulsifiers on the characteristics of beta-carotene nanoemulsions prepared by high pressure homogenization. *Food Technology and Biotechnology*, 47:336–342, 2009.

[229] Kabalnov, A. Ostwald ripening and related phenomena. *Journal of Dispersion Science and Technology*, 22:1–12, 2001.

[230] Kabalnov, A. Effects of micellar pseudophase on Ostwald ripening in emulsions. *Abstracts of Papers of the American Chemical Society*, 206:49–COLL, 1993.

[231] Kabalnov, A. and Shchukin, E. Ostwald ripening theory – applications to fluorocarbon emulsion stability. *Advances in Colloid and Interface Science*, 38:69–97, 1992.

[232] Taylor, P. Ostwald ripening in emulsions. *Advances in Colloid and Interface Science*, 75:107–163, 1998.

[233] Thomson, W. 4. on the equilibrium of vapour at a curved surface of liquid. *Proceedings of the Royal Society of Edinburgh*, 7:63–68, 1872.

[234] Lifshitz, I. M. and Slyozov, V. V. The kinetics of precipitation from supersaturated solid solutions. *Journal of Physics and Chemistry of Solids*, 19:35–50, 1961.

[235] Wagner, C. Theorie der Alterung von Niederschlägen durch Umlösen (Ostwaldreifung). *Zeitschrift für Elektrochemie, Berichte der Bunsengesellschaft für physikalische Chemie*, 65:581–591, 1961.

[236] Weers, J. G. *Chapter 9 Molecular Diffusion in Emulsions and Emulsion Mixtures*. The Royal Society of Chemistry, 1998.

[237] WALSTRA, P. Principles of emulsion formation. *Chemical Engineering Science*, 48:333–349, 1993.

[238] Acosta, E. Bioavailability of nanoparticles in nutrient and nutraceutical delivery. *Current Opinion in Colloid & Interface Science*, 14:3–15, 2009.

[239] Anton, N. and Vandamme, T. F. The universality of low-energy nano-emulsification. *International journal of pharmaceutics*, 377:142–147, 2009.

[240] Leong, T. S. H., Wooster, T. J., Kentish, S. E., and Ashokkumar, M. Minimising oil droplet size using ultrasonic emulsification. *Ultrasonics sonochemistry*, 16:721–727, 2009.

[241] Pouton, C. W. and Porter, C. J. H. Formulation of lipid-based delivery systems for oral administration: Materials, methods and strategies. *Advanced Drug Delivery Reviews; Lipid-Based Systems for the Enhanced Delivery of Poorly Water Soluble Drugs*, 60:625–637, 2008.

[242] Maali, A. and Mosavian, M. T. H. Preparation and application of nanoemulsions in the last decade (2000–2010). *Journal of Dispersion Science and Technology*, 34:92–105, 2013.

[243] Gupta, A., Eral, H. B., Hatton, T. A., and Doyle, P. S. Nanoemulsions: formation, properties and applications. *Soft Matter*, 12:2826–2841, 2016.

[244] Prakash, R. and Thiagarajan, P. Nanoemulsions for drug delivery through different routes. *Res J Biotechnol*, 2:1–13, 2011.

[245] Gutierrez, J. M. et al. Nano-emulsions: New applications and optimization of their preparation. *Current Opinion in Colloid & Interface Science*, 13:245–251, 2008.

[246] Freitas, S., Rudolf, B., Merkle, H., and Gander, B. Flow-through ultrasonic emulsification combined with static micromixing for aseptic production of microspheres by solvent extraction. *European Journal of Pharmaceutics and Biopharmaceutics*, 61:181–187, 2005.

[247] Freitas, S., Hielscher, G., Merkle, H., and Gander, B. Continuous contact- and contamination-free ultrasonic emulsification – a useful tool for pharmaceutical development and production. *Ultrasonics sonochemistry*, 13:76–85, 2006.

[248] Yin, L. J., Chu, B. S., Kobayashi, I., and Nakajima, M. Performance of selected emulsifiers and their combinations in the preparation of beta-carotene nanodispersions. *Food Hydrocolloids*, 23:1617–1622, 2009.

[249] Chu, B. S., Ichikawa, S., Kanafusa, S., and Nakajima, M. Preparation and characterization of beta-carotene nanodispersions prepared by solvent displacement technique. *Journal of Agricultural and Food Chemistry*, 55:6754–6760, 2007.

[250] Sole, I. et al. Nano-emulsions preparation by low energy methods in an ionic surfactant system. *Colloids and Surfaces A-Physicochemical and Engineering Aspects*, 288:138–143, 2006.

[251] Sole, I., Maestro, A., Gonzalez, C., Solans, C., and Gutierrez, J. M. Optimization of nano-emulsion preparation by low-energy methods in an ionic surfactant system. *Langmuir*, 22:8326–8332, 2006.

[252] Sole, I. et al. Nano-emulsions prepared by the phase inversion composition method: Preparation variables and scale up. *Journal of colloid and interface science*, 344:417–423, 2010.

[253] Sonneville-Aubrun, O., Simonnet, J., and L'Alloret, F. Nanoemulsions: a new vehicle for skincare products. *Advances in Colloid and Interface Science*, 108:145–149, 2004.

[254] Wang, L., Li, X., Zhang, G., Dong, J., and Eastoe, J. Oil-in-water nanoemulsions for pesticide formulations. *Journal of colloid and interface science*, 314:230–235, 2007.

[255] Wang, L., Mutch, K. J., Eastoe, J., Heenan, R. K., and Dong, J. Nanoemulsions prepared by a two-step low-energy process. *Langmuir*, 24:6092–6099, 2008.

[256] Wang, L. et al. Formation and stability of nanoemulsions with mixed ionic-nonionic surfactants. *Physical Chemistry Chemical Physics*, 11:9772–9778, 2009.

[257] Lamaallam, S., Bataller, H., Dicharry, C., and Lachaise, J. Formation and stability of miniemulsions produced by dispersion of water/oil/surfactants concentrates in a large amount of water. *Colloids and Surfaces A-Physicochemical and Engineering Aspects*, 270:44–51, 2005.

[258] Izquierdo, P. et al. The influence of surfactant mixing ratio on nano-emulsion formation by the pit method. *Journal of colloid and interface science*, 285:388–394, 2005.

[259] Shinoda, K. and Saito, H. Stability of o/w type emulsions as functions of temperature and hlb of emulsifiers – emulsification by pit-method. *Journal of colloid and interface science*, 30:258–&, 1969.

[260] Pey, C. M. et al. Optimization of nano-emulsions prepared by low-energy emulsification methods at constant temperature using a factorial design study. *Colloids and Surfaces A-Physicochemical and Engineering Aspects*, 288:144–150, 2006.

[261] Porras, M., Solans, C., Gonzalez, C., and Gutierrez, J. M. Properties of water-in-oil (w/o) nano-emulsions prepared by a low-energy emulsification method. *Colloids and Surfaces A-Physicochemical and Engineering Aspects*, 324:181–188, 2008.

[262] Sajjadi, S. and Jahanzad, F. Nanoparticle formation by highly diffusion-controlled emulsion polymerisation. *Chemical Engineering Science*, 61:3001–3008, 2006.

[263] Sajjadi, S. Nanoemulsion formation by phase inversion emulsification: On the nature of inversion. *Langmuir*, 22:5597–5603, 2006.

[264] Sajjadi, S. Formation of fine emulsions by emulsification at high viscosity or low interfacial tension; a comparative study. *Colloids and Surfaces A-Physicochemical and Engineering Aspects*, 299:73–78, 2007.

[265] Fernandez, P., Andre, V., Rieger, J., and Kuhnle, A. Nano-emulsion formation by emulsion phase inversion. *Colloids and Surfaces A-Physicochemical and Engineering Aspects*, 251:53–58, 2004.

[266] Rallison, J. The deformation of small viscous drops and bubbles in shear flows. *Annual Review of Fluid Mechanics*, 16:45–66, 1984.

[267] Oh, D. H. et al. Effect of process parameters on nanoemulsion droplet size and distribution in spg membrane emulsification. *International journal of pharmaceutics*, 404:191–197, 2011.

[268] Vladisavljevic, G. and Williams, R. Recent developments in manufacturing emulsions and particulate products using membranes. *Advances in Colloid and Interface Science*, 113:1–20, 2005.

[269] Vitale, S. and Katz, J. Liquid droplet dispersions formed by homogeneous liquid-liquid nucleation: "the ouzo effect". *Langmuir*, 19:4105–4110, 2003.

[270] MILLER, C. Spontaneous emulsification produced by diffusion – a review. *Colloids and Surfaces*, 29:89–102, 1988.

[271] Aryanti, N., Williams, R. A., Hou, R., and Vladisavljevic, G. T. Performance of rotating membrane emulsification for o/w production. *Desalination*, 200:572–574, 2006.

[272] El-Din, M. R. N. and Al-Sabagh, A. Preparation of water-in-hexane nanoemulsions using low energy emulsification method. *Journal of Dispersion Science and Technology*, 33:68–74, 2012.

[273] Panagiotou, T. and Fisher, R. Improving Product Quality with Entrapped Stable Emulsions: From Theory to Industrial Application. *Challenges*, 3:84–113, 2012.

[274] Graves, S., Meleson, K., Wilking, J., Lin, M. Y., and Mason, T. G. Structure of concentrated nanoemulsions. *The Journal of chemical physics*, 122:134703, 2005.

[275] Jafari, S. M., Assadpoor, E., He, Y., and Bhandari, B. Re-coalescence of emulsion droplets during high-energy emulsification. *Food Hydrocolloids*, 22:1191–1202, 2008.

[276] Jafari, S. M., He, Y., and Bhandari, B. Production of sub-micron emulsions by ultrasound and microfluidization techniques. *Journal of Food Engineering*, 82:478–488, 2007.

[277] Jafari, S. M., He, Y., and Bhandari, B. Optimization of nano-emulsions production by microfluidization. *European Food Research and Technology*, 225:733–741, 2007.

[278] Qian, C. and McClements, D. J. Formation of nanoemulsions stabilized by model food-grade emulsifiers using high-pressure homogenization: Factors affecting particle size. *Food Hydrocolloids*, 25:1000–1008, 2011.

[279] Meleson, K., Graves, S., and Mason, T. Formation of concentrated nanoemulsions by extreme shear. *Soft Materials*, 2:109–123, 2004.

[280] Fryd, M. M. and Mason, T. G. Time-dependent nanoemulsion droplet size reduction by evaporative ripening. *Journal of Physical Chemistry Letters*, 1:3349–3353, 2010.

[281] Lee, S. J. and McClements, D. J. Fabrication of protein-stabilized nanoemulsions using a combined homogenization and amphiphilic solvent dissolution/evaporation approach. *Food Hydrocolloids*, 24:560–569, 2010.

[282] Abismail, B., Canselier, J., Wilhelm, A., Delmas, H., and Gourdon, C. Emulsification by ultrasound: drop size distribution and stability. *Ultrasonics sonochemistry*, 6:75–83, 1999.

[283] Bondy, C. and Söllner, K. Quantitative experiments on emulsification by ultrasonic waves. *Transactions of the Faraday Society*, 32:556–567, 1936.

[284] Anna, S., Bontoux, N., and Stone, H. Formation of dispersions using "flow focusing" in microchannels. *Applied Physics Letters*, 82:364–366, 2003.

[285] Squires, T. and Quake, S. Microfluidics: Fluid physics at the nanoliter scale. *Reviews of Modern Physics*, 77:977–1026, 2005.

[286] Torza, S., Cox, R. G., and Mason, S. G. Particle motions in sheared suspensions xxvii. transient and steady deformation and burst of liquid drops. *Journal of colloid and interface science*, 38:395–411, 1972.

[287] Kentish, S. et al. The use of ultrasonics for nanoemulsion preparation. *Innovative Food Science & Emerging Technologies*, 9:170–175, 2008.

[288] Sutradhar, K. B. and Lutful, A. M. Nanoemulsions: increasing possibilities in drug delivery. *European Journal of Nanomedicine*, 5:97–110, 2013.

[289] Shah, P., Bhalodia, D., and Shelat, P. Nanoemulsion: A pharmaceutical review. *Systematic Reviews in Pharmacy*, 1:24–32, 2010.

[290] Strydom, S. J., Rose, W. E., Otto, D. P., Liebenberg, W., and de Villiers, M. M. Poly(amidoamine) dendrimer-mediated synthesis and stabilization of silver sulfonamide nanoparticles with increased antibacterial activity. *Nanomedicine-Nanotechnology Biology and Medicine*, 9:85–93, 2013.

Index

https://doi.org/10.1515/9783110640250-006

www.ingramcontent.com/pod-product-compliance
Lightning Source LLC
Chambersburg PA
CBHW081529220326
41598CB00036B/6377